P9-BJK-872

The Perfect Machine

The Perfect Machine

BUILDING THE PALOMAR

TELESCOPE

Ronald Florence

HarperCollins*Publishers*

HarperCollins books may be purchased for educational, business, or sales promotional use. For information please write: Special Markets Department, HarperCollins Publishers, Inc., 10 East 53rd Street, New York, NY 10022.

FIRST EDITION

Designed by George J. McKeon
Insert designed by Barbara DuPree Knowles

Library of Congress Cataloging-in-Publication Data

Florence, Ronald.
    The perfect machine / Ronald Florence.
        p. cm.
    Includes index.
    ISBN 0-06-018205-9
    1. Reflecting telescopes—California—Palomar, Mount—History. 2. Mount Wilson and Palomar Observatories—History. I. Title.
QB88.F55 1994
522' .29794'98—dc20   93-46374

94 95 96 97 98 ❖/RRD 10 9 8 7 6 5 4 3 2

For my son, Justin

How minute are our instruments in comparison with the celestial universe!
—EL KARAT, CAIRO, ELEVENTH CENTURY

Starlight is falling on every square mile of the earth's surface, and the best we can do at present is to gather up and concentrate the rays that strike an area 100 inches in diameter.
—GEORGE HALE, 1928

For I can end as I began. From our home on earth we look out into the distances and strive to imagine the sort of world into which we were born. Today we have reached far out into space. Our immediate neighborhood we know intimately. But with increasing distance our knowledge fades ... until at the last dim horizon we search among ghostly errors of observations for landmarks that are scarcely more substantial. The search will continue. The urge is older than history. It is not satisfied and it will not be suppressed.
—EDWIN HUBBLE, 1936

Russell Porter's 1936 conception of the two-hundred-inch telescope.

Contents

Illustrations follow page 278.

1

April 1921

The two men met on the platform of the Southern Pacific depot in Los Angeles. In their wool suits and stiff collars, they might have been mistaken for commercial travelers starting out on a week's run to peddle their wares—at least until they bought tickets straight through to Washington, D.C. Transcontinental travel was enough of a novelty to turn heads in 1921.

The area around the depot looked like the backgrounds of the Mack Sennett Keystone Kops comedies, which were filmed nearby. Elegant homes abutted vacant lots. Trolley cars ran down streets lined with palm and eucalyptus trees, past incongruously empty pastures. Signs of construction were everywhere. Civic boosters, buoyed by preliminary reports of the 1920 census, were already bragging that Los Angeles had passed San Francisco in population to become the fastest-growing city in the United States.

At the depot the men exchanged pleasantries and agreed to share a compartment, but neither said much to the other, letting the flurry of boarding passengers, conductors checking tickets, baggage handlers, and porters fill the silence until the train left. Outside the Pullman window they watched cart after cart being wheeled down the platform to the dining car, laden with food, linen, and menus. Prohibition had eliminated the wine list and the prospect of a bar car, but even in the infancy of transcontinental travel, the railroads knew that leisurely meals served by porters in starched white jackets were one way to fill the long hours of the journey.

The trolley cars had a nine-story terminal at Sixth and Main, but railroad passengers didn't count for much in the City of the Angels. Eager to match the splendor of the great eastern stations, like the grand concourse of New York's Penn Station, where departing passengers on the famed Limited trains were greeted by a stationmaster in tails, top hat, and white gloves, Los Angeles had long campaigned to get the Southern Pacific, Union Pacific, and Atchison lines to build a grand Union Station as part of a grandiose civic center plan. But by the 1920s Southern California already seemed hell-bent on the automobile. The ambitious plans of the tire, gasoline, and auto companies were open secrets, and the railroads, unsure of the future of passenger traffic, resisted municipal entreaties. Eastbound passengers in Los Angeles had to settle for a cluttered platform in a dilapidated depot, where the train pulled away with no more ceremony than an everyday local, and the engineer had to ride his whistle to clear the way over a series of annoying grade crossings along busy Alameda Street. Automobiles had already wrought havoc in Los Angeles.

The train didn't pick up speed until it emerged from the built-up downtown area. By then the travelers could see billboards advertising canary farms, artificial pools for fishing, stands selling fried rabbit, dogs at stud, grass-shack eating huts, psychic mediums, vacant-lot circuses, storefront evangelicals, bicycles for rent, and frogs for sale. California was like no other economy in the world. Real estate companies sprang up and disappeared overnight. Where most cities had restaurants, Los Angeles had cafeterias. Signs advertised the eternal glories of Forest Lawn and the Hollywood Pet Cemetery. There were stores that sold pet caskets, as well as dog beauty shops and dog restaurants. Los Angeles had more telephones and automobiles than any city of comparable size in the United States. Its divorce and suicide rates were double the national average.

Some attributed the madness to Hollywood, a boomtown even by California standards. The studios had started making movies in Los Angeles only a half-dozen years before and were already one of the biggest industries in the state. All over the country people eagerly followed the antics, orgies, fame, depravity, scandals, and successes of the stars. Thousands flocked to California every week in pursuit of fame, stardom, or quick money in the movie business. Street-corner vendors were already hawking maps with addresses of the stars.

Although they were interested in stars and, like the moviemakers, had been drawn to California by the weather and skies so clear that the nearby mountains seemed to reach up and touch the heavens, the two men who boarded the train at the Southern Pacific depot hadn't come to California because of the movie industry. They were astronomers, from what were then the two most advanced observatories in the world: Heber Curtis, from the Lick Observatory on Mount Hamilton, near San Jose; and Harlow Shapley, from the Mount Wilson Observatory, above Pasadena. For years they had been engaged in an exchange of articles in the professional journals, criticizing each other with the remarkable vehemence that scientific journals seem to encourage. To a layperson who tried to wade through the articles, the spirited exchange about "globular clusters" and "spiral nebulae" might have seemed a tempest in a teapot. But this tempest had ultimately gathered national attention, finding its way onto the pages of newspapers that normally favored murders, adultery, and political scandal to serious science. At the end of their journey across the country, the two men were scheduled to debate at the annual meeting of the National Academy of Sciences, in the presence of an august audience that—and this was one detail the newspapers never failed to mention—would include the famous professor Albert Einstein.

From the sprawling Los Angeles Basin, a refuge ringed by mountains, the train climbed through the San Bernardino Mountains toward the Cajon Pass. Beyond the pass the open desert stretched for hundreds of miles, its bleakness punctuated only by chaparral, cactus, and the occasional mountain that pierced the clear air. The desert mountains were strange geological formations, isolated from the dangers of earthquake, remote enough to be untainted by the urban light pollution that even then affected most astronomical observatories. Their peaks thrust up into air so still that stars were pinpoints against the black heavens.

The great California observatories, like Lick and Mount Wilson, were themselves isolated by traditional standards. Before they were built most astronomical observatories were located more for convenience—usually on the campuses of urban universities—than for optimum *seeing*, the term astronomers use to describe the stillness and transparency of the atmosphere. Yet, compared to the remoteness of the desert peaks that Shapley and Curtis saw from the train window, the Lick and Mount Wilson observatories were almost urban. In 1921 the desert peaks, like Palomar Mountain, were so isolated by the lack of regular roads that only the Indians really knew them. Some of the mountains were sacred to the Indians: Legend had it that they were the homes of the Spirits—the departure point for souls journeying to the heavens. Shapley and Curtis could only imagine what the night sky would be like from a desert mountaintop.

The National Academy of Sciences often had distinguished speakers at its annual meetings. In the interest of widespread appeal the topics were generally selected from some aspect of what today would be labeled "applied" science. Reports of scientific developments with immediate and direct practical applications were a sure way to attract the attention of the newspapers, and potentially of the wealthy benefactors on whom the academy was dependent.

Early in 1921 Charles G. Abbot, secretary of the academy, began organizing a program of speakers for that year's annual meeting. George Hale, a distinguished solar astronomer and the director of the Mount Wilson Observatory, as well as an officer of the academy, proposed that a Hale Lecture, named in honor of his father, might occupy one evening. The topic he suggested, relativity, was a new and quite fashionable subject in scientific circles, especially since the newly famous Professor Einstein was scheduled to make his first visit to the United States that spring.

Only two years before, a much-publicized research expedition to Principe Island, off Africa, led by the well-known British astronomer and cosmologist Sir Arthur S. Eddington, had measured the deflection of starlight during a solar eclipse. When the measurements confirmed the predictions of Einstein's theory of gravitation, *The Times* (of London) called Einstein's work a "revolution in science." The sudden newspaper publicity transformed the shy former inspector at the Swiss Patent Office in Berne into a world celebrity.

Despite Einstein's fame and the pages that had been devoted to his theories in the newspapers and magazines, Abbot responded that a talk on relativity would be incomprehensible to a majority of the members of the academy, who came from all branches of science and weren't necessarily familiar with Einstein's writings. As an alternative Hale suggested a topic that some of the California astronomers had been debating in the scientific journals. On the basis of new sky surveys, some astronomers were convinced that the Milky Way, our own galaxy, was only one of many "island universes" in the heavens—an assertion many other astronomers found incomprehensible, untenable, or unreasonable. The primary evidence for the debate had come from Lick Observatory and Hale's own Mount Wilson Observatory in California, but well-known European astrophysicists like Eddington and James Jeans had also written about the island universes debate. To an astronomer like Hale it was a hot topic.

Abbot responded that he was afraid the members of the academy also wouldn't be interested in island universes. He refrained from telling George Hale that the only astronomical topic of widespread interest was Percival Lowell's search for canals on Mars, or that to much of the public astronomers were the butt of jokes: boring old men with long beards who spent hours at the eyepieces of their telescopes, scribbling inscrutable numbers and making sketches that meant little to anyone else. Not too long before, a traveler had noted that one reason Westerners considered "the Chinese such barbarians is on account of the support they give to their Astronomers—people regarded by our cultivated Western mortals as completely useless. Yet there they rank with Heads of Departments and Secretaries of State. What frightful barbarism!"

In lieu of astronomers Abbot suggested that they get a speaker on a topic like medical progress in treating wounded soldiers. With the war fresh in memory, it would be a good response to the publicity recently generated by the antivivisectionists. Hale dismissed the topic as too pedestrian. The two men corresponded until Abbot, realizing that they were running out of time before the meeting, acquiesced and fired off telegrams to invite young Harlow Shapley from the Mount Wilson Observatory and Heber Curtis of the Lick Observatory to speak on "The Scale of the Universe." Abbot wasn't the first to discover that George Hale, a mild-mannered, self-effacing man, who looked almost cherubic in his tiny wire-rimmed glasses, usually got his way.

Harlow Shapley accepted the invitation to speak immediately. A young man from Missouri who still wore his straight black hair slicked back in a rural "hick" style, Shapley was in a hurry, eager to make his mark. When, in his third year at the University of Missouri, he fell in love with a woman named Martha Betz, he told her, "Listen, I'm a busy man. If you want any more letters from me you will have to write my language." Shapley's language was the Gregg shorthand system, which they used to save time even after they married.

He had been a reporter on small-town newspapers before he went to the University of Missouri in the hope that a little more education would get him a position on a bigger newspaper. Shapley later claimed that he had chosen astronomy as an undergraduate major because there was no school of journalism and he couldn't pronounce the first discipline he came across in the college catalog, *archaeology*. His professor at Missouri was Frederick Seares, who recommended him for a prestigious graduate fellowship at Princeton and later helped him obtain the plum of a first appointment at Mount Wilson after he received his doctorate. Shapley's ready sense of humor and boisterous horselaugh took the edge off his unconcealed ambition, though some older astronomers thought him too eager to take credit.

At Mount Wilson, Shapley poured his energies into both his primary responsibilities, assisting other astronomers, and his own program of research, quickly building a reputation for what some called boldness and others saw as hasty conclusions. At the time bachelor astronomers at Mount Wilson were paid ninety-five dollars a month and were provided with a room at the "Monastery" on the mountain. Shapley, who had married just before he arrived at Mount Wilson, got a munificent $135 a month to cover rent in Pasadena and the cost of getting himself up and down the mountain. Hale, the director of the Mount Wilson Observatory and Shapley's boss, offered him $250 for expenses on the trip to Washington, but Shapley would have gone even if he had had to pay for the trip from his meager salary. "We had already resigned ourselves to poverty," he wrote.

For an ambitious young astronomer the debate in Washington, and the opportunity to address a wide audience, was the chance of a lifetime. George Hale had recently nominated Shapley for the prestigious position of director of the Harvard College Observatory, a remarkable coup for a man so young. A good showing in the debate could make or break his chances for the appointment.

Heber Curtis, the principal critic of Shapley's new ideas, was skeptical of the proposed debate. Curtis was from an older, more restrained generation, gentlemen scientists wary of publicity, public forums, and young men with half-cocked ideas who put unseemly ambition ahead of methodical science. Taciturnity, an understated senior-common-room style, and an instinctive mistrust of faddish notions were badges of distinction for serious scientists. Curtis finally agreed to participate, but only after some negotiation, insisting on the topic proposed in Abbot's original telegram, "The Scale of the Universe," and objecting to any title that would exclude his own research interests in favor of Shapley's. The evening at the academy was scheduled as a symposium rather than a formal debate, but once the newspapers heard that Einstein would be present, they began ballyhooing the forthcoming evening as the greatest scientific debate since the trial of Galileo.

Neither man knew what to expect. Both had written primarily for scientific journals and spoken mostly to audiences of professional astronomers, a small world in 1921. Tempting though it was to discuss their qualms and apprehensions, they agreed when they first boarded the train that it would be best if they did not talk about the upcoming debate. And so, day after day, they rode on, avoiding the subject on their minds in favor of small talk or the books and notebooks they had brought in their briefcases.

The transcontinental railroad tracks were still a novelty, a tenuous tie to the remote coasts. In an age before the telephone was widespread, when radios were not yet in every home, when only businessmen in a hurry and the War Department reporting casualties used the telegraph, and when the network of highways had just begun to reach out from the cities, that thin line of railroad tracks was the only thread tying the country together. Across many routes travelers could ride for most of a day without seeing any settlements except tiny hamlets, watering stations, and mail stops along the tracks.

The America they crossed was a land of small farms, producing not only the crops they sold but their own milk, eggs, meat, and vegetables. Rural folk made do. In the summers families enjoyed the bounty of the land. In the winters they drew from the larder or the root cellar. Except for store-bought dresses and suits for special occasions, or the rugged ready-made garments that were becoming available in the catalogs, they wore homemade clothing, buying fabric and notions from country stores, catalogs, or itinerant peddlers. News about technology came from the Sears catalog, ubiquitous reading material in outhouses across America. In 1921 it featured .22 caliber rifles for $4.25; an upholstered, curly-backed rocker for $5.95; women's middy blouses for $0.98; and a treadle-powered sewing machine for $29.95.

Farmers worked the land with draft animals. Only in the cities had auto exhausts replaced manure as the hazard of the streets. Speed limits in most cities were still twenty miles per hour. A few who happened to live near the railroad tracks could watch the speedy trains bridging the land; to most the whistles of the trains were as remote as the contrails of jet planes to a later generation.

Nights were quiet time. The wireless wasn't in many homes yet, although Westinghouse had broadcast early results of the elections in November. Victrolas were a luxury that plain folk considered showing off. It wasn't unusual for a family to spend a summer evening outside, on the porch or in the yard, sitting on rockers or swing benches, staring at the stars. With no city lights, no highways with nightly columns of trucks and cars, and electric power unavailable beyond the fringes of the cities, families could enjoy the glories of dark skies that revealed the Milky Way not as an occasional lucky sight but as a regular evening spectacle.

The stars were a nightly wonder. Some accepted the canonical explanations of the Bible and thought of the heavens as one more impenetrable miracle of Creation. Others contented themselves with the thought that pretty soon scientists, using those big new telescopes out in California, would know what it was all about.

From the vast prairie land of the Midwest and the sharecropped farms of the Mississippi River Valley, the travelers rode on into the industrial belt of the eastern states, the largest single concentration of heavy industry in the world. The United States prided itself on superlatives—the most railcars of coal extracted from a mine in a day, the most tons of steel produced, the most feet of rail rolled. Corporations, armed with their new public relations departments, eagerly joined the chorus of hyperbole, issuing press releases to announce the largest electrical network ever built, the biggest turbine, the largest milling machine.

Some Europeans saw the American habit of superlatives as a sign of collective insecurity, but to Americans there was a comfort in the concrete symbols of achievement. From the lonely farmers on the boundless prairies, to the factory workers of the mill towns, to the men of untold wealth who were not ashamed to describe themselves as capitalists, Americans held up industrial might as a challenge and a response to the alleged sophistication, cosmopolitanism, and grandeur of Europe. The United States was on a roll. Business was booming. The smokestacks were going full-time. Although no one bragged about it, the United States could also claim the smokiest skies and dirtiest rivers in the world. What another generation would see as threats to health and the future were symbols of progress and prosperity in 1921.

In this America of the biggest, the grandest, and the greatest, science was on its way to a new, elevated status. Already hucksters, journalists, teachers, and advertisers were cavalierly tossing off the claim that "Science tells us" or "Science teaches us" as a preëmptive answer to arguments. Einstein had not yet visited the United States, but already his name had entered the common vocabulary as a synonym for genius. The mysteries of the General Theory of Relativity were widely touted as the most important scientific discovery of the century. A myth circulated that only twelve men in the world could understand the theory, but the alleged limits of comprehension didn't stop editors and soapbox orators from extolling the importance of relativity, tossing off a casual $E = mc^2$, or announcing that "there are no absolutes, everything is relative" to prove that they too were part of the great age of science.

Yet the intellectuals and poseurs who revered Einstein were a tiny minority of the American public. For much of the country formal science was too abstract, so obscure that it was somehow un-American. In 1914 a congressman questioning a witness at an appropriations hearing said: "What is a physicist? I was asked on the floor of the House what in the name of common sense a physicist is, and I could not answer."

Even the august National Academy of Sciences enjoyed less than universal prestige. Andrew Carnegie typified the American reaction when he dismissed a request for funds for the academy: "Oh," he said. "That's just one of those fancy societies."

Americans had their own science. To the ordinary folk of Sinclair Lewis's *Main Street,* science meant know-how, the ability to make cars, vacuum cleaners, electric irons, light bulbs, radios. America was the country that could build anything. Americans believed that they had won the Great War in the shipyards and mills and factories as much as the trenches, and few doubted that there was any problem of science that couldn't also be solved by the same commonsense engineering that brought invention after invention out of the laboratories of Thomas Edison and car after car out of the factories of Henry Ford.

In 1921 few Americans had ever heard of Harlow Shapley or Heber Curtis. Most would have named Edison as the greatest living scientist. But the United States was a land of newspaper readers, and the newspapers had discovered the art of turning the commonplace into the kind of stories that readers demanded. A mine cave-in that killed seventy men earned a brief mention in the paper; a single man trapped in a mine was a story that could be developed and enhanced to hold readers for days. A good murder trial could hold them for months. The National Academy of Sciences wasn't a usual newspaper beat, but then Albert Einstein in the audience wasn't the usual lead. If there was ever a science story that would get readers, this was it. On April 26, 1921, the newspapers promised, at the annual meeting of the National Academy of Sciences, held in the central hall of the Smithsonian Institution in Washington, in the august presence of Professor Albert Einstein, the most basic questions about our universe would be answered.

By the second day of the journey, the travelers on the train were weary. The steady click-clack of the bolted rails, reassuring the first day, was monotonous. The view through the miasma of black smoke from the soft coal that fueled the engine was no longer exciting. The panorama outside the windows, hour after hour of wheat or rice or woods or bottomlands, became tedious. Those who hadn't prepared for the journey with reading material or games were soon bored.

Heber Curtis, an amateur classicist as well as an astronomer, had brought Latin and Greek texts with him. Reading the classics was a gentleman's avocation. Hour after hour he would sit with a familiar, leather-bound volume open on his lap, as if he were in a club chair in the Atheneum. Harlow Shapley was fascinated by Curtis's choice of reading material. He had been educated a generation later, when the classics had already faded in high school and college curricula.

There was no room for extraneous reading in Shapley's life. He thought of himself as a modern man, with a modern education. His avocation when he couldn't work on astronomy was nature studies. He was an amateur naturalist, but he approached nature as he approached astronomy problems, carrying a notebook with him wherever he went. If he couldn't be near the observatory and its instruments, he cataloged the species of insects or plants he found, writing notes as methodically as his logbook of observation runs on the sixty-inch telescope at Mount Wilson.

On one nature walk in California he had stumbled on a colony of ants, scurrying to and from their nest. Shapley timed how fast they were moving. What factors determined the speed of their travel? he asked in his omnipresent notebook. He gathered enough data to hypothesize that the ants' speed of travel was determined solely by the ambient temperature. Armed with a theory, Shapley needed data. Wherever he went he would search out an ant colony and accumulate more measurements to bolster his theory.

Less than a day from Washington, on the east side of Birmingham, Alabama, the train broke down, close enough to the city that the passengers could still see the smoke-darkened skies from the steel mills. The conductors made the rounds of the cars, reassuring the passengers, but as the day went on and the train stood still under the broiling sun, the passengers grew restless and hot inside the cars. Curtis took one of his classical texts and lay down in the shade to read. As he read he could see Harlow Shapley—notebook, stopwatch, and thermometer in hand—chasing through the jasmine and the new kudzu vines that had been planted to control erosion, in pursuit of a colony of ants.

For a while, the Scale of the Universe seemed far away.

2

Washington

Washington was a sleepy town in 1921, more like the capital of a small state today than the full-time capital of a great nation. Congressmen and senators spent most of the year in their home districts, commuting to congressional sessions of limited duration. The staffs of Congress and the president each numbered a few people. It would take a dozen years before there would be a telephone on the desk in the Oval Office. The British Foreign Office classified the city as a semitropical hardship location. Even the press wasn't there in droves yet: Washington politics weren't considered important enough to attract permanent press bureaus or hordes of lobbyists.

April was one of the better months, before the oppressive heat and humidity of summer. The hundreds of cherry trees around the Tidal Basin, a gift only nine years before from the mayor of Tokyo, were in bloom, a welcome relief from the dank mosquito infestations that had once marked the area. Relations with the Japanese weren't as friendly as they had been in 1912, but most Americans, after the experience of the war to end all wars, weren't interested in other countries.

The headlines on the newspapers at Union Station were depressing. Warren Harding had replaced the ailing Woodrow Wilson, who had spent the last years of his presidency sequestered in the White House. Wags who had speculated whether Mrs. Wilson or Colonel House was running the Wilson White House now wondered whether *anyone* was running the government. The "Red scare" was in full swing. In the pages of the *Dearborn Independent* Henry Ford attacked what he called the "International Jews"; the revived Ku Klux Klan blamed the woes of the nation on the triad of Jews, Roman Catholics, and blacks; police chiefs like William Francis Hynes in Los Angeles sent squads of officers to break up union and leftish meetings; and almost everyone seemed willing to take a swipe at the Industrial Workers of the World, the Wobblies.

Fortunately there were diversions from the pall of politics. Babe

Ruth, who had pitched and played occasional outfield for the Boston Red Sox, was in his second year with the New York Yankees and proving he was worth the astonishing $125,000 they had paid to get him. Man o' War was the Babe Ruth of the racetrack, and handsome, charming Jack Dempsey seemed equally unbeatable in the boxing ring.

In the bookstores the talk was of F. Scott Fitzgerald's daring *This Side of Paradise*. Readers turned down the corners of the pages on which one of Fitzgerald's heroines confessed: "I've kissed dozens of men. I suppose I'll kiss dozens more," or, "Oh, just one person in fifty has any glimmer of what sex is. I'm hipped on Freud and all that, but it's rotten that every bit of real love in the world is ninety-nine percent passion and one little *soupçon* of jealousy."

A few adventurous women had started wearing short-sleeved or sleeveless dresses in the evening, sometimes showing their knees and stockings rolled below the knee. A risqué few even smoked in public and went out in the evenings without corsets because, as the whispered saying had it, "Men won't dance with you if you wear one." They were the fringe exception, the radicals who attracted sensational press and wagging fingers from the guardians of morality. Still, it wasn't hard to imagine that before long there would be bathing-beauty contestants in skin-tight suits with naked legs, cheek-to-cheek dancing, people getting "blotto," and necking and petting in parked cars—exactly the stuff the moralists most feared.

The National Academy of Sciences didn't even have a building of its own in 1921, which was why its annual meeting that year was scheduled to be held in the strange, turreted, brick castle of the Smithsonian Institution in the middle of the empty mall that ran from the Capitol to the Potomac. On the evening of April 26 a steady stream of motorcars drove up to the sheltered portico of the castle. The founders had modeled the institution after the long-established academies of Europe. They wisely stopped short of the formal dress that might have evoked protests of outrage from those who would be sure to insist on American plainspun. In France or England plumes and sashes were de rigueur. The men who came to the annual meeting of the American academy—science was not yet a proper pursuit for a woman—dressed in dark wool suits for the occasion. Even science was supposed to be democratic in America.

The ticket George Hale had gotten for Shapley entitled him to a seat at the head table, among the notables. He sat next to W. J. V. Osterhout of the Botany Department at Harvard, but the banquet had been served before Shapley had a chance to talk about his nature studies. They were still eating when the speeches began.

Stylized elocution was fashionable in 1921. A parade of long-winded speakers followed one another to the podium, first to honor the Prince of Monaco for his support of oceanographic studies, then to praise the achievements of a bureaucrat named Johnson, who had devoted his life to hookworm control. To keep his own nervousness in check, Shapley silently cataloged the speeches: "Johnson the Scientist," "Johnson the Operator," "Johnson the Man." Out in the audience he could see heads nodding off.

Einstein was at one end of the head table, next to the secretary of the Netherlands Embassy, there to accept a prize on behalf of the Dutch scientist Pieter Zeeman. During one of the speeches Einstein leaned over to whisper something to the Dutchman. Reporters later rushed to ask what Einstein had said. He said, the Dutchman reported with a grin, "I have just got a new Theory of Eternity."

Finally it was time for the much-publicized symposium. Shapley came to the podium first. He had never before addressed a large audience of nonspecialists.

As Shapley looked around the room, one of the few faces he could recognize was that of his mentor at Princeton, the legendary Henry Norris Russell, the dean of American astronomy. Russell was a shy and formal professor, given to strolling the Princeton campus with his cane in hand, brushing aside students in his way and addressing even his best graduate students as if they were servants. For all his formality, Russell was an inspiring teacher, and he had been profoundly appreciative of the superb graduate student who had come his way. As Russell put it years later: "I had this struggle with darkening at the limb of an eclipsing binary. All these observations had to be worked over; it looked hopeless, and then the good Lord sent me Harlow Shapley."

Shapley's diligence and success studying eclipsing binary stars at Princeton earned him the prized postdoctoral appointment at the Mount Wilson Observatory, in the hills above Pasadena, California—then the home of the world's largest telescope. Shapley, brashly self-confident, was sure "that I could do something significant at Mount Wilson if the people there gave me a chance. . . . My desire, almost from the first, was to get distances."

Distances—how far away the various objects in the heavens were from our vantage point on the earth—seem an obvious question for the astronomer. In 1914, when Shapley came to Mount Wilson, there were few convincing answers. From our perspective, in an era of powerful telescopes on earth and in space, and after a remarkable revolution in the sciences of astronomy and astrophysics, it is astonishing to realize how limited man's understanding of cosmology was earlier in our own century. When Shapley arrived at Mount Wilson astronomers could pinpoint the location of objects within a few arc seconds,* but

---

* The astronomer's measurements of angular distances are like the latitude and longitude used for earthbound navigation. The 360 degrees of a circle are each divided into sixty arc minutes, and each arc minute is divided into sixty arc seconds. The arc second is 1/1,296,000th of a circle.

they had few tools or techniques to determine the distance to the objects they saw in the night sky. Kepler and Newton had provided the mathematics to calculate the orbits of planets, and refined observations made it possible to calculate the distance to the planets with remarkable precision. But even the most sophisticated observatory equipment presented objects beyond our solar system to the astronomer as they appeared to the casual observer: like pinpoints of light on the inside of a great black sphere overhead—as if they were all at the same infinite distance away. Without a method of determining distances, the myriad objects the astronomer could see or photograph in his or her telescope were effectively a two-dimensional frieze.

The methods we use to measure distance on earth are useless for astronomical distances. We obviously can't use a tape measure or yardstick. We can't scale the size of a familiar object the way a hiker estimates the distance across a valley by comparing the apparent size of known objects like a fellow hiker, because stars appear as pinpoints of light in even the most powerful telescopes. Triangulation—calculating distance to a remote object by measuring angles to the object from two widely separated points—is an inviting technique, but in 1914 no equipment on earth had the resolution to measure the parallax of a star from two points on earth. Even the longest baseline available to an earth-born observer—the span of the earth's orbit around the sun—is tiny compared to the distance of the closest stars. Measurements taken six months apart show a parallax shift of the star against the background of other stars only for the closest stars.

Many astronomers reluctantly accepted the limitations of the available technology. The energy of astronomers went into the laborious and unrewarding task of cataloging data: measuring positions, spectra, and apparent brightness of stars. Columns of numbers accumulated at observatories; generations of women scribes tested their vision on the tables of copperplate numbers. The data would all, *someday*, be invaluable, the secrets to understanding the most basic questions of cosmology—as soon as someone figured out how to use it.

Shapley chose the problem of distances precisely because it was a bold enough problem to make a mark in the world of astronomy. To his good fortune, just about the time he came to Mount Wilson, there was an unexpected breakthrough in the techniques of astronomy from what many in the world of early-twentieth-century science would have thought the least likely source—a woman.

At Harvard the "computers" who worked long days and nights calculating and tabulating observational data, were women, hired by Edward Pickering, the longtime director of the Harvard College Observatory, for twenty-five to thirty-five cents per hour. They worked with quill pens and black ink, writing long lists of figures in neat script, without corrections. Pickering had turned to women out of exasperation with a male assistant. Declaring that even his maid could do a better job, he hired Williammina P. Flemming, a twenty-four-year-old Scottish immigrant, to assist him. She stayed for a total of thirty years and before long was in charge of an entire staff of women assistants. Flemming, a divorced mother, supported herself on the meager wages Pickering paid. Other computers were college graduates interested in science. The prevailing social notion of "separate spheres" for men and women left them no room in the observatories and laboratories.

Henrietta Swan Leavitt came to the Harvard College Observatory from the Society for the Intercollegiate Instruction of Women, the forerunner of Radcliffe College, in 1895. Pickering stacked glass photographic plates from Harvard's Southern Station observatory in Peru in front of her and told her to look for variable stars, stars that cycled in brightness.

It was tedious work. Star by star Leavitt compared glass photographic plates of the same area of sky until she found a speck that was darker or fainter than it had been on a plate taken earlier or later. With enough plates of the same area of sky, and records of when the plates were exposed, she could measure the period of the variable stars—how long it took to cycle from maximum to minimum brightness. After years of laborious measurements Leavitt had cataloged more than 2,400 Cepheid variable stars, named after the constellation Cepheus, where they were first discovered.

Not content just to catalog the data, Leavitt searched for a correlation between the intrinsic brightness of the Cepheid variables and the period of their cycle. In 1908 she tried plotting the logarithm of the period of variable stars in the Small Magellanic Cloud, and hence at the same distance, against their apparent brightness. The data on her graph fell on a straight line: Cepheids a thousand times as luminous as our sun completed their bright-dim-bright cycle in three days; Cepheids ten thousand times as luminous as our sun took thirty days to complete their cycle. By 1912 she had graphed enough data to publish an article, "Periods of Twenty-five Variable Stars in the Small Magellanic Cloud," in the Harvard College Observatory *Circular*. After the article appeared Pickering ordered her not to pursue the subject further. His attitude mirrored what was then a widespread viewpoint: A lady of science's place was in the back room, writing columns of numbers, not in the observatory or the scientific journals.

Another piece of the puzzle fell into place when the Danish astronomer Ejnar Hertzsprung used studies of proper motions (a form of triangulation using the long baseline of the sun's motion through space over a period of decades) to determine the average magnitude of a typical Cepheid variable in the Milky Way. Independently, Henry Norris Russell determined the absolute magnitudes of thirteen Cepheids in the Milky Way. Shapley was still Russell's graduate student at the time.

When Shapley saw Hertzsprung's article in *Astronomische Nachrichten*, he realized that he might have his cosmic ruler. The Cepheids Leavitt had charted were all in the same star group, the Small Magellanic Cloud, which is visible only in the Southern Hemisphere. Shapley observed that if the stars on her charts were at roughly the same distance from earth, the differences in apparent brightness she had measured must indicate differences in the intrinsic luminosity of the stars. With the reasonable assumption that variable stars everywhere in the universe were the same, he took his extrapolation one step further: If all variables of comparable period, anywhere in the heavens, had the same intrinsic luminosity, then by measuring the apparent brightness of a variable star and comparing it to stars of comparable period on Leavitt's chart, he could calculate the relative distance of the new star.

The mathematics, and the concept that "faintness means farness," were invitingly simple. The apparent brightness of an object is inversely proportional to the square of its distance from the observer: When an object is twice as far away, it appears one-fourth as bright. If Shapley found a Cepheid variable of the same period as one of the sample on Henrietta Leavitt's charts, with an apparent luminosity one-fourth that of her sample star, his star was twice as far away as hers. The question was where to look for these "yardstick" stars. Where could he find Cepheid variables far enough away to help him construct a theoretical model of the heavens?

Before he had gone off to his new job at Mount Wilson, Shapley had taken a trip to Yale, Brown, and Harvard. At Harvard, Solon I. Bailey, interim director of the Harvard College Observatory, told Shapley, "I have been wanting to ask you to do something. We hear that you are going to Mount Wilson. When you get there, why don't you use the big telescope to make measures of stars in globular clusters?"

It was an intriguing suggestion. Even in a powerful telescope, globular clusters are faint, mysterious objects. On a photographic plate the centers of these clusters of thousands or millions of stars look like a stellar gridlock, as if the great mass of stars at the heart of the cluster were in physical contact with one another. Depending on the equipment used, the clarity of the atmosphere on a given night, and the observer's mood, clusters sometimes seem as though they can be resolved into agglomerations of thousands of individual stars; other times the image, especially of faint clusters, is too nebulous to appear much more than a luminous blob.

Astronomers had cataloged globular clusters for centuries, wondering what secrets they held. How far away were they? Were they part of our local galaxy, the Milky Way? The difference for Shapley was that he was going to study these clusters with the largest working telescope in the world, the great sixty-inch reflector on Mount Wilson, which had been finished in 1908.

As a junior man Shapley's principal assignment at Mount Wilson was to assist a more senior astronomer with traditional studies of star colors and magnitudes. On the nights when he was allowed to do his own research, Shapley used the big telescope to search for Cepheid variable stars in globular clusters. The nights were cold, the controls on the telescope were balky, and the long exposures put a premium on the astronomer's skill of bladder control. Shapley spent so many hours examining plates that he discovered a new asteroid, which he named after his newborn daughter Mildred.

Even on the sixty-inch telescope, the most modern research instrument, astronomy was a physically demanding science. Astronomers work when the objects they need are "up," the weather is clear, and the seeing is good—three conditions that rarely come together on a balmy summer evening. Metal can fatigue and crack from the cold, lubricants and even the ink used to write notes in the logbook freeze, and human efficiency falls.

The painstaking work paid off as Shapley began to identify Cepheid variable stars in the globular clusters. Of the approximately one hundred globular clusters visible in the Northern Hemisphere, he identified Cepheid variables in a dozen. By comparing the brightness and period of these stars to stars of comparable period on Henrietta Leavitt's graph, he could extrapolate the relative distances to the globular clusters.

Each step in Shapley's research required a leap of extrapolation. The variable stars on Henrietta Leavitt's graphs cycled between bright and dim in a few days; the stars Shapley was studying took weeks to complete their cycle. There was no evidence to indicate that all variable stars he and she measured were *not* the same sort of star, so Shapley extrapolated the relationship Leavitt had plotted to include his own observations, even though the much longer periods of variation he measured put the intrinsic luminosities of his stars off the end of her chart. Leaps of reason and data are a necessity of astronomy. The paucity of available information forces astronomers to assumptions that might seem outrageously bold in sciences with a surfeit of experimental data.

He derived distances to a dozen globular clusters—an important first. In July of Shapley's first year at Mount Wilson, he showed his initial results to J. C. Kapteyn, perhaps the best known cosmologist of his day. Shapley's distances were so enormous, compared to the scale of contemporary models of the universe, that Kapteyn suggested Shapley recheck his observations and calculations.

Shapley persisted. In January 1918 he reported a breakthrough to Arthur Eddington, announcing that the consequences of his studies extended not just to clusters but to the entire galactic system: "With startling suddenness and definiteness, they seem to have elucidated the whole sidereal structure." Buoyed by his findings, Shapley began publishing his results. The articles followed each other so quickly that by the 1918–19 issues of the *Mount Wilson Contributions*, fully half the articles were by Shapley.

The results of these first efforts were so satisfying that Shapley went a step further. In each nearby globular cluster, he isolated the most luminous stars, stars that the astronomers identified as red giants and supergiants, and compared their apparent brightness to the brightness of the Cepheid variable stars. When he had a large enough sample to feel confident of his calibration of the absolute magnitude of the giant stars, he began using the giant stars as yardsticks to estimate the distance to faint, distant globular clusters where he could not resolve Cepheid variable stars but could resolve red giant and supergiant stars.

Shapley's total sample for these extrapolations was only a few stars. But even a small sample was enough to begin measuring the universe, if he could assume that stars of similar spectra*—stars that reflected or absorbed the same colors of light—were in fact similar, whether they were relatively close to us or in some distant corner of the observable universe. Shapley also assumed that the relationship of brightness to period that Henrietta Leavitt had discovered in the Cepheid variable stars she had studied in the Magellanic Clouds would apply equally to variable stars throughout the observable universe. Finally, he assumed that space was essentially empty, that there was no absorption of light from distant objects by interstellar dust or gases. Skeptical critics shook their heads as they tallied up the assumptions, but Shapley's results were too exciting to ignore.

Eager to have everyone understand his arguments, Shapley assumed little from his audience at the symposium. He did not define a light-year until the seventh page of his nineteen-page script, and he devoted the last three pages to descriptions of an intensifier he had developed to photograph faint stars, a subject that had little bearing on his argument, though it might impress those members of the audience who were associated with the Visiting Committee of the Harvard College Observatory, where he was a candidate for the directorship.

---

* The spectrograph, a glorious discovery of nineteenth-century German science, is the primary research tool of the astronomer. Using a fine grating or prism, the astronomer can spread the light of the distant object into a spectrum, like a rainbow. Each chemical element, when heated, emits light of specific colors (wavelengths). The spectra that the astronomer records from distant objects include bright lines and black strips at various wavelengths, telltale signatures of the emission or absorption of specific wavelengths of light. By comparing the spectrum of a distant object with reference spectra, the astronomer can determine the presence or absence of chemical elements in the distant body. Spectra also enable the astronomer to classify types of stars and other celestial objects by color and, by measuring shifts from the expected position of various lines on the spectra, to determine the motion of distant objects.

Shapley's total evidence was meager—he had had only a few years to work sporadically on the big sixty-inch reflector at Mount Wilson, and had only observed for three months on the new one-hundred-inch reflector that had just come into service—but if you accepted his initial assumptions, the subtle arguments cascaded one upon the next with compelling logic, and with the elegant simplicity that so often characterizes good science. In Shapley's model the globular clusters outline the extent of our galaxy, the Milky Way, and it is a big one. The diameter of Shapley's universe was 300,000 light-years, or 19×10^{17} miles (19 with 17 zeros after it)—approximately ten times as large as the cosmological models that prevailed when he began his work.

His huge new universe made man seem very small indeed. But the consequences of Shapley's argument went deeper. In his observations he found that the clusters were not scattered evenly around the observable heavens, as one would expect if the sun were at the center of the universe. Instead, the clusters appeared to be concentrated in the area of the constellation Sagittarius. To Shapley the implications were obvious: "One consequence of accepting the theory that clusters outline the form and extent of the galactic system, is that the sun is found to be very distant from the middle of the galaxy. It appears that we are not far from the center of a large local cluster or cloud, but that cloud is at least 50,000 light-years from the galactic center."

In other words the center of our galaxy, the Milky Way, was tens of thousands of light years away, in the direction of Sagittarius. Copernicus and then Galileo had demonstrated that the sun, not the earth, was the center of our solar system. Shapley had taken Copernicus one step further to argue that our sun was not the center of the universe, but only a perfectly ordinary, second-class star, somewhere out toward the edge of the galaxy. His new universe made man seem very small indeed. If his evidence and calculations were correct, Shapley had revolutionized astronomy.

Not everyone was convinced he was right.

Heber Curtis was old enough to be Shapley's father. For his official portraits he posed next to the telescopes at the Lick Observatory in a coat and tie, the usual observing attire in an age when science could still be a relaxed and gentlemanly pursuit. His trimmed mustache and stiff collar seemed appropriate for a man who had studied classical languages as an undergraduate at the University of Michigan and who was still as comfortable reading a Greek or Latin text as the daily newspaper. Curtis was also a classicist in his astronomy, a staunch believer in the tradition of incremental observation. He was a patient and careful observer who had put in his time on balky telescopes on cold nights. He had been trained in the old tradition, his astronomy studies devoted as much to practical optics as to the newer physics of particles and waves.

By temperament Curtis was a skeptic. He became the chief critic of Shapley's articles, answering them with a vehemence that reflected a reaction to the perceived arrogance of an "upstart" going off half-cocked, and the rivalry between the Lick and Mount Wilson Observatories, as much as intellectual disagreement. To Curtis, Shapley's theories weren't necessarily wrong; he awarded the Scottish verdict: *Not proved*. Curtis had made the long trip to Washington to make sure those theories didn't get more credit than they deserved.

Following Shapley to the podium, Curtis was in the position of a reviewer. He had no new cosmology of his own to present. His job was to challenge Shapley's thesis by questioning the logic and the evidence, to convince the audience that while the arguments might be intriguing, the evidence was too thin, the hypothesis stretched too far. Curtis argued that Shapley's sample of only eleven stars was too small to determine the average brightness of stars in clusters, that the dispersion of magnitudes of the stars in the sample made any calculation of an average from this data suspect, and that the techniques Shapley had used to "smooth" his data before plotting it were also suspect. Curtis's own plot of Shapley's data came out not a smooth curve but an essentially meaningless scatter plot. "It would seem," he said, "that available observational data lend little support to the fact of a period-luminosity relation among galactic Cepheids."

Curtis also attacked Shapley's effort to use other stars as distance guides. Citing his own studies, Curtis showed that the average magnitude of stars in the neighborhood of the sun is *less* than the brightness of the sun. Since there is no evidence that giant stars predominate in the clusters, he argued, if we accept the proposition of uniformity throughout the universe, then the average stars in clusters must also be dwarfs, smaller than the sun. And if the stars in Shapley's sample were dwarfs rather than giants, they could not be as far away as Shapley calculated. By Curtis's reasoning the distance to the clusters Shapley had observed, and the diameter of our galaxy, were about one-tenth those determined by Shapley, or the same modest dimensions that had prevailed among astronomers before young Shapley came along.

Curtis was a deft speaker. He made his presentation without notes, after Shapley had read his long report. But he was in the unenviable position of having to debunk exciting, potentially pathbreaking work, and he was speaking not to astronomers who might share his respect for details or his skepticism about Shapley's data, but to generalists. Curtis's dry arguments weren't the pinprick he needed to burst Shapley's balloon.

As Curtis reached the conclusion of his remarks, it seemed almost as though he were abandoning the subject of the debate. He turned to his own studies of the spiral nebulae, mysterious wispy structures, like pinwheels in the sky, the brightest of which are barely visible through binoculars or a small telescope. Little was known about the spirals.

They looked so unlike any other class of celestial object that some believed they were entire galaxies, separate "island universes," comparable in scale to our own Milky Way. In proposing the symposium for the annual meeting of the National Academy, George Hale had originally suggested "island universes" as a topic.

Less than a century before, Lord Rosse in Ireland had used a huge telescope with a seventy-two-inch mirror to study and sketch spiral nebulae. The size of the telescope, and the long hours Lord Rosse put into his observations, had given credence to his claim that the spiral nebulae could be resolved into individual stars. If there were individual stars in these spirals, they might be "island universes," separate from but perhaps equal in scale to the Milky Way. The existence of other galaxies outside the realm of our own but as large as the Milky Way, would raise havoc with Shapley's scale for the galaxy.

In 1898 James E. Keeler had begun a survey of spiral nebulae at the Lick Observatory. Although his telescope, the Crossley reflector, had a mirror just half the diameter of the one Lord Rosse had used, the carefully figured glass mirror provided higher resolution than the speculum (polished metal) mirror in Lord Rosse's telescope. Keeler also had the advantage of using photographic plates to record images of the spirals. A photographic emulsion can accumulate faint light over a long exposure, building up an image over a period of hours from an object that might be invisible or barely visible to the human eye. A few of the nebulae Keeler photographed, like M31 in Andromeda and M33 in Triangulum, appeared incredibly large in plates taken on the Crossley. Unless they were fundamentally different from all other observable spirals— and there was no reason to believe they were—the tiny apparent size of the thousands of other spirals observed with the reflector suggested that they must be at great distances.

Heber Curtis, who took over Keeler's survey of the spirals, had concluded that the spiral nebulae were fundamentally different from every other form of celestial object:

> Grouped about the poles of our galaxy, they appear to abhor the regions of greatest star density. They seem clearly a class apart. *Never* found in our Milky Way, there is no other class of celestial objects with their distinctive characteristics of form, distribution, and velocity in space. . . . The evidence at present available points strongly to the conclusion that the spirals are individual galaxies, or island universes comparable with our own galaxy in dimensions and in number of component units.

If Curtis was right—if the spirals were indeed individual galaxies, comparable in scale to our own Milky Way—then Shapley's model, with the globular clusters marking the edge of the Milky Way at distances on the order of one hundred thousand light-years, would have placed the spiral nebulae, as separate galaxies, at what every astronomer in 1921 would have argued were impossible distances.

In his concluding remarks Curtis relaxed his tone from the language of his formal presentation. After presenting a brief summary of the evidence on spirals, he turned to Shapley. Where, he asked, do the spirals fit in your scheme? Are they part of this enormous grand galaxy you've drawn in your model? Or are they, as the evidence would seem to indicate, separate "island universes," comparable to our own galaxy, and at distances far beyond the limits of our own galaxy?

Curtis's questions caught Shapley unprepared. Although Shapley had read the literature about spirals in the journals, he wasn't ready for the give-and-take of a spontaneous debate on the subject. Less than a month before, he had written to his mentor, Henry Russell, that he would not say much about spirals because "I have neither time nor data nor very good argument." In his talk he had kept to his word, never mentioning the spiral nebulae or the island universe theory. But on the floor of the symposium, before the distinguished audience, Shapley could not wave Curtis's questions aside as irrelevant.

And, although Shapley hadn't studied the spirals himself, a colleague of his, Adrien van Maanen, had been working on spirals at Mount Wilson since 1912. Van Maanen was charming, a bachelor, and as Shapley put it, "society." He and Shapley had become good friends.

Van Maanen had been measuring the rotation of individual spiral nebulae by comparing photographs of the nebulae taken five years apart. He used an instrument called a stereocomparator, which compares two plates by using a movable mirror to blink rapidly from one plate to the other. If one of the thousands of images on the two plates is in a different relative position, the human eye will catch it when the mirror flips. This was the same technique Clyde Tombaugh would use at the Lowell Observatory in Arizona to search for the still-undiscovered planet beyond Neptune. Van Maanen spent so much time studying the plates of spirals on the instrument at Mount Wilson that a stern warning note was posted: DO NOT USE THIS STEREOCOMPARATOR WITHOUT CONSULTING A. VAN MAANEN.

His patience seemed to pay off. Van Maanen reported measurements of large rotations in the spiral nebulae, so large that if the spiral nebulae were at a distance great enough to be outside the Milky Way—at least a Milky Way of the dimensions Shapley proposed—the spiral nebulae would have been spinning faster than the speed of light, a proposition that anyone at the Smithsonian that night understood to be absurd, even without amplification by Professor Einstein from his seat at the head table. For Shapley, van Maanen's evidence was persuasive. He had written to his friend, "Congratulations on the nebulous results. Between us we have put a crimp in the island universes, it seems,—you by bringing the spirals in and I by pushing the galaxy out. We are indeed clever, we are. It is certainly nice of those nebulae to have measurable motions."

"I don't know what they are," Shapley answered to Curtis's question about the spiral nebulae, "But according to certain evidences they are not outside [our galaxy]." As proof Shapley first cited a supernova in the Andromeda Nebula, which had been discovered on a plate taken at the Lick Observatory in 1885. If the Andromeda spiral were outside the Milky Way, he argued, that single exploding star had equaled the light of millions of stars—an idea he presented as absurd. Shapley then cited van Maanen's data in support of his own view of the size of the universe.

The respected Henry Norris Russell stood up to support Shapley's position. But Russell's support was too little, too late. No one else had ever produced measurements comparable to van Maanen's. The rules of science are unforgiving: Results that cannot be reproduced are suspect.

Heber Curtis took the podium again, suddenly relaxed, ready for repartée. His question to Shapley had subtly but significantly shifted the subject of the evening. Without van Maanen's data, Curtis pointed out, Shapley's model of the universe collapsed on the question of the spiral nebulae. Of van Maanen's data he said, radiating confidence and pausing for effect: "There are some observations that are not worth a damn, and others that are not worth a damn. In my opinion, two damns are no better than one damn."

The room burst into laughter. Curtis didn't need to say more as the guffaws of the audience erased all that had gone before. The newspaper reporters, most of whom hadn't really done their homework or followed the details of the presentations that night, treated the laughter as a verdict. In the popular press, the great symposium on the scale of the universe, the biggest science story of the decade, was decided by a one-liner.

If the newspapers awarded Curtis a knockout, the astronomers in the audience remained divided. Some were content to ignore Shapley's findings, either because they had been persuaded by Curtis's arguments or because the small sample of data from a newcomer wasn't enough to rescale the universe. Others found Shapley's elegant method and clever use of evidence too compelling to ignore.* Predictably, the participants split their decision. Curtis thought that his approach had

---

* Shapley had made some fundamental mistakes. He assumed that Cepheid variables in the globular clusters did not differ from those in the Milky Way. In fact, the Cepheids in the globular clusters belong to a different stellar population and are approximately two magnitudes fainter at a given period than classical (Population I) Cepheids. At the same time Shapley underestimated the

probably been too technical for the audience, but that overall, "the debate went off fine in Washington, and I have been assured that I came out considerably in front." Shapley was more circumspect in his claims: "I think I won the debate from the standpoint of the assigned subject matter."

Some were disappointed that the "scientific debate of the century," instead of revealing definitive new truths about the universe, had turned out a draw. But for one man at the Smithsonian that evening, that very inconclusiveness was the symposium's most important outcome. George Hale—director of the Mount Wilson Observatory, an officer of the National Academy of Sciences, a distinguished solar astronomer, and the promoter of the original topic for the symposium—diplomatically refused to take a public position on the big questions of the evening. When pressed, Hale answered in statesmanlike fashion that the evidence was intriguing but not yet persuasive.

The important word for George Hale was *evidence*. As an experienced astronomer and the director of a major research facility, he knew that the frontiers of science would always have vague borders. The issues at the symposium had been argued among astronomers for as long as men had gazed at the sky. What made it different now was that Harlow Shapley and Heber Curtis were tantalizingly close to being able to *answer* those questions. Astronomy was at the dawn of one of those breathtakingly exciting moments in the history of a science when scientists were ready to explore a whole new level of inquiry.

As the two men presented data from the most powerful telescopes in the world, George Hale thought about astronomers extending their observations and their cosmologies even further, to the very edge of the universe, to the beginning of time.

Hale was no novice with telescopes. He had already shepherded through construction three great telescopes, each larger than any before it. The next step, in his mind, was not just an increment over those machines, but a leap in technology and design, an instrument orders of magnitude larger and finer than any that had ever been built. He had enough experience with telescopes to know that to design and

---

magnitude of the Milky Way Cepheids because he did not make provision for the dimming of the apparent brightness of stars due to absorption of light by interstellar material. With striking luck the two mistakes cancelled each other out, and his estimates of the size of the Milky Way, and of the position of the sun within the Milky Way, were not far off.

He had been flat wrong about the spiral nebulae. Looking back years later, Shapley wrote, "I consider this a blunder of mine because I faithfully went along with my friend Van Maanen and he was wrong on the proper motions of galaxies—that is, their cross motions. . . . I stood by Van Maanen." Shapley had made a classic scientist's mistake: citing the work of a colleague and friend on trust.

build a machine on that scale and to those specifications would require an unprecedented national engineering and scientific effort: the coordination of the talents and efforts of hundreds of scientists, engineers, designers, artists, and craftsmen all over the country; the cooperation of the largest corporations, universities, and research institutions in the land; an appropriation of funds larger than any that had ever been made in support of a single scientific instrument; the extension of optics, metallurgy, control systems, and large-machine-construction technology to limits few had ever imagined possible; and a coordination of facilities and individuals that had never been attempted.

Anyplace but the America of the 1920s, it would have been an inconceivable project. The war that had humbled and exhausted Europe had been one more challenge for the United States—a challenge, like the Panama Canal or the Brooklyn Bridge, that could be solved with American resources, might, and spirit. While Europe licked its wounds, exhausted, enveloped in still-unsolved questions of diplomacy and hegemony, Americans talked of damming the great rivers of the West, bridging the Golden Gate at the entrance to San Francisco Harbor, building skyscrapers and superliners that would dwarf the achievements of an earlier era.

Even in that optimistic America of the 1920s, the machine Hale had in mind would press the limits of technology, stretching the confidence of a cocky nation perilously close to hubris. But science is compelling, the promise of timeless answers irresistible. And George Hale, as the secretary of the National Academy of Sciences had learned, was not a man to take no for an answer.

3

The Worrier

Even as a boy, growing up in Chicago, George Hale worried too much. The family firm, Hale Elevators, had profited handsomely from the construction boom after the Great Chicago Fire, and the Hales lived well. A few split branches in the family tree, including the divorce of George's maternal grandparents, caused a stir in their time, but quiet money, a reputation for generosity, and a splendid town house gave William Hale and his family a secure position in Chicago society.

George was a sickly child. The doctors could find no explanation for his chronic stomach trouble, backaches, and fainting spells, so George's mother, a semi-invalid who confined herself to dark rooms because of paralyzing headaches, prescribed her own medicine, a regimen of reading—Homer, *Robinson Crusoe*, *Don Quixote*, and Grimms' *Fairy Tales*. George acquired, or inherited, both her love of literature and her unpredictable, terrifying headaches.

When his mother retired to her dark room, George liked to experiment with tools and instruments. His brother and sister would discover him lost in a world of his own. When they spoke, he didn't answer. If they made a loud noise, he seemed not to hear it. To those who had never seen the concentration of a scientist absorbed in his work, George's behavior was mystifying. He didn't seem an ordinary boy. On a trip to London, George's friend bought magic tricks; George spent his allowance on an expensive spectroscope.

Each new passion became an obsession: When he became interested in microscopy, his father bought him a fine Beck binocular microscope. For months microscopy excluded all other interests, until it was replaced by the next passion. George's father worried about his son. What would become of a boy with no focus, no plans or ambition, who preferred puttering alone to sports and friends, whose dabbling in one science after another seemed to lead nowhere?

Then George discovered astronomy. As far as anyone in the family could remember, it started when he read Jules Verne's *From the Earth to the Moon.* Soon nothing else mattered. He devoured books and articles about astronomy, visited observatories, interviewed astronomers and telescope makers, planned his own astronomy journal, and fired off a flurry of letters to general interest magazines, offering to write articles. He was thirteen years old.

Before his fourteenth birthday the enthusiastic young astronomer got up the nerve to call on Sherburne Wesley Burnham, a quiet Chicago court reporter by day and obsessed amateur astronomer by night. Burnham let young Hale assist him in his painstaking observations of double stars. One evening he told Hale about a secondhand four-inch telescope, built by the well-known optician Alvan Clark, that was available at the right price. It was the first of many telescopes that George decided he must have.

This time George's father refused to indulge what seemed yet another whim. But George argued and pleaded until William Hale ultimately relented and surprised George with the Clark telescope in time for George to observe a transit of Venus on December 6, 1882. George no sooner had the telescope when he started planning to equip it with a spectrograph and other ancillary devices. He dreamed of bigger and better telescopes and his own fully equipped observatory.

He dutifully trudged off to MIT for college study, but found the courses in his major of physics dull. Arthur Noyes's course on qualitative chemistry hinted at the excitement of scientific research, but George was marking time in Cambridge. The only activity he looked forward to each week was on Saturdays, when the famed Edward Pickering allowed the eager young man from MIT to work for ten hours as a volunteer at the Harvard College Observatory. Astronomy, George had already decided, would be his future.

When George was close to graduating, William Hale, reluctant to see his son plunge headlong into a future with so few options, tried to divert him with the enticement of a large block of stock and a directorship in a new building the elder Hale was putting up in Chicago. George countered with his own proposal for his future: He would continue his research in solar astronomy and marry his childhood sweetheart Evelina Conklin, whom he had met on summer vacations at his grandmother's home in Madison, Connecticut. In the end George got his way, but the strain of what he called the "interview" with his father left George prostrated with a splitting headache and nervous indigestion—the same symptoms that had invalided his mother for as long as he could remember. For George they were a harbinger of things to come.

Poor Evelina, who had known George only on summer holidays, had no idea what was in store for her. They got married two days after George graduated from MIT. For a honeymoon they took the traditional trip to Niagara Falls, but only as a stop on a trip across the country to California, where he planned to visit the Lick Observatory.

When they paused in Chicago on their way to California, George's mother wrote of her newlywed son: "I wish he cared a *little* more for Society, but now he cannot be induced to make calls or do anything in that line that is not *absolutely* necessary. He is as absorbed in his studies as his father in business—otherwise a model son."

In 1875 James Lick, a wealthy Northern California businessman, asked a friend to witness the signing of his will. Lick's proposed bequests included a marble pyramid larger than that at Cheops, which he wanted erected on the shore of San Francisco Bay; giant statues of his father, his mother, and himself on North Beach; a home for old ladies; and on Market Street in San Francisco, a telescope larger and more powerful than any other in the world. His draft will included funds to endow each bequest, from five hundred thousand dollars for the telescope to three thousand dollars to provide for a previously unacknowledged illegitimate son.

As far as anyone could recall, Lick had never seen a major astronomical telescope. There is no record of where or how he stumbled on the notion of the glory of astronomical discovery. But the explanation for his bequest may not be difficult.

James Lick was one of many who had prospered in the real estate booms that followed the California gold rush. "O this California," one transplanted New Englander wrote home during that wild era, "what a madness there is about it." There were easy fortunes to be made. Businessmen and real estate moguls gambled on the next boom; those who played their cards right emerged with sudden and tremendous wealth. Blasted in the press as greedy parvenues and robber barons and excluded from both eastern and transplanted western society as nouveaux riches, these wealthy Californians were left to hope that by leaving suitable endowments behind, they would be remembered by future generations for something other than their moneymaking. What better symbol of open-mindedness, scientific dedication, and vision than a large telescope, permanently named after its donor? Built on a massive foundation of granite, under a majestic dome, literally and symbolically reaching out to the heavens, a telescope would be a glorious memorial for a man eager to transform a quick fortune into an eternal monument.

Lick's friend, David Jackson Staples, tried to persuade him that the traffic on Market Street in San Francisco would disturb an instrument as delicate as an astronomical telescope, and that the location would certainly be too foggy to be useful. Lick was reluctant to abandon the location where people would be able to see his great telescope, so a battery of lawyers was called in to negotiate with Lick. They were as skillful at getting the money as Lick had been at making it. When they finished, Lick had bequeathed seven hundred thousand dollars to the Regents of the University of California for "a powerful telescope, superior to and more powerful than any telescope ever yet made." There were few constraints on the bequest, and the grant was generous enough to allow the planners of the new telescope free rein to design any kind of instrument they wanted.

Their first decision was whether to use a lens or a mirror to gather the light. Galileo's telescope had been a refractor, using lenses to bend, or refract, the light from distant objects. Newton, a century later, used a reflector, with a parabolic mirror to gather and focus the light. Each had advantages and problems, and, as with so many technologies, over the years the pendulum of astronomical preference swung back and forth.

By the late nineteenth century, refractors—the familiar telescope built with an objective lens at one end of a long tube and an eyepiece at the other—had been the telescope of choice for many years, especially in the United States. The disaster of a huge reflector attempted in Australia in the 1850s had stopped reflector building for decades, and there were several American opticians who had built fine refractors, but few who had ground large mirrors. A refractor, which excels at high-magnification study of planets and nearby objects, and at photometric and statistical measurements, also fit into the observation programs then prevailing at most observatories. The Lick Trust decided on the safe course of a refractor and ordered two thirty-six-inch-diameter glass disks, each larger than any piece of glass ever cast, from Feil & Cie. of Paris. France had been for many years the only country with glass technology and experience adequate to cast large glass disks of the required clarity and consistency.

The objective lens of a refractor is normally built of two elements, one of flint glass, the other of crown glass, figured and sandwiched together to function as a single lens. The combination of two different formulations of glass is an effort to counter or correct chromatic aberration, the varying refraction or bending of light of different colors. Without correction of the aberration, it would be impossible to focus a sharp image of a star.

The flint-glass element for the Lick telescope was cast successfully, but the initial effort at a crown-glass blank cracked when it was packed for shipment. The Feil brothers worked for two years to cast a second satisfactory crown blank and ultimately went into bankruptcy before the elder Feil came back, took charge, and in 1885 shipped a satisfactory disk to be figured for the Lick telescope.

The mounting of the big telescope was entrusted to the firm of Warner & Swasey of Cleveland, Ohio. Worcester Reed Warner and Ambrose Swasey, both born on New England farms, met as apprentices and worked at Pratt and Whitney in Hartford, Connecticut, before starting their own business designing and manufacturing fine machine tools. Swasey's skills were designing and building machine tools; Warner was a production line expert and amateur telescope

maker. Their first effort at a telescope mount, for Beloit College in
Wisconsin, proved so successful that a Warner & Swasey mount
became the mark of a fine telescope.

They were the high bidders on the mounting for the Lick telescope, but with an obviously superior design that the Lick Trust
accepted. The mounting for a large refractor has to hold and point the
long, heavy tube of the telescope with exacting precision and move it
around a polar axis for east-west motion parallel to the axis of the
earth, and a declination axis for north-south motion, so that the telescope follows the apparent movement of celestial objects as the earth
turns during the night. The only practical design for a refractor was
one that supported the long tube in the middle, which meant that as
the telescope rose and fell in altitude from the horizon to the zenith,
the eyepiece would swing in a large arc up and down. To allow for the
motion the telescope had to be mounted on a tall pedestal, with ladders, a lift, and a movable floor provided to make the working end of
the telescope accessible to observers and for the installation of cameras, spectroscopes, and other devices. To achieve the rigidity required
for astrophotography with a structure this large was a considerable
challenge of machine design and construction.

The third big decision about the telescope was location. This was
in some respects the biggest innovation of the project. In a bold move
the new observatory was sited not on a university campus in a large
city, which had been the norm for telescopes, but on remote Mount
Hamilton, a coastal peak near San Jose, far from the loom of light pollution of the cities and high enough for the night atmosphere above
the observatory to be still and clear. Carrying the components of a
huge, high-precision optical device up the steep paths to the mountaintop proved a greater challenge than anyone had anticipated, and it
took longer to finish the observatory than even the pessimists prophe-
sied. Lick died in 1879 without ever seeing the completed instrument
that would bear his name. His body was interred beneath the pedestal
of the telescope.

The telescope saw first light later that year and was an immediate
success. Astronomers from eastern universities, hearing of the number
of cloud-free nights, the quality of the seeing, and the light-gathering
power of the new telescope, clamored for invitations to the observatory.

Hale heard about the Lick telescope from his mentor Sherburne
Burnham in Chicago, but it was not until his honeymoon trip that he
had a chance to visit the famed site. Access was by a rugged trail, six
hours and 366 hairpin turns by horse-drawn stage from San Jose; the
peak, even with the telescope and housing facilities, was bleak. But the
skies were all that Burnham had described. Evelina was bored in California, eager to leave after a short visit, and frustrated when Hale

extended their stay on the mountain. She didn't realize that George Hale had found his destiny.

Hale declined an offer to stay and work at Lick. He had already planned a career as an independent astronomer, his father had agreed to build him a private observatory in Chicago, and the equipment had been ordered: an excellent twelve-inch refractor from the fine firm of John Brashear, with a mounting from the same Warner & Swasey shop that had built the mounting for the Lick telescope.

When they returned to Chicago, George and Evelina moved in with George's mother, now a recluse in her darkened upstairs room. George spent all his time at his new observatory, leaving Evelina to care for his mother. Evelina proposed that they set up housekeeping on their own, but George's mother wouldn't hear of it. Other people, she said, would think she had thrown them out.

About the time that George's Kenwood Observatory was almost finished, William Harper, the president of the new University of Chicago, was taking advantage of the seemingly limitless backing of Rockefeller money to recruit the finest talent in every field to his faculty. He proposed that Hale and his observatory join the University as the nucleus of an astronomy department. Put off by Harper's aggressiveness, Hale turned down the offer.

Harper could be as stubborn as George Hale. He negotiated with George's father, and after he persuaded the famed physicist Albert Michelson, the first American recipient of a Nobel Prize in physics, to accept the chair of physics at the new school, he used the prestigious appointment as leverage to strike a deal that would give George Hale a year to evaluate a university appointment before his father gave the Kenwood Observatory and $25,000 to the University of Chicago. A clause of the agreement specified that the University would subsequently raise $250,000 for a larger observatory facility. William Hale was a good enough businessman to expect a proper return on his investments, including those he made in his son's career.

George was so absorbed in his work at the observatory that he begged off almost all social engagements and other distractions. It was months before Evelina could drag him away for a vacation at Lake Saranac, New York. Even on holiday George couldn't relax away from his telescope. He finally abandoned the pretense of fly fishing to sneak off to Rochester and a meeting of the American Association for the Advancement of Science. There he overheard Alvan Clark talking about a pair of glass blanks—the largest ever cast—which were in his shop.

Clark explained that shortly after the opening of the Lick Observatory, a group of supporters of the University of Southern California, anxious that Southern California not be outdone by anything in Northern California, had made plans for an even larger telescope. They organized their publicity before their funding and made sure that

Scientific American reported that although the new telescope would be only one-ninth greater in diameter than the Lick telescope, its light-grasp would be one-fourth greater and that "the existence of a large city on the moon would readily be detected by the telescope." They ordered two forty-inch glass blanks from the Paris firm of M. Mantois—there was still no American firm that could pour large glass castings. The blanks were successfully cast and shipped to Alvan Clark, who had built George Hale's first telescope, to be ground and figured. Warner & Swasey got the contract to build a mounting for the new telescope.

Before Clark began grinding the disks, the crash of 1893 popped the real estate bubble in Southern California. The businessmen who had pledged funds for the telescope decided that they had other priorities more important than beating Northern California in a telescope race, and the huge glass disks, unground and unpaid-for, languished in Alvan Clark's Massachusetts shop.

Hale excitedly pulled Clark aside. What, he asked, would it take to figure those disks into the objective lens for a large refractor? Clark gave him a rough estimate, and they discussed details of mounting, housing, and siting a telescope that large. From that day Hale was a man possessed. He had always dreamed of bigger and better instruments. Now he would build the biggest and best telescope in the world. He would site it in a fully equipped observatory, with laboratories right on the premises, with instruments like no others that had ever been built. All he needed, he calculated, was three hundred thousand dollars.

In the 1890s it was a considerable but not impossible sum. George Hale had grown up with wealth, and he knew there were many men in Chicago who could afford it. The Hale name gave him a ready introduction, and he wasn't embarrassed to make his pitch for a telescope. For months he made the rounds of offices and homes of Chicago society, proposing his venture to anyone who would listen, his eyes sparkling with enthusiasm as he described his proposed observatory and what it could accomplish for science. Astronomy, he told anyone willing to listen, was ready for a revolution.

Hale did his best to explain that the old astronomy, men looking through telescopes to sketch what they saw, was exhausted. His proposal was something entirely different, an observatory equipped with the most modern laboratories and facilities, darkrooms, and spectroscopes. But no matter how enthusiastic his pitch, he found no takers. Times had changed, potential donors told him. Money was tough, the climate was wrong, this wasn't the moment. For George Hale it was a good lesson in the vagaries of fund-raising.

Finally, on a tip from a mutual friend, Hale approached the streetcar magnate Charles Tyson Yerkes. Friends called him "Yerkes the Boodler." The Boodler liked the idea of a telescope with his name on it. When Hale promised that the telescope would be the largest in the world, bigger than the one at the famed Lick Observatory, Yerkes liked the idea enough to call in the press. "Here's a million dollars," he was quoted as saying in the *Chicago Tribune*. "If you want more, say so. You shall have all you need if you'll only lick the Lick."

Yerkes' farsightedness and vision—always good terms for the donor of a telescope—dimmed considerably as the time came around to make good on his commitments. When he realized that the observatory would be at a remote site, and that much of the funding would go to laboratories and other facilities that were far less flashy and less likely to attract favorable publicity than the big telescope, Yerkes balked. Hale, relentless, cajoled Yerkes to follow through on his pledges, pressured contractors to get the work done, negotiated with local and university officials, and mediated between the perfectionists who would fiddle with the lenses and machines forever and the astronomers eager to use their new facility. The strain of the project, especially the battles with Yerkes, took their toll. Hale began suffering recurrent headaches, sometimes bad enough to keep him home in bed.

Friends, noticing his nervousness and anxiety, urged him to go easy. The optician John Brashear, who knew Hale from years of dealings on optical equipment, wrote, "You have a big responsibility on your hands . . . the only thing I beg you to look out for, *don't overwork yourself.* . . . delegate all the work you can. Save yourself for that—which you can *do better than anyone can do for you.*"

George Hale was twenty-four years old.

Despite the headaches Hale got the telescope and the observatory built. Yerkes, at what was then a remote site on the shores of Lake Geneva, eighty miles north of Chicago, emerged the most complete observatory in the world. The great forty-inch refractor, with its Clark lenses and Warner & Swasey mounting, was considered a sufficient engineering marvel to be exhibited at the 1892 World Exposition in Chicago.

In spite of routine winter temperatures at the observatory of –20°F, Hale attracted extraordinary optical and astronomical talent to Yerkes. For a period he could boast having the best observers and the best glass grinders in the world together under the domes and in the laboratories. Those who had seen the toll the construction took on George Hale urged him to settle down to the promising career of director of the great observatory. But even before Yerkes was dedicated, George Hale had a new idea.

In his own studies, with solar telescopes, Hale used a spectrograph to study the chemical composition of the sun. By identifying lines in the spectrogram that corresponded to the emission or absorption of particular materials, he could identify the presence of various elements in the sun—almost as accurately as if he had a sample of solar material in a laboratory. Knowing the chemical composition of the sun and the solar atmosphere, astronomers could begin to ask what chemical or atomic processes were at work to create energy and light. What, Hale asked, if the same techniques that he applied to the sun could be applied to distant stars? It would be a whole new field, astrophysics, a discipline devoted to trying to determine what the stars and other celestial bodies outside our solar system were made of and what processes created the enormous energy in them. Once astronomers and physicists made some headway on those questions, they could take on the even grander field of cosmology, which tried to understand how the universe was put together, to discern the size, shape, structure, and origin of the cosmos.

The answers to those questions would demand telescopes far more powerful than even the great Lick and Yerkes instruments. Hale's new dream was a huge new telescope, somewhere out in the clear air of California, where cloudless mountaintop skies provided night after night of good seeing, and where a facility could be all but immune to light pollution from urban illumination. The telescope Hale had in mind would not only be even bigger than the great Yerkes, but it would turn the circle from refractors, like the Lick and Yerkes telescopes, back to reflectors, like the telescope Newton had once used, relying on a mirror instead of a lens to gather and focus the faint light from distant objects.

The technology of refractors was temporarily exhausted. It might have been possible to cast and grind larger glass lenses than the forty-inch-diameter disks in the Yerkes refractor, but the sheer weight and fragility of the enormous glass disks, which can be supported only at the edges, and the engineering of the long tube, which must rigidly hold and point the lenses, had reached their limits. In France a refractor with lenses close to sixty inches in diameter was built and displayed at the Paris Exhibition of 1900, but it was not successful as a telescope. Even if a bigger refractor could be built, a reflector had many arguments in its favor for astrophysics research.

The light of a star or other distant object goes through the objective lens of a refractor. Because light of various colors bends, or refracts, differently as it goes through the glass, a refractor is not achromatic; the lens forms a series of images of different wavelengths. Only some of these effects can be corrected. A reflector, by using the surface of a mirror to focus the light, avoids this problem. The optics of a big reflector are also easier to grind and polish. Instead of four surfaces of glass to grind, figure, and polish, two on each of the elements that are sandwiched to make up an objective lens, a reflector requires only a single optical surface, the face of the primary mirror.

Finally, the physical mounting of a refractor, with its long, rigid tube supporting the objective lens at one end and an eyepiece or instrument at the other, presents difficult engineering problems as the instrument gets larger. The long tube must be balanced on its equatorial mounting, and for the telescope to reach all areas of the sky, the mounting must be on a tall pillar. When the telescope is pointed toward targets of low altitude, the eyepiece is high off the floor, out of reach of the observer or his cameras. At Lick and Yerkes this problem was solved by having the entire floor of the observatory rise and fall around the fixed telescope mounting pillar. An early accident with the moving floor was another hint that refractors were approaching their technical limits.

Because the light can be bounced back up the tube from the primary mirror to a secondary mirror, in effect folding the focal path of the telescope, a reflector can be built with a relatively short tube, short enough in most instances for the telescope to be mounted in a movable fork with its pivot point close to the primary mirror. Without the weight of a heavy lens to support at the far end of the tube, the reflector can use an open tube, resulting in a lighter structure and greatly simplifying the construction of the instrument.

The reflector is also more versatile than a refractor. The eyepiece, or more typically for a large instrument, the cameras or spectrograph, of a reflector can be mounted at one side of the high end of the tube, in what is called the Newtonian position, after Newton's early design. The light from the primary mirror is deflected to the Newtonian focus with a small diagonal mirror suspended inside the telescope. It is also possible to bounce the light from a secondary mirror back through a hole in the center of the main mirror so that cameras and other instruments can be mounted at the base, or supported end of the tube at what is known as the Cassegrain focus. With additional mirrors, the light can be directed to a fixed Coudé position of extreme focal length in a separate temperature and humidity-controlled room. Finally, if the telescope is big enough, the light can be deflected through the hubs of the declination axis, to Nasmyth foci on either side of the telescope. The different foci, each with different focal lengths, add up to increased versatility for the reflector.

When George Hale began thinking about a new telescope, the arguments for a reflector weren't only theoretical. In 1895, the same year that the lenses for the great refractor at Yerkes were finally finished, Edward Crossley of Halifax, England, presented the thirty-six-inch Calver-Common reflector to the Lick Observatory. The new reflector was overshadowed in publicity by the larger telescope at Yerkes. While Yerkes' telescope dominated the press, the mirror of what came to be known as the Crossley reflector was quietly refigured and a new mounting built for photographic work. Keeler, the director of the Lick Observatory, used the Crossley, the first large reflector in the United States, to reveal an immense number of spiral nebulae that had never before been recorded.

The Crossley was an awkward telescope to use, with a stiff mount that required a kick from time to time to get it to behave, but so many spiral nebulae could be photographed, or even seen visually with the telescope, that it raised disturbing cosmological questions. Heber Curtis's continuation of this study provided the background for his contributions to the great debate in Washington.

Hale kept abreast of the work with the Crossley reflector at Lick. The incentive of following up on that work, and a program to extend the detailed spectrographic studies Hale had made of the sun to distant stars, was a compelling agenda for a big reflector. And when the headstrong young director of Yerkes Observatory got an idea in his head, there was no stopping him. Before the forty-inch refractor at the Yerkes Observatory was dedicated, Hale persuaded his father to contribute the funds to have a sixty-inch glass blank—the largest piece of glass the French foundries could mold in a single pour—cast for a giant reflector. William Hale made it clear that his gift was seed money; he would pay for the glass blank and nothing more. It would be up to George to find the money elsewhere to have mirror and other optical surfaces ground, to mount the telescope, and to build an observatory. George accepted the challenge.

The Saint-Gobain glassworks in France successfully poured the disk. It was annealed—heated, then gradually cooled over a period of months to avoid strains in the glass from rapid cooling—and shipped to Yerkes. Although funding to complete the project was nowhere in sight, Hale had a colleague at Yerkes, George W. Ritchey, begin grinding the mirror.

Ritchey had little academic training, but he had built several telescopes as a student at the University of Cincinnati, and in 1888 he set up a laboratory at home in Chicago before coming to work for Hale at the Kenwood Observatory. From Kenwood he moved on with Hale to Yerkes. Among astronomers Ritchey was known for his fierce concentration and what one colleague called "the temperament of an artist and a thousand prima donnas." Ritchey would sometimes spend hours on a single photograph, setting and resetting the focus until it was exactly right, waiting for the perfect seeing conditions, then concentrating so intensely on guiding the fine motions of the telescope that an explosion nearby would not have distracted him. The resulting photograph would be an artistic masterpiece—except when Ritchey, lost in his concentration, neglected to record the date, time, or sky conditions, so that the plate was useless for scientific purposes.

Though hard on colleagues, Ritchey's perfectionism was ideal in the optics laboratory. He was delighted to seal himself off for hours, even days, at a time, allowing no one near his project, ruling over his domain as an absolute tyrant while he patiently ground, then polished the disk.

While Ritchey began grinding the sixty-inch disk at Yerkes to the optical shape the future telescope would require, George Hale pounded the pavements again, making his pitch for funds to build the sixty-inch telescope. In 1901 he persuaded John D. Rockefeller to visit Yerkes Observatory. The usual show for VIPs was to mount an eyepiece on the telescope and point it to familiar objects for the entertainment and enlightenment of the visitor, but clouds covered the sky most of the day and evening of Rockefeller's visit. Instead Hale took Rockefeller to the optical laboratory where Ritchey was working on the sixty-inch mirror. Ritchey showed off the procedures and tests used to figure and test a mirror. In one test Ritchey put his finger on the glass surface for a minute, then demonstrated that the distortion of the surface from the heat of his finger could actually be measured. Rockefeller was fascinated by the sensitivity of the test and asked to see it again. But interest wasn't commitment, and Rockefeller wasn't willing to fund Hale's telescope. Nor, it appeared, was anyone else. Hale ran out of pavement to pound. The days of big telescope bequests seemed to be over.

Then, in 1902, Andrew Carnegie announced his gift of $10 million to establish the Carnegie Institution of Washington, "to encourage investigation, research and discovery in the broadest and most liberal manner, and the application of knowledge to the improvement of mankind." Hale knew one member of the board of directors of the new institution, the legendary Elihu Root, secretary of war in Theodore Roosevelt's cabinet. Root's father, Oren Root, was a mathematician and astronomer at Hamilton College in upstate New York. Hale was convinced that Root had persuaded Carnegie to give the money.

He wasn't entirely wrong. Carnegie's original plan was to donate the money to a quasi-public body that would be under the authority of the president and Congress. President Theodore Roosevelt liked the idea. Nicholas Butler, the president of Columbia University, argued against it. At ten o'clock one night, when their discussions at the White House hit an impasse, Butler suggested that they call Elihu Root, famed for his cool intellect, to solicit his views. Root, openly grumpy about being telephoned so late, came over to the White House, read through the proposal and asked the president, "What damn fool suggested this idea?" With that Carnegie decided to establish an independent body, the Carnegie Institution, to administer his bequest. Root was a charter member of the board and later became vice chairman, then chairman.

Ever thorough, Hale prepared his pitch to the new organization by reading Carnegie's writings. Carnegie had publicly expressed his admiration for Lick's telescope bequest: "If any millionaire be interested in the ennobling study of astronomy—here is an example which could well be followed, for the progress made in astronomical instruments and appliances is so great and continuous that every few years a new telescope might judiciously be given to one of the observatories upon this continent, the last always being the largest and best, and certain to carry further and further the knowledge of the universe and our relation to it here upon earth." To George Hale that sounded like an invitation.

Hale launched an appeal to Carnegie with a barrage of letters, photographs, recommendations from other astronomers, and exhibits of what a big reflector could do compared to even the large refractors at Lick and Yerkes. Edward Pickering, director of the Harvard College Observatory, who had been chosen to chair the Advisory Committee on Astronomy for the new Carnegie Institution, asked Hale to join the committee. Hale was thrilled at the invitation, but even a position as an insider didn't help with his pet project. Like many philanthropists Carnegie preferred to initiate programs, rather than finish programs started by someone else.

The first substantial astronomy grant the Carnegie Institution made was to support J. W. Hussey of the Lick Observatory in an expedition to search for possible sites in the United States, Australia, or New Zealand for "a southern and solar observatory." Hale's own work was in solar astronomy, and when Hussey set off in 1903, Hale followed his reports of the excellent atmospheric conditions in Southern California. One site Hussey visited was a remote mountain rising above the desert roughly halfway between Los Angeles and San Diego. "Nothing prepares one for the surprise of Palomar," Hussey wrote.

> There it stands, a hanging garden above the arid lands. Springs of water burst out of the hillsides and cross the roads in rivulets. The road is through forests that a king might covet—oak and cedar and stately fir. A valley where the cattle stand knee deep in grass has on one side a line of hills as desolate as Nevada; on the other side majestic slopes of pines.

The observing conditions at a station some thirty miles southeast of Palomar were so bad that Hussey didn't bother taking his telescope and measuring equipment to Palomar. The mountain had no regular stage or telephone and could be reached only by a road laid out with a 10 percent grade and some steeper portions; it was too remote for an observatory in 1903. But Hale would long remember the description, and Hussey's raptures of the beauty of the spot and the clarity of the seeing. On the basis of Hussey's reports, the Carnegie Institution chose Wilson's Peak, close to Pasadena, as the site for a solar observatory and, Hale hoped, the sixty-inch telescope.

Even with a site selected for the future observatory, the partly ground sixty-inch mirror languished in the basement optics laboratory at Yerkes, waiting for money to finish figuring the mirror and to build a mounting for a telescope. In 1904 the Carnegie Institution gave Hale $10,000 for Mount Wilson, and even that was only a token grant, barely enough to keep Hale's hopes up. He paid $27,000 out of his own funds to keep the work going. It wasn't until the marine biologist

Alexander Agassiz decided not to accept a $65,000 per year grant for two years of research from Carnegie that Hale got a substantial grant for the sixty-inch telescope.

By then, the grinding of the disk had already gone as far as it could in the optical lab at Yerkes. The funds from the Carnegie Institution let Hale appoint Ritchey director of new optical and mechanical labs that were being constructed in Pasadena, close to what Hale hoped would be the site of the telescope on Mount Wilson. Hale himself moved to Pasadena to be near the project. In California he began a friendship with an attractive young woman named Alicia Mosgrove, who may have become his mistress. Hale was a discreet and private man. Evelina Hale, who later knew Miss Mosgrove, seems not to have known about George Hale's relationship with her.

The optical laboratory in Pasadena was built with what were, for the time, unusual precautions. To keep dust off the grinding surfaces, the walls and ceiling were shellacked, and all air entering the room was filtered. During the final polishing operations the painted cement floor was kept wet, and a canvas screen was suspended over the mirror surface to protect it from airborne dust. Double windows and special heating equipment kept the shop and the disk at a steady temperature.

The sixty-inch glass disk was eight inches thick and weighed a ton. The turntable that held the blank was cushioned with two thicknesses of Brussels carpet, the looped threads serving as a spring mounting. For the first time carborundum was used as the abrasive to grind the glass. Carborundum, which had been invented in 1898 and first made at the Niagara Falls Elstree Works, was six times as effective at cutting as the emery powder it replaced, which reduced the time for rough grinding the shape of the sixty-inch disk from years to months. The quick-cutting abrasive also raised the price of an error. Ritchey's reaction to that possibility was to become even more protective of his new lair than he had been in the optics lab at Yerkes. When the new optical lab was finally in operation, only Ritchey, dressed in a surgeon's cap and gown, was allowed through the door.

Month after month he worked on the mirror, first grinding the disk to a rough spherical shape, then reshaping it to a parabola, and finally polishing the disk to the final optical surface. In spite of the extraordinary precautions, one morning in April 1907, while the mirror was being polished to its final figure, the surface was found covered with scratches. The cause was never discovered, but the scratches were serious enough that the mirror had to be reground, delaying the completion and putting new pressures on George Hale's strained budget for the telescope.

It wasn't just the mirror that raised the ante for the new project. The initial reports on the Mount Wilson site had been favorable. Hale hiked up the mountain himself to test the quality of the seeing, even climbed trees on top of the mountain with a small solar telescope, hoping the slight additional elevation would escape the effects of ground heat on the optics of the telescope. He scribbled notes that might be thought unusual for an astronomer:

> Hay about 2c a lb. at top.
> Grain about 2c a lb. at top.
> Burros need about 100 lbs. a week of hay and grain together.
> Burro cost about $25 with saddle, pack saddle, paniers, rope.
> Basset & Son have 4 year lease of everything, road etc.

The conditions on the mountain were rough on men and equipment. Before a road adequate for tractors and dollies was finally built, components of the early solar telescopes and the heavy mounting for the sixty-inch telescope had to be carried up the narrow trail on the back of a donkey, mule, or man. On one early trip an unbroken burro rolled over with a valuable spectrograph prism on his back, destroying it. It was the price they paid for a good site.

It was 1908 before the sixty-inch telescope was ready. Ritchey had polished the great mirror for four years. The mountings, cast at the Union Iron Works, were almost lost in the San Francisco earthquake and fire. To reduce the friction on the bearings, the fork that held the telescope tube was mounted on a ten-foot-diameter float in a tank of mercury. The drive gears, marked off with a finely graduated circle fixed to the polar axis, had been ground to a precise shape and polished with jeweler's rouge. To keep the mirrors of the telescope at a steady temperature, the dome was fitted with a canvas screen on a skeleton framework, and the mounting of the telescope was encased in blankets during the day. The shutters of the dome, fitted so they were almost airtight, were kept closed until shortly before sunset each evening. These precautions were dismissed as extreme by astronomers at eastern universities, until tests demonstrated that the mirror retained its figure during the temperature changes from day to night so well that it was optically as perfect at midnight at the site as when it had been tested in the temperature-controlled shop.

First light with the new telescope was in December 1908. To their delight the astronomers were able to obtain "perfectly round" star images of 1.03 seconds of arc—phenomenal resolution by comparison to other large telescopes. These images required exposures of eleven hours, during which the guiding mechanism of the telescope had to be corrected by hand. By 1910, with improved photographic emulsions, exposures of only four hours yielded images of stars of magnitude twenty—an improvement of several magnitudes over all previous telescopes.

As the work on Mount Wilson progressed, Hale resigned from

Yerkes and was appointed director of the new Mount Wilson Observatories, a division of the Carnegie Institution of Washington. The new laboratories and offices on Santa Barbara Street in Pasadena were the most modern in the world. With the lure of the solar telescopes on the mountain and the new sixty-inch reflector, Hale assembled a talented cast of astronomers at Mount Wilson. But even the sixty-inch reflector on Mount Wilson—the biggest telescope in the world and the first to be built with a mounting and guidance system of the precision and temperature stability that long photographic and spectrographic exposures of distant, faint objects would require—wasn't enough to satisfy George Hale.

4

The Whirligus

The competition for bigger and better machines was the spirit of the times. The years when Hale was guiding the construction of the forty-inch refractor at Yerkes, then the sixty-inch reflector at Mount Wilson, were the years of the great dreadnaught-building contest between Great Britain and Germany, an arms race as frightening to contemporaries as the nuclear arms race would be to a later generation. One country would launch a great battleship, then the other would counter with an even bigger battleship, extending the technology of armor, ballistics, and explosives until they provoked yet another generation of ships with larger displacements and more guns. Cartoonists and editorial writers goaded the competition, as the huge ships became symbols of national pride.

A parallel competition took place in architecture and technology. In France, Gustave Eiffel's tower temporarily won the race for greater heights, but architects and developers in the United States were already dreaming of taller buildings. In Britain and the United States, Isambard Kingdom Brunel and John Roebling raced to build longer and taller bridges than the world had ever known. The Panama Canal, thought an impossible engineering feat after Ferdinand de Lesseps's catastrophic failure, was nearing completion by American engineers, as important as a symbol of the triumph of American engineering as for military strategy and commerce of a nation with coastlines on two oceans. New York was building subways and massive aqueducts; the railroads were conquering mountain passes; aircraft engineers were building bigger and better flying machines.

George Hale had no competition in the telescope race, but each achievement of technology, each leap further into the cosmos that a big telescope provided, fueled the dream of reaching still further. Even before first light on the sixty-inch telescope, Hale approached a hardware merchant in Los Angeles, John D. Hooker, with the idea that he donate the money for the mirror of a still-larger telescope. Hale had bought hardware from Hooker's company for the observatory, and he knew that investments in the oil industry had made Hooker a very wealthy man.

When George Hale called on Hooker in 1906, at his spacious two-story colonial home on West Adams Street, with its grand, wicker-furnished porches overlooking the gardens, Hooker had already lived in Los Angeles for twenty years, long enough to establish a reputation as a community leader and philanthropist. Still, he was flattered to be called on by a distinguished scientist; that same year Hale had been offered and had declined the presidency of MIT and a chance to become the secretary of the Smithsonian Institution. When the enthusiastic Hale described what an 84-inch telescope could do, and the attention that the astronomical discoveries it would make would draw to the telescope and its donor, Hooker asked how much the mirror would cost. Hale said $25,000. Within a few days Hooker pledged $45,000, on the condition that the new telescope that would bear his name be the largest in the world.

Though for different reasons, both the donor and the fund-raiser were convinced that bigger was better. To be sure that it wouldn't be topped by any competitor, Hooker and Hale later agreed to increase the size of the proposed telescope to one hundred inches, a nice round number.

The trustees of the Carnegie Institution accepted the gift from Hooker with the understanding that they would not commit to mount the mirror—that if further funds were not found to build a telescope, the mirror, and the project, might languish. The order to pour the disk went out in September of 1906 to the Saint-Gobain glassworks, the only foundry that would even attempt to cast and anneal a one-hundred-inch-diameter glass blank.

Even Saint-Gobain, which had successfully poured the 60-inch blank, did not have crucibles large enough to melt the five tons of glass required for the new disk. They calculated that the mold would have to be filled with three successive pours of plate glass of one and a half tons of glass each—a technique that had never been tried. After preliminary trials Saint-Gobain produced a glass disk 101 inches across and 13 inches thick, weighing more than four and one-half tons. It was the largest plate glass casting ever poured.

The disk was annealed for a full year before Hale received notice that the blank had been successfully cast. Three months later, when the blank was shipped to Hoboken, New Jersey, the New York papers called it the most valuable single piece of merchandise ever shipped. From Hoboken the disk was reshipped to New Orleans and then brought overland to Pasadena, where it arrived on December 7, 1908, the very day the 60-inch telescope was set in place on Mount Wilson.

Hale and Ritchey watched impatiently as layer after layer of packing materials were stripped off the disk. Finally the crating was gone, and the two men got their chance to examine the immense mass of glass.

They were appalled at what they found. There were heavy sheets of bubbles between the layers, where the glass from successive pours had not fused completely. Preliminary tests with a light source and a polarimeter (a device to measure strains in the glass) indicated that the long annealing process had resulted in partial devitrification, a breakdown of the internal structure of the glass. The disk appeared to have lost both strength and rigidity. Hale and Ritchey agreed that it was unlikely that the disk would take, let alone retain, a good surface figure. After hearing Ritchey's verdict, Hooker announced that he would not pay for the blank.

Hale called on Hooker to urge patience and to persuade the businessman that if he withdrew his commitment of funds it would never be possible to build the great telescope. Hooker refused to receive him. Hale came back again and again, presenting his card to Hooker's butler. Each time Hooker refused to see him, and Hale was left to make small talk with Mrs. Hooker. Mrs. Hooker collected books and supervised the plantings, fashioning her gardens in what she called a "palazzo style." With his slight mien and oval glasses, George Hale radiated a boyish energy. He was well read and willing to talk about Mrs. Hooker's hobby of collecting fine book bindings in the Italian style. He asked her to teach him Italian. Unknown to Hale, Hooker was fiercely jealous of his wife. He got angrier with each visit Hale made.

Hale estimated that he would need another five hundred thousand dollars from the Carnegie Institution or other sources to build the telescope. He kept up his rounds of calls, but no one wanted to pour money into a project that seemed to reach too far. The pressures of the search for funds, pulling at Hale from every direction, took their toll in renewed excruciating headaches.

Hale began confiding in Harold Babcock, a gentle and understanding fellow astronomer who had recently joined the Mount Wilson staff. He told Babcock about his "terribly hard dreams," in which he would sometimes get up in the middle of the night and try to climb the picture frames on the wall. His letters to his wife hinted at an inchoate anxiety. Others became aware of Hale's increasing nervousness and urged him to rest. He refused. With one telescope barely finished and another seemingly stillborn, there was too much to do.

The Saint-Gobain glassworks agreed to bear the loss for the first one-hundred-inch disk and pour another. Ritchey was sent to Paris to discuss the arrangements, and before long Saint-Gobain built a new furnace and annealing oven. They shipped smaller disks to Pasadena for testing. These blanks would be used for the secondary mirrors of the telescope, and from the preliminary tests they seemed excellent.

During the summer of 1909 the reports of the canals Percival Lowell thought he saw on Mars filled the newspapers with speculation about the possibility of life on the Red Planet. The Mount Wilson astronomers had little patience for the debate about Martians, but during the close approach of Mars that summer, Hale turned the sixty-inch telescope on Mount Wilson—still the largest in the world by a substantial margin—toward Mars, took photographs with red-sensitive emulsions, and concluded that the widespread public belief in canals (fueled by publicity from Lowell and Giovanni Schiaparelli, an Italian planetary observer who had coined the label *canali*) was nothing more than wishful imagination. To Hale and the other astronomers, there were more important tasks for this and the next big telescope.

Within months news arrived that a new one-hundred-inch disk had been cast at Saint-Gobain and consigned to a manure pile, which would generate and hold a steady temperature, for annealing. Now all they could do was wait. The news of the new casting was the perfect background for a visit to Mount Wilson by Andrew Carnegie. Everyone knew that Carnegie's personal blessing would go a long way toward assuring the completion of the telescope.

There had been working facilities at the top of the mountain for a decade when Carnegie visited, but Mount Wilson was still a rugged site, at least when compared to the refined eastern observatories. A tin-roofed machine shop and library on one side of the mountain were joined to the other facilities by a narrow balcony and walkway over a one-thousand-foot sheer drop. The site planned for the one-hundred-inch telescope was across a gorge that would have to be spanned with a bridge.

The famed philanthropist arrived at the peak wearing a long fur coat, which on his ample girth gave him the appearance of a great bear. Carnegie, veteran of thousands of VIP tours, was ever the curious observer, asking questions, taking mental notes on everything he saw. He seemed bored by the huge fixed solar telescopes that studied the sun during the day, and the great sixty-inch reflector that studied the heavens at night. Between peeking through the eyepieces and listening to the descriptions of instruments and research projects, Carnegie's gaze kept drifting to the stands of ancient pine trees on the mountaintop, some of them 120 feet tall and 17 feet in diameter. Hale had to call his attention back to the telescopes.

Hale was careful not to make a direct appeal to Carnegie. He had met the great philanthropist in New York and Washington and knew that Carnegie liked to distance himself from the nitty-gritty of his eleemosynary projects. Presenting a proposal required a delicate diplomacy—enough hard sell to be sure Carnegie understood the importance of the work at Mount Wilson and the need for a larger telescope, interlaced with a subtle appeal to Carnegie's oft-stated preferences for what he considered productive science with a chance of changing the world. Hale did a good job that day. Although Carnegie made no commitment, he seemed impressed with Hale and the work at the observatory and apparently enjoyed himself at dinner in the Monastery. Calling the astronomers' residence the Monastery was only half facetious. There were no women observers, and observatory policy did not allow women to spend the night on the mountain.

Reading the tea leaves of Carnegie's remarks and gestures was an arcane and demanding science. Hale was optimistic, but optimism couldn't build a telescope. And while he waited nervously for signs, the strain of overwork and compulsive worry about progress in France and Pasadena took their toll. Hale tried to work, juggling his own research, the administration of the observatory, and his efforts to get the new telescope under way, but the headaches and exhaustion wouldn't go away. His sunken, gray-rimmed eyes and pained expression made his suffering obvious to all who met him. In the language of the day, he had "a bad case of brain congestion and exhaustion."

Nervous exhaustion was a commonly diagnosed affliction in the first decade of the twentieth century. Some attributed it to the technological millenialism that the humorist Robert Benchley summed up in a book title: *After 1903—What?* For Benchley the big question of the day was: "Are we living too fast for our nerves?" Whatever the cause, Theodore Dreiser, Theodore Roosevelt, Frank Norris, William and Henry James, Edith Wharton, and others less celebrated or chronicled suffered symptoms of exhaustion, depression, nervous prostration, dyspepsia, lack of appetite, insomnia, and apathy, sometimes complicated by hysteria, invalidism, or hypochondriasis. It was a peculiarly American affliction, according to New York neurologist George Beard, who coined the fancy label *neurasthenia*. The symptoms were real and often incapacitating, but some sufferers found consolation in identifying neurasthenia as a mark of intelligence, sensitivity, and spirituality, a disease of intellectuals and leaders in a world that was changing too fast.

The cures ranged from patent medicines for the poor, like Dr. Hammond's Nerve and Brain Pills, to expensive rest cures. Teddy Roosevelt and Frederic Remington went west to seek a cure in the wide open spaces. Elihu Root and Theodore Dreiser went to Muldoon's spa, run by a former wrestler and boxer who imposed a regimen of militaristic discipline and vigorous exercise. The stodgy Mrs. Peniston in Edith Wharton's *The House of Mirth* followed George Hale's mother's prescription: Lying down was her "panacea for all physical and moral disorders."

Alas, none of the cures would work for George Hale. He was too much the scientist to trust patent medicines, already as far west as he could travel, and too busy to take time off for a rest cure. His physician, Dr. McBride, urged him to leave Pasadena for a sustained period of complete rest. Hale refused. He was at a crucial point in his own solar research, the sixty-inch telescope was just coming into full service, and the one-hundred-inch project was at a critical juncture. The most he would agree to was to cut back a little. In 1910 the Solar Union scheduled an international meeting at Mount Wilson, the first time a large academic congress was held at the observatory. Hale was too exhausted to attend more than a garden party, a single dinner, and one brief meeting on the mountain. When the astronomers prepared to look through the sixty-inch telescope, Hale's pride and joy, he had to retire to bed with a painful headache.

He finally agreed to leave Pasadena and Mount Wilson for a much-needed rest. He chose Europe, far from the day-to-day pressures of the observatory.*

Hale was at the peak of his career as a solar astronomer and the director of the leading observatory of the world. In Europe he was feted wherever he went, elected a member of the Royal Society of London, and toasted at receptions at the Paris Observatory, at a stag dinner at Prince Rothschild's, and at the International Association of Academies. He spent a day with the distinguished Dutch physicist H. A. Lorentz and received an honorary doctor of science degree at Oxford in the company of Oliver Wendell Holmes and Earl Grey, governor-general of Canada. He visited Saint-Gobain and was cheered by the progress on a new mirror. When he came home he felt rested, ready for work again.

Then the news arrived that the second one-hundred-inch disk had broken during annealing. The Saint-Gobain glassworks admitted it was unlikely that they could produce a disk better than the one they had shipped. There were no other options for the telescope, so Hale suggested another look at the first disk. He recalled that when he had studied the rejected blank in Pasadena, the bubbles were not close to the surface. He questioned whether it was possible that despite their threatening appearance, the bubbles might not interfere with the formation of an optical surface. He fired off telegrams ordering more tests. Arthur L. Day of the Geophysical Laboratory at the Carnegie Institution declared after examining the blank that the layer of bubbles strengthened rather than weakened the disk.

Hooker and Ritchey protested. Hooker thought he was investing in immortality when he offered the funds for the telescope. He had been promised that the instrument with his name on it would be the largest and most perfect instrument in the world. Now, he demanded a release from all further obligations if he paid the $45,000 in full.

Hale called on him, urging patience, but Hooker was adamant. He

* Escaping to Europe to cure nervous exhaustion was another American custom. In the James family a trip to Europe was the "reward" for whichever child was most ill. Henry often campaigned for as long as a year with hypochondriac symptoms to finagle the trip for himself over his brother William and sister Alice.

wanted no part of a telescope built from a flawed disk. Without his funds the project threatened to unravel, and with it the hope for a new era of astrophysics and cosmology.

The new round of squabbles with Hooker was too much for Hale's frail health. Ever since the interview with his father before he graduated MIT, he had grown accustomed, in times of stress, to a painful ringing sensation in his ears and the ghastly headaches. Now, suddenly, the symptoms were accompanied by an uncontrollable and frightening vision: A little man appeared, as if in a dream, advising and berating Hale about the decisions of his life. The medical vocabulary of the time said his neurasthenia had progressed to nervous prostration. We would call it a nervous breakdown.

The doctors told Hale that that he had to get away again. Hale chose Egypt, via Menton on the Riviera. He had long been a close friend of James Breasted, the Orientalist on the University of Chicago faculty who was trying to record the history of the Nile Valley through hieroglyphs, and Breasted had suggested an itinerary. Once they were abroad Evelina censored his letters, trying to keep the worries from California at bay. But in each city Hale sneaked away to the telegraph office to send off queries on the progress of the telescope.

No matter how far he traveled, Hale wasn't alone. From Egypt to Rome, wherever he went, the little man appeared in visions, telling Hale that what he was doing wasn't important, that the books he was reading didn't matter, that his plans were nothing. Though George Hale did not confide in his wife, Evelina could see that her husband's situation had worsened. She wrote to Walter Adams, the acting head of the Mount Wilson Observatory: "I wish that glass was in the bottom of the ocean."

It was when the symptoms were at their worst, the frightening visits of the little man coming almost daily, that Hale—on a train from Menton to Genoa—spotted a tiny notice in a copy of *Le Petit Niçois*, a French Riviera newspaper, that Carnegie had given another ten million dollars to the Carnegie Institution of Washington. When they arrived he hurried to buy *Corrière della Sera*, which confirmed the report and mentioned Carnegie's special interest in Mount Wilson and the one-hundred-inch telescope. Hale knew what it meant: Finally the one-hundred-inch telescope was funded.

Ritchey still voiced doubts about the disk, but Hale wired him to start grinding the preliminary shape into the mirror. Ritchey took even more extreme precautions for the grinding shop than those that had been taken with the sixty-inch disk. Because even slight variations in the temperature of the shop would affect the results of any testing of the surface, Ritchey and Hale decided that the final figuring and testing would be done only in the summer months. The winter months would be devoted to rough-grinding and work on the auxiliary mirrors. Ritchey ordered electric fans to circulate the air in the optical laboratory, to prevent temperature stratification of the air. With this regime figuring the surface of the mirror from spherical to paraboloid took one year of steady work. Then they stopped to prepare a sixty-inch-plane-silvered mirror that could be used for optical tests at the focus of the big mirror, before they would polish the surface to the final figure. Ritchey insisted on working alone.

Hale sailed home in 1911. The doctors were cautious, sending him off to a sanatorium in Maine for rest and a regimen of wood sawing, copious good food, rests on the floor with a billet under his spine, daily massage, medicine nine times a day, lectures on the subconscious, and self-hypnosis exercises. When he finally returned to Pasadena, Dr. McBride limited his schedule to working from nine to twelve-thirty in the mornings, and warned him that he would never again be able to return to his previous routine.

Doctor's orders or not, Hale couldn't relax. Even Andrew Carnegie urged: "Pray show your good sense by keeping in check your passion for work, so that you maybe spared to put the capstone upon your career, which should be one of the most remarkable ever livd [*sic*]." Hale watched nervously as Ritchey ground and polished the one-hundred-inch blank. Ritchey had never been an easy man. Now, with the world watching his progress on the priceless blank, he became even more intractable, raging at anyone who approached his inner sanctum to inquire about progress on the disk.

Ritchey had his own ideas about telescope design. At his home laboratory he had been experimenting with using a deep, fast primary mirror and a complex hyperboloid curve in the secondary mirror. The resulting design would produce a telescope with a large field of sharp focus and a shorter tube design that could be mounted in a fork like the sixty-inch telescope. For the careful photographs Ritchey liked to take of deep-space objects, his design would be a boon. But a wide field of sharp focus was not necessary for the spectrographic studies that would form a substantial portion of the work of the Mount Wilson telescopes. Given the difficulties of fund-raising, Hale and Adams would not support experiments on Ritchey's new designs.

Ritchey, chafing that his ideas were ignored, approached Hooker privately for money to set up his own shop. Hale, outraged that a colleague would try to divert Hooker's funds from "the benefit of the Observatory as a whole" and the "advancement of science" began distancing himself from Ritchey, drawing up a new contract that limited Ritchey's privileges at the observatory. When Ritchey later advertised his telescope-making business and ordered special stationery imprinted "Professor G. W. Ritchey" with the Mount Wilson Observatory address, the relationship turned bitter. Ritchey and Hale both claimed credit for an edge-support mechanism used for testing of the disk. Rumors began to circulate that Ritchey sometimes had strange attacks when he was alone with the disk, and that he was an epileptic, subject to unpredictable seizures. He was relieved from any further work on the mirror.

Ritchey was furious. When another optician was given the task of the final figuring and polishing of the mirror, Ritchey announced that the disk was fundamentally flawed, that it had a "strong" and a "weak" diameter at right angles to each other and would never work no matter who figured it. Despite Ritchey's predictions the opticians kept polishing, gradually bringing the entire surface of the one-hundred-inch-diameter disk to the required optical figure. A deviation of one-millionth of an inch would show up on the optical tests, and ultimately in the quality of the images the telescope produced. The only prescription was more polishing and more testing.

Months of polishing the surface stretched into years. The continued crises over the new telescope brought an end to the temporary remission of Hale's dreaded symptoms. The familiar pattern returned: first a ringing in his ears and an agonizing headache, then physical exhaustion, a tingling in his feet, frantic excitement, insomnia, indigestion, spastic colitis, and the sensation that his mind was whirling out of control. The little man who had tormented him in Egypt came back, a dreaded, unwelcome companion offering unsolicited advice. There seemed to be no escape.

The early twentieth century was not an age of medical candor. Hale told no one outside a few close confidants the extent of his torments. When he did confide in a friend, like Harold Babcock, he called the collection of symptoms "Americanitis," his term for the madness and impatience that Charlie Chaplin later portrayed in *Modern Times*. Other times Hale called the symptoms, onomatopoeically, the "whirligus." Whatever he called it, Hale was powerless in the face of its tortures. He tried everything, even thanking the elf and promising to follow its advice. The whirligus would always come back.

The relapse in 1913 sent Hale back to Dr. Gehring's sanatorium in Maine. Gehring's diagnosis was that prolonged use of one part of Hale's brain had allowed poisons generated in a displaced large intestine to accumulate. He labeled the condition *ptsosis*, and prescribed sawing wood, subconscious exercises, and light massage of the intestinal area. The treatment helped, temporarily relieving the driving headaches and sending away the whirligus. But by summer Hale needed yet another trip abroad to get away from the pressures of the observatory and the unfinished telescope. Hale's life had become a seesaw, brief moments of relative peace alternating with the tormenting headaches and uncontrollable visions.

During one of the clear moments, in 1914, Hale interviewed an energetic young astronomer in New York. The astronomer came to the meeting primed to discuss the latest astronomical research, especially the variable stars on which he had done considerable work of his own. To his surprise Hale asked him about music, opera, drama—anything but astronomy. George Hale knew that a candidate who came with formidable recommendations from Henry Norris Russell would be a capable observer and astronomer; what he needed to ascertain was whether the man was the right sort of fellow, whether he could function well in the atmosphere of close collegiality that Mount Wilson required. The young astronomer was Harlow Shapley, who would soon be using the sixty-inch telescope for his research on globular clusters.

Through sheer force of will, Hale fought back the horrible symptoms often enough to carry on, recruiting a steady stream of first-rate astronomers for the observatory and dealing with the institutional and administrative problems of a major research institution. When the components of the huge new telescope proved too heavy to go up the mountain on mules, special trucks were bought for the treks up the mountain. The daily trips of the one- and three-ton trucks tore up the road and required regular labor teams to keep the crude path passable. Crisis after crisis came up, and with World War I beginning to draw on American resources, Hale could not turn down the invitation of the National Academy of Sciences that he assist in the organization of a National Research Council, which took over many of the laboratories and facilities at Mount Wilson for war-related research.

George Ritchey, though relieved as optician in charge of the grinding of the mirrors and other optics for the one-hundred-inch telescope, remained at the observatory, and other astronomers still chafed at his sanctimonious and sometimes sadistic attitude. His perfectionism was more than mildly annoying, not only in the time he would spend on the big telescopes, taking one *perfect* exposure (and, as often as not, forgetting to note the parameters that would make the plate useful for scientific purposes), but in his procrastination of any writing, so that collaborations involving him fell impossibly behind schedule. One night he helped a visiting French astronomer, Henri Chrétien, use the sixty-inch telescope for the first time. Ritchey loaded the telescope's plate holder and Chrétien spent several hours guiding the telescope, perched precariously at the eyepiece, in freezing temperatures, pushing buttons on a paddle to hold a star image steady in the crosshairs. When the exposure was done, Chrétien knew he had done a good job and was eager to develop this, his first plate on the famed telescope. Then Ritchey told him that it was a blank, that he had not been willing to risk a good photographic plate on a novice.*

Other astronomers, especially Walter Adams, who served as acting

---

* Chrétien later said that the exposure had only been forty-five minutes. The experience seems to have left no scars: In the 1920s Chrétien collaborated with Ritchey on the telescope design that was to bear their names.

director of the observatory during Hale's absences, were outraged by Ritchey's sadistic and authoritarian manner. By the middle of the war, Ritchey, who was in charge of war production at the optical shop, began signing his correspondence "Commanding Officer, Mount Wilson Observatory."

He rarely passed up an opportunity to give the staff of the optical laboratory or visitors to the observatory his views on the one-hundred-inch-telescope project. He had spent long enough with the disk, he said, to know that it changed shape with even slight changes in position, and that the layers of air bubbles from the three separate pours had fatally weakened the glass. Ritchey's success in figuring the mirror for the sixty-inch telescope weighted his predictions. Doubts about the new telescope were widespread.

5

First Light

Their hundred inch reflector, the clear pool,
The polished flawless pool that it must be
To hold the perfect image of a star.
And, even now, some secret flaw—none knew
Until to-morrow's test—might waste it all.
Where was the gambler that would stake so much,—
Time, patience, treasure, on a single throw?
The cost of it,—they'd not find that again,
Either in gold or life-stuff! All their youth
Was fuel to the flame of this one work.
One in a lifetime to the man of science,
Despite what fools believe his ice-cooled blood,
There comes this drama.
*   If he fails, he fails*
Utterly.
    —ALFRED NOYES, WATCHERS OF THE NIGHT

It was 1917 before the telescope was ready for first light.

The weather was clear and cool when the group left Pasadena for the drive up to Mount Wilson, the kind of November evening in Southern California that condenses enough moisture to require windshield wipers on a clear evening. Pasadena is a flat city, with streets laid out in a rectangular grid. Were it not for the palm trees that line the main streets, the stucco exterior walls, and red tile roofs, it could pass for a midwestern city.

The mountains arise abruptly at the northern edge of the city. As the party drove up the rugged nine-mile path from the base of Mount Wilson, through patches of scrub oak, sagebrush, and black-cone fir, layers of fog closed in, first scattered, and then so dense that driving was difficult. Only at the summit, at 5,700 feet, did they break through the fog that obscured the light of the sprawling Los Angeles basin below.

The city's pain was the astronomers' gain: The frequent fogs below the summit, and the inversion layer that trapped pollutants in the Los Angeles Basin, were a blessing on Mount Wilson. The fog scattered and occluded the light of the city below, and the inversion stabilized the atmosphere. From the peak the sky overhead was filled with stars, pinpoints in a celestial dome of black velvet. The stillness of the atmosphere left the star images stable, without the twinkle that inspires poets and frustrates astronomers. Mount Wilson has among the best seeing of any observatory site in the world.

That cold November night in 1917, twenty men walked across the narrow wooden drawbridge to the one-hundred-inch telescope. The dome was immense, dwarfing the dome of the sixty-inch telescope in the distance. A table had been set up so each of them could sign the logbook that W. P. Hoge, night assistant on the sixty-inch telescope, used to record the evening's events.

The great telescope loomed above them, a seventy-five-foot-high erector set of riveted black steel girders and huge brass gears. An astronomical telescope is an impossible combination of the scale of a battleship and the precision of a microscope. The heart of the telescope was a fraction of an ounce of silver, the coating of the great mirror that had been polished to as perfect an optical shape as the opticians could achieve. Everything else—the whole huge structure of girders, gears, bearings, and drive mechanisms—was there to cradle and aim that ounce of silver.

The mounting had been fabricated by the Fore River Shipbuilding Company in Quincy, Massachusetts—a division of Bethlehem Steel more accustomed to building naval gun turrets than precision optical devices. Sections of the telescope tube were so large they had to be shipped around Cape Horn to Los Angeles Harbor. Assembled, the instrument weighed one hundred tons. The clock mechanism alone required a ton of bronze castings, one and a half tons of iron castings, a two-ton driving weight, a seventeen-foot driving wheel, and a maze of hand-machined gears, each wider than a man's reach. Altogether the telescope, dome, and shutters needed thirty electric motors to move them.

The great one-hundred-inch-diameter mirror was mounted at the bottom of an open tube of riveted steel, eleven feet in diameter and more than forty feet long. The tube pivoted at its center in a heavy steel yoke; motors turning hand-machined gears rotated the tube on the pivots to direct the telescope to objects higher or lower in declination. The steel yoke was suspended at each end on huge floats in mercury bearings and precisely aligned with the axis of the earth, so that as the telescope turned synchronously with rotation of the earth, the heavens would seem to stand still. The massive yoke was designed by Francis Pease, an astronomer at the observatory, in the so-called English style, with the north end closed. The price of the great rigidity was that the telescope could not be lowered enough to be aimed at objects around the north celestial pole. This telescope would never see Polaris.

The auxiliary mirrors, eyepieces, spectroscopes, and other instruments that would record the light of faint objects were mounted at the opposite end of the tube from the mirror, all in precise alignment with the primary mirror. A deviation of a fraction of a millimeter in the alignment would degrade the image. The tiniest wobble in the mounting would make the image too unsteady for photography or spectroscopy. Astronomical telescopes are unforgiving instruments.

The twenty men on the mountaintop that night included astronomers, machinists, electricians, and carpenters. Hale had also invited the poet Alfred Noyes, in the hope that he might capture and record the majesty of the occasion of first light. Hale deliberately did not invite the press. Hale had a scientist's skepticism of journalistic oversimplification and sensationalism. Noyes captured Hale's fears of the press:

> As for the stars, if seeing them were all,
> Three thousand million new-found points of light
> Is our rough guess. But never speak of this.
> You know our press. They'd miss the one result
> To flash "three thousand millions" round the world.

Once the sun dropped below the horizon, the only illumination inside the dome was dim red night-lights. They too would be turned off when the actual observations began. Through the open shutters of the huge dome, the visitors could see the sky, punctuated with uncounted pinpoints of light. Even for the experienced astronomers, who had spent hundreds of nights on mountaintops with the big telescopes, it was an inspiring sight. Mountaintop observatories create the feel of a cathedral, with the heavens as their ceiling.

If it worked, the new telescope would almost triple the light-gathering ability of the sixty-inch telescope Shapley had used for much of his work. But in 1917 no one could say for sure whether a telescope as large as the new one-hundred-inch reflector would ever achieve its theoretical resolution of faint and distant objects. The effective resolution of a large telescope is limited by the turbulence of the earth's atmosphere. Mixed air of varying density, which produces the twinkling of stars, leads to irregular refraction under magnification: Stars appear as blurred images instead of pinpoints of light. The larger the lens or mirror of a telescope, the more light rays from widely separated paths are united in a single image, and the more sensitive the instrument becomes to the tremors of the atmosphere. Even at a site like Mount Wilson, with its superb seeing, there were many who thought that the sixty-inch telescope was already pressing against the limits.

Although the new telescope was intended for use almost exclusively for photographic and spectrographic work, an eyepiece was mounted that evening so they could visually test its resolution and light-gathering ability. When the sky was dark enough, Walter Adams pressed buttons on the control panel to swing the great telescope around toward Jupiter. The others gathered around the base of the telescope as Hale was given the privilege of the first look through the eyepiece. He climbed the ladder to the eyepiece, high above the concrete floor, stared for a moment, then came down the ladder without saying a word.

Adams, an astronomer and experienced optician, went next. He couldn't believe what he saw through the eyepiece. Instead of a single image of Jupiter, there were six or seven overlapping images in the eyepiece. "It appeared," Adams later wrote, "as if the surface of the mirror had been distorted into a number of facets, each of which was contributing its own image." If that was the best the telescope could do, it was worthless for astronomical work.

Hale and Adams looked at each other, wondering if the predictions of doom for the big telescope had come true. Had telescope building reached its limit? Had they labored for most of a decade—raising the funds and painstakingly building the giant machine—for nought?

Then someone recalled that the workmen on the mountain had left the dome open during the day. Maybe the sun shining on the mirror during the day had heated it enough to distort the images. The astronomers waited around. Every fifteen minutes one of them climbed the ladder to check the eyepiece, until someone suggested that a watched pot doesn't boil.

Adams and Hale walked glumly from the new dome down to the Monastery, the multiple images of Jupiter fresh in their minds, Ritchey's predictions of failure hung in the air like a curse. Already astronomy in the United States had divided into two camps: While the California astronomers worked on building bigger and better instruments, astronomers at some of the eastern universities, with limited access to good viewing sites, argued that astronomy didn't need bigger or fancier instruments. What was needed, they said, was more and deeper analysis of the data that was already available, accumulated in thousands of painstakingly compiled ledgers at the observatories. Ritchey's assessment of the new mirror added fuel to the fires. Outside the Mount Wilson offices and optical laboratories, there was no shortage of doubters, who had joined Ritchey in predicting failure for the machine.

Hale and Adams agreed to meet at the telescope at three in the morning for another look.

In his room at the Monastery, Hale didn't even undress. He couldn't sleep and couldn't concentrate on the mystery he had brought with him. He knew the new telescope wasn't without problems. Some were the results of deliberate compromises. Francis Pease's design for the mounting provided great rigidity, and the mercury flotation systems on the pedestal bearings provided smooth motions, but the support beam at the end of the closed cradle made it impossible to lower the telescope to point at the celestial pole. To insulate the telescope from changes in the air temperature outside, the dome had been fabricated of a double thickness of thin sheet steel, with an insulating air space between. Still, it didn't take extensive calculations to determine that the immense plate-glass mirror would be slow to adjust to changes in the ambient temperature.

At two-thirty in the morning, earlier than they had planned to meet, Hale walked back to the dome. Walter Adams showed up too, confessing that he also couldn't sleep. By then Jupiter was out of reach in the West. They chose a bright star for their second test. The night assistant didn't identify the star in the logbook.* After the earlier observations of Jupiter, they probably assumed that the logbook would be short-lived.

Again they slewed the telescope around, and again Hale climbed the ladder to take the first look through the eyepiece. This time he came away from the eyepiece with a broad smile on his face. For the first time in years, his impish eyes sparkled. "With his first glimpse," Adams remembered, "Hale's depression vanished." Adams took a look for himself. The image of the star stood out in the eyepiece as a single, sharp point of light, dazzling in its brilliance. Within hours everyone on the mountain had come over to take a look through the eyepiece.

Spirits were high, but even with the most spectacular images anyone had ever seen in a telescope eyepiece, the results of the evening had been a mixed success. The long, cool hours of the night had been enough to let the mirror resume its normal figure, but further tests confirmed that it took twenty-four hours to cool the mirror of the telescope ten degrees Celsius. During the cooling period the telescope was useless for accurate observations, which meant that sudden changes in the weather on the mountain would severely limit its use. The addition of a cold-water pipe system behind the disk, an effort to maintain the temperature of the mirror, didn't really help because the night temperatures on Mount Wilson were unpredictable. The only way the telescope could be used without the disasters of that first night was to keep the dome tightly closed all day, with the mirror housed in a cork-lined insulating chamber. And even when a complex routine was estab-

---

* Adams later said it was Vega, and in her biography of Hale, and again in her history of Palomar, Helen Wright repeats Adams's account. Vega is a summer star. At three in the morning in early November, Vega would be below the horizon. A more likely candidate would have been Capella, in the constellation Auriga. Hale's diary said only, "With Alfred Noyes to Mountain. First observations with 100"—Jupiter, Moon, Saturn."

lished to limit the thermal instability of the mirror, it became clear during the later testing that the definition of the telescope fell off in certain inclinations. The one-hundred-inch was a temperamental machine.

Yet for all its teething problems, the telescope put to rest the doubts that had raged about whether the atmosphere, even at a site like Mount Wilson, was steady enough to permit a large telescope to reach its limits of resolution. Within months observers were using the telescope to reach out to distant objects too faint to resolve on the sixty-inch telescope. As bugs in the mirror supports and temperature stabilization procedures were gradually worked out, the telescope produced significant results. Exposures for images and spectrograms of distant objects that had required several nights on the sixty-inch could be completed in a single session on the one-hundred. On a good night, when the mirror settled down and behaved itself, the telescope could resolve faint objects well beyond the reach of the sixty-inch telescope. Astronomers queued up for time on the machine. The limits of the universe, the elusive edge that tantalizes astronomers, had been pushed back.

Mount Wilson now had two great telescopes in addition to the solar telescopes. It was unquestionably the center of astronomical research, attracting the best and brightest young astronomers, a must station on the tour of visiting scientists. Yet, whatever the successes of the great telescope, Hale could never forget that first night and the horrors of that first glance through the eyepiece. Even if the one-hundred had been a perfect instrument, with no teething problems, George Hale was never a contented man. He had always been impatient. In the opening paragraph of some autobiographical notes he compiled, he wrote, "I was impatient to make rapid progress: as my father used to say, I wanted 'to do it yesterday.'"

Building telescopes was a Herculean task. For ten years the Hooker telescope had been George Hale's obsession, one hundred tons of iron and steel and glass held together by his compulsive worrying. It had left Hale's life swinging wildly between the moments of clarity and the quicksand of the whirligus.

Most men would have quit then.

6

Waiting

By the mid-1920s George Hale was spending much of his time in his private solar laboratory on Holladay Road in San Marino, not far from the Huntington Library, which he had helped establish. In 1921, the year of the great debate in Washington and three years after the one-hundred-inch telescope went into service, the visits of the tormenting demon and the incapacitating nervous breakdowns had become so frequent that he had had to give up the directorship of the Mount Wilson Observatory.

Set back from the road on a quiet side street, the solar lab is an attractive stucco building. The lintel over the mahogony doors, a stone bas-relief of the sun from a Theban tomb, bridges Hale's interests in solar astronomy and Egypt. Another bas-relief, over the fireplace in the paneled library, shows the pharoah Akhenaton riding a chariot surrounded by the sun and planets. It's only a few steps from the library to the research instrument, a spectroheliograph designed by Hale, with a movable coelostat mirror on a tower to follow the sun and reflect the light through a series of mirrors to the console. The electrical controls, including racks of hand-wound relays, were made by Jerry Dowd, the electrician from the Mount Wilson Observatory. The laboratory was equipped with a workshop, a spacious library, and a private study. With the blackout curtains drawn, Hale could be alone with his thoughts or with the demons. For a man in retirement, it was an ideal laboratory. He could pursue his studies of astronomy undisturbed by the trivia of administration.

But George Hale refused to retire, and try as he might, he could not drive away his demons. He tried a different sanatorium, Dr. Riggs's, in Stockbridge, Massachusetts, where instead of woodchopping and hypnosis, the regime was weaving and walks. He tried more trips abroad, visiting Egypt in the midst of the Tutankhamen excavations. No one expected a miracle cure.

His secretary, Miss Gianetti, wrote frequent letters explaining that

Hale was temporarily unavailable for meetings or to reply to correspondence. He was still secretary of the National Academy of Sciences and active in the effort to get the academy its own building in Washington, so that annual meetings and symposia would not have to be held in borrowed quarters like the Smithsonian. He and his old MIT professor Arthur Noyes led a group that transformed the Throop School in Pasadena into a polytechnic institution. Recruiting first-class faculty, like Noyes himself to head a chemistry department, the ambitious effort to build a first-rank scientific and engineering school succeeded faster than anyone would have expected. By 1920 the small school had renamed itself the California Institute of Technology. When the new name was announced at a campus meeting, a student shouted "Hooray for Caltech!" The nickname stuck.

A year later Noyes and Hale succeeded in recruiting the famed physicist Robert Millikan to join the Caltech faculty. Although he never took the title of president, Millikan took over the executive direction of the school to such a degree that it was known to many as "Millikan's school." By 1923, when he was awarded the Nobel Prize in physics, Millikan had developed the fine art of convincing Southern Californians and foundations that it was a privilege to contribute endowment and grant frunds to Caltech. "Just imagine," Wilhelm Röntgen said. "Millikan is said to have a hundred thousand dollars a *year* for his researches!"

The generous funding and the prestigious names of Millikan, Hale, and Noyes soon attracted a faculty of first-rate appointments. Before the decade was out, Caltech achieved the impossible, vaulting into competition with eastern universities that had spent centuries building their faculties, facilities, and prestige.

For George Hale the special appeal of the California Institute was that it would provide laboratories and faculty in physics and astrophysics as a resource for the Mount Wilson Observatory. Hale had predicted, years before, when he established the *Journal of Astrophysics*, that new research and theory in astronomy would draw heavily from developments in physics, from Einstein's work on gravity to the theoretical work on particles and waves in Europe. At the same time physicists and mathematicians with no formal training in astronomy were drawn to the problems of cosmology and astrophysics. As astronomers began to explore the basic processes at work in stars and other celestial objects, physicists saw a grand laboratory in which they could explore the most basic phenomena of nature. Between the new faculty and laboratories at Caltech, and the facilities of the Mount Wilson Observatory, George Hale was in the midst of exciting developments. The 1920s were heady days for astronomy.

Not long after the great debate in Washington, and after Henry Norris Russell turned down the position, Harlow Shapley was appointed director of the Harvard College Observatory. For years Harvard, with its southern station in Peru and its voluminous records of observations recorded by generations of "computers," had been the leading institution in American astronomy. Shapley would miss the access to the powerful telescopes at Mount Wilson, but his ambition had clashed often enough with others at the observatory that there was little love lost at his departure. One astronomer recalled, "I have never seen a quicker mind, a more agile sense of humor, or a more complete absence of what usually passes for humility." Walter Adams, who had been virtually in charge during Hale's frequent absences for illness or other commitments, thought that Shapley's methods with Cepheids were "not new and that he has never given the credit where it belongs."

At Harvard, Shapley had enough to keep him busy without worrying about what his former colleagues thought. The famed Harvard College Observatory was equipped with ancient telescopes and instrumentation, much of it located where urban growth had made serious observation impossible. The famed computers and their accumulations of ledgers were no substitute for state-of-the-art observation facilities. Shapley engaged Fecker & Co. of Ohio to explore the design of a large modern telescope for Harvard, while Shapley familiarized himself with the foundations. He soon became a regular visitor to the Rockefeller Foundation, willing to talk astronomy or to give his views on proposals from other observatories. The prestige of his position at Harvard, along with his experience at Mount Wilson, his fame after the great debate in Washington, and his ready access to reporters, made him a force to be reckoned with in American astronomy. Others were ready to take his place at Mount Wilson.

Edwin Hubble, like Harlow Shapley, was born in rural Missouri. He attended the University of Chicago on a scholarship and distinguished himself as a student of math and astronomy and a letterman in track and basketball. It is tricky to sort fact from myth in Hubble's early life. He liked to give the impression that he had been a good-enough boxer to fight an exhibition with Georges Carpentier, then the reigning light-heavyweight champion, and that promoters had asked him to fight as a "white hope" against Jack Johnson, the black heavyweight champion, but there is no record that he ever fought professionally. He did accept a Rhodes scholarship to Oxford, where he read law at Queen's College. He returned to the United States in 1913. He liked to tell people that he had passed the Kentucky bar and practiced law, but the only records have him teaching and coaching basketball at a high school in Indiana, just across the Kentucky line. Teaching bored him. According to a friend, the astronomer Nicholas Mayall, Hubble "chucked the law for astronomy" because "astronomy mattered."

Hubble reënrolled in the University of Chicago in 1914, and went to Yerkes to get his Ph.D. in astronomy. Service in the infantry in France interrupted his studies, but by 1919 he had arrived at Mount Wilson, sporting a newly acquired English accent, a tweed jacket, and a pipe. He was eager to use the famous one-hundred-inch telescope.

From the beginning Hubble's interest was the nongalactic nebulae, including the spirals, those wispy pinwheels in the heavens that had formed the basis of Heber Curtis's arguments in the "great debate." In his Ph.D. dissertation Hubble argued that even if the current data were inconclusive, astronomers should proceed on the assumption that the "white" nebulae are galaxies or island universes—Heber Curtis's position at the debate—because the assumption that they are within our stellar system leads to a full stop in research. At first Hubble was careful not to take a formal position on the Shapley-Curtis debate. "There appears to be a fundamental distinction between galactic and nongalactic nebulae," he wrote. "This does not mean that the latter class must be considered as 'outside' our galaxy."

In 1917 Ritchey had identified a nova—an exploding star—in an old photograph of a spiral nebula in the constellation Cygnus. Heber Curtis, at the Lick Observatory, then found plates from the Lick archives that showed novas in other spiral nebulae. These newly discovered novas were so much dimmer than novas in the Milky Way—the relative luminosity was as if a one-hundred-watt bulb were viewed at a distance of miles instead of inches—that if they had the same intrinsic luminosity they would have had to be very distant indeed. The novas were a powerful argument for island universes.

Hubble began a program of searching for novas in distant nebulae, counting nebulae for their space distribution, and photographing a wide range of nebulae to derive a morphology. He began his work at Mount Wilson with the sixty-inch telescope. His first night of observing was marked by seeing that the Mount Wilson astronomers rated as extremely poor. Hubble, used to the conditions at Yerkes, came back from the darkroom jubilant. "If this is a sample of poor seeing conditions," he said, "I shall always be able to get usable photographs with the Mount Wilson instruments."

Night after night he exposed plates on the big reflector. Day after day he studied the plates. The telescope drive turned the telescope on its polar axis to compensate for the rotation of the earth, but to make a successful plate, the observer had to track a guide star within the field of the exposure, using electrical hand controls to make tiny movements in the position of the telescope that would compensate for atmospheric effects, the flexure of the telescope tube and mount, and the slight quirks of the telescope-guiding mechanism. As the telescope slewed to different portions of the sky, Hubble had to contort his body on a precarious perch to keep the guide star in the crosshairs of the eyepiece. The winter nights were cold enough to freeze his tears to the eyepiece. The exposures were long enough to test his bladder control.

By morning his body would be a bundle of cricks. Lack of sleep and hours staring at the glass plates gave him headaches.

In 1923 Hubble began using the one-hundred-inch telescope to search for novas in the spiral nebula M33 in Triangulum and M31, the Andromeda Nebula. These are the only spiral nebulae visible to the naked eye, and he assumed that they were the closest to the earth. Lacking a wide-view camera to survey the entire nebulae for likely targets, he selected what he hoped would be promising areas of the two nebulae and photographed them repeatedly with the narrow field of view of the one-hundred-inch telescope. The relative luxury of the many nights of time he was allowed on the telescope meant that Hubble could take plates on the darkest nights with the best seeing, hoping the combination of perfect conditions would reveal the faint images he sought.

Each plate covered only a small region of sky, but the light-grasp of the big telescope, and the long exposures, revealed thousands of stars. "Critical tests made with the 100-inch reflector," he wrote, ". . . show no difference between the photographic images of the so-called condensations in Messier 33 and the images of ordinary galactic stars." The images he was resolving were fainter than photographic magnitude eighteen, dim enough to require long exposures on the regions of the nebulae where the stars were far enough apart from one another that they could be resolved. The resulting plates had to be scrutinized star by star in the search for images that had not appeared in the same place or at the same brightness on a previous plate of the same region. After months of exposures Hubble shifted to a more sensitive photographic emulsion and finally began to find novas. In all he studied more than 350 plates of the Andromeda Nebula.

On a plate from the night of October 5–6, 1923, Hubble marked off three apparent novas in M31 (Andromeda). On later examination he concluded that one of the three was not a nova at all but a Cepheid variable star. He extrapolated its luminosity and period from the formula and tables Shapley had used and concluded that the star had to be at an extreme distance from the Milky Way. He had found evidence that the Andromeda spiral was indeed an "island universe," a separate galaxy like our own Milky Way. It was a staggering discovery: If the Milky Way was one of many galaxies, the universe was immense, bigger than even Shapley had imagined.

Hubble did not publish his findings immediately. His earlier study of law had made him cautious about evidence. He went back and reëxamined earlier plates to confirm his results. He switched from Andromeda to a cluster of stars named NGC6822 in the rich star clouds of Sagittarius. A British veteran of the Lick Observatory named Charles Perrine had studied the cluster from Argentina and concluded that it was *not* part of the Milky Way. Hubble identified fifty Cepheid variables in NGC6822. If Shapley's Cepheid yardstick was correct, the cluster was deep in the cosmos, clearly an "island universe," distinct from the Milky Way. "The principle of the uniformity of nature," Hubble wrote, "seems to rule undisturbed in this remote region of space." He leaked his findings to a few astronomers but waited to present them in a formal paper.

The perfect opportunity arose at the 1924 joint meeting of the American Astronomical Society (AAS) and the American Association for the Advancement of Science (AAAS), in Washington, where a prize of one thousand dollars, approximately six months' salary for Hubble, was being offered for the best scientific paper. Henry Norris Russell, who knew of Hubble's findings, arrived late at the conference and found that Hubble not only wasn't there but hadn't submitted a paper.

"Well," said Russell, "he is an ass."

Russell and a small group decided to draft a telegram to Hubble, urging that he send his new results so Russell and Shapley could draft them into a paper. Russell and Joel Stebbins, secretary of the AAS, were on their way to the telegraph desk when they spotted a large envelope from Hubble, addressed to Russell. They returned to the group, holding up the paper and joking that they had gotten a very quick reply to their telegram.

The next day, New Year's Day 1925, Russell read Hubble's paper, announcing the discovery of Cepheid variable stars in the Andromeda galaxy. By then Hubble had also discovered Cepheid variables in M33. By Shapley's yardstick, which mapped the extent of the Milky Way at one hundred thousand parsecs,* these two spiral nebulae were so distant that they were clearly outside our galaxy. Hubble and the one-hundred-inch telescope had settled the celebrated question of the great debate. Astronomy would never be the same.

Success begets ambition. Hubble's discovery of Cepheid variables in the Andromeda galaxy, like Shapley's earlier work, and Hale's early work on magnetic fields in sunspots, had certified Mount Wilson as the leader in astronomical research. While other observatories were playing catch-up, searching for funds to upgrade existing facilities, in

---

* Astronomers use a variety of measurements for distances that are far too large to express conveniently or usefully in terrestrial units like the mile or kilometer. The AU, or astronomical unit, used for relatively small regions like the solar system, is the mean distance from the earth to the sun, or approximately 93 million miles. The LY, or light-year, is the distance light travels in one year, approximately six trillion miles. The parsec, sometimes abbreviated pc, is approximately 3.26 light-years. Kiloparsecs (kpc), one thousand parsecs, and megaparsecs (Mpc), one million parsecs, are used for extreme distances.

the offices and library on Santa Barbara Street in Pasadena, and at the dining table in the Monastery on Mount Wilson,* the astronomers had begun to talk of the possibilities of an even bigger telescope, powerful enough to penetrate to the depths of the universes that had been revealed by the one-hundred-inch telescope. George Hale was a regular at the talks: Big telescopes were a topic he couldn't resist.

At Hale's request, in 1921 Francis Pease prepared a preliminary drawing of a three-hundred-inch telescope. By 1925, after many revisions to his sketches, Pease had built a wooden model of his tentative design. Rumors of the plans leaked out, and Lester Markel of the *New York Times* fired off telegrams asking for details of the design and confirmation of a tentative price tag of $12 million. Bethlehem Shipbuilding, which had done some of the fabrication work on the one-hundred-inch, wrote to offer its facilities to build the new telescope. Harlow Shapley, eager to know what was in the works, wrote a friendly query from Harvard.

All the telegrams and letters were answered with evasions and denials. Pease's drawings and model weren't a working plan. They were a blown-up derivation of the sixty-inch telescope on Mount Wilson, with rudimentary concessions to the problems of a larger machine.

There were two problems with any plan for a bigger telescope. The first was that there were no funds available to build it and no prospect to raise those funds. And even if at some later date Yerkes or Hooker had miraculously come forward with the funds needed for a new telescope, it wasn't clear that the technology existed to build such a machine. As a machine gets bigger, the engineering demands grow exponentially. The volume and mass of a machine increase not in linear proportion to the increase in linear dimensions, but by the *cube* of the increase. A three-hundred-inch telescope wouldn't be 5 times as massive as a sixty-inch, but closer to 125 times. Pease's early drawing shows the domes of the sixty- and one-hundred-inch telescopes superimposed on the dome of the new instrument. They look like toys in the corner.

Casting the mirror of the one-hundred-inch telescope had pressed plate-glass technology to the limit. The largest glass furnaces in the world had required three pours to produce enough plate glass for the one-hundred-inch mirror. There weren't enough furnaces in France to melt the glass for a three-hundred-inch mirror all at once, and even if

* The discussions at the Monastery table are legendary. By tradition the astronomer observing on the one-hundred-inch telescope took the place of honor at the head of the dinner table and was granted the privilege of guiding the discussion for the evening. Hubble was famed and feared for his habit of arriving for dinner fifteen minutes early, looking up an article on some obscure topic in the eleventh edition of the *Encyclopaedia Britannica* in the library and channeling all discussion during dinner to the obscure topic about which he and he alone could be expected to know anything.

a mirror that huge could be poured, it would require an astronomer's working lifetime to anneal a plate-glass mirror that immense.

A mounting to support and point a mirror that large presented a comparable challenge. At the longest focus, the focal length of the telescope would be more than two hundred feet. A skyscraper of comparable dimensions sways inches; the tower of a suspension bridge may sway twenty feet. The telescope would have to be rigid enough to maintain the alignment of that long shaft of light on a spectrograph slit just one-thousandth of an inch wide, even as the telescope itself moved to follow the sidereal motions of the object across the heavens.

The sixty- and one-hundred-inch telescopes could be floated on mercury bearings, to provide low-friction precision motion. But there wasn't enough mercury available to support the volume of the telescope Pease had sketched. Even a freshman engineering major could calculate that the huge roller bearings Pease included in his drawing would entail so much friction that vibration-free movement of the machine would be impossible.

Although the engineers and astronomers were aware of these limitations, the reporters refused to understand that Pease's sketches weren't meant to be a working model. They were a pipe dream, a concept the astronomers batted around the way managers of sports teams fervently discuss the dream team they know they can never put together. As enticing as the huge machine was, Hale and his colleagues at Mount Wilson had enough experience building large telescopes to know that every step of the process would be several orders of magnitude more difficult, more time-consuming, and more expensive than anyone had ever anticipated.

Still, the dream tantalized the astronomers and cosmologists. For even as Hubble and others were expanding the known realm by careful use of the one-hundred- and sixty-inch telescopes, theoreticians in astrophysics and cosmology had begun raising questions about the very structure of the universe that cried out for investigation, demanding observations deeper into space than any existing telescope could reach.

In the years after the publication of his General Theory of Relativity in 1915, Einstein and others began to explore the cosmological implications of his revolutionary theory of gravity. One conclusion of Einstein's theory was that the universe could not be static: It had either to be expanding or contracting. Yet there was no observational evidence for this novel and disturbing idea. For centuries, with increasingly sophisticated and precise instruments, astronomers had measured the proper motions and radial velocities of stars. All the data, gathered with the finest equipment and the most careful analysis, suggested that the stars wandered more or less randomly through space. Einstein concluded that there was something wrong with his theory and included a term in the field equations of gravity that he called the cosmological constant, or Λ, which was designed to make the radius of the universe hold steady with time. The cosmological constant was a "fudge factor," with no known physical basis. Einstein readily admitted that it was introduced "only for the purpose of making possible a quasi-static distribution of matter, as required by the fact of the small velocities of the stars."

In fact the inelegant fudge factor didn't even accomplish its avowed purpose. The Russian mathematician Aleksandr Friedmann discovered that Einstein had made an algebraic error in introducing the term. Friedmann threw out the cosmological constant. Without Λ, a universe according to the General Theory of Relativity had to be continually expanding or contracting.

The same year that Einstein had reluctantly introduced the cosmological constant, an American astronomer named Vesto Slipher published a remarkable paper with evidence that the constant might not be needed after all. Slipher was on the staff of Percival Lowell's observatory in Flagstaff, Arizona, where Lowell pursued his much-publicized efforts to prove there were canals on the surface of Mars. Lowell gave Slipher the assignment of taking spectra of spiral nebulae, in the hope that measurements of the rotation of the nebulae would prove that the spirals were the early stages of solar systems like our own, a theory Laplace had advanced years before and to which few astronomers other than Lowell subscribed.

Using a new spectrograph with a very fast lens on the twenty-four-inch telescope at the Lowell Observatory, Slipher found the rotations Lowell sought—which we now know are the spiral motion of the billions of stars that make up each spiral nebulae. Slipher also found that the spectral lines of most of the spiral nebulae were shifted toward the red end of the spectra. The only reasonable explanation for this displacement was a Doppler shift, the same phenomenon we notice in the pitch of a car horn as the car drives away from or toward us. As the car approaches, the pitch of the horn seems to rise; it seems to drop as the car drives away. The same shifts in color or wavelength occur to the light emitted by a body when it is moving away from or toward the observer. Astronomers had long made use of Doppler shifts to measure stellar velocities. In fact it had been Doppler shift measurements, which asserted the randomness of the motions of stars in the Milky Way that had prompted Einstein to put the cosmological constant into his equations in the first place.

But the velocities of the spiral nebulae Slipher measured were not random. Twenty-one of the twenty-five spirals for which he had spectra were displaced toward the red, indicating that they were flying away from each other and from the earth. The displacements were enormous, suggesting that the spirals were moving at speeds that had never before been measured.* By 1924 a trio of astronomers—Carl Wirtz in Germany, Knut Lundmark in Sweden, and Ludwik Silberstein in the United States—had assembled evidence that the smallest and presumably most distant spiral nebulae were receding from the Milky Way faster than the nearer ones. As provocative as this material was, the data were marginal. The velocities the astronomers claimed to have measured seemed too great to believe.

But to Edwin Hubble, who had been able to use the Cepheid variable stars he had resolved in the Andromeda and Triangulum spiral galaxies as a yardstick to estimate the distance to those galaxies, Slipher's observations suggested a whole new line of inquiry. If these galaxies were actually receding, the universe was expanding—each galaxy or island universe moving away from the others at incredible velocities. The cosmological implications were exciting, but Hubble was not a man to shoot from the hip. What was needed was observations, data. He proposed an extensive program to measure the red shifts of galaxies within range of the one-hundred-inch telescope. His program would require a demanding schedule of observations, far more than one man could ever manage. By 1928 Hubble had a talented assistant and, ultimately, collaborator to aid him in his observations.

Milton Humason had been at Mount Wilson long before Hubble arrived. He was originally hired as a mule packer, trekking material up the steep path to the observatory. After years on the path he was promoted to busboy in the dining room and then to observatory janitor. Humason's eighth-grade education didn't stop him from asking questions of the observers. "I decided the only way to stay awake at night was to get interested in what was going on," Humason said. "So I did." Shapley and some of the others were so impressed by Humason's intelligence that they urged his promotion to night assistant on one of the smaller telescopes. Over the years Humason was promoted to minor observational assignments on the six-inch and then the ten-inch telescope on Mount Wilson. Finally Hubble asked Humason to assist him with observations on the one-hundred-inch telescope.

Traditionally there had been a chasm of education and status between the observatory staff and the observers. Observers were professors of astronomy, usually with Ph.D.s from eastern universities. They brought tweed jackets, pipes, common-room-sherry habits, haughty accents, and the latest research to Pasadena and Mount Wilson. The observatory staff were onetime farmers, truck drivers, and

* The exceptions, the few spirals that were not receding, we now understand, were nearby galaxies that are gravitationally bound to the Milky Way and therefore do not participate in the expansion.

mechanics, men who were comfortable living year round on a lonely mountaintop where the winter temperatures ranged from painful to unbearable and where the isolation drove many men to drink despite the observatory bans. Humason, who had come up through the ranks, remained one of the boys. He chewed tobacco, played poker and billiards, and went fishing with the night assistants on his days off. He was also a patient and talented observer. Before long he was doing the bulk of the actual observation work on the big telescope. Hubble, his ever-present pipe clenched in his teeth, pored over the plates.

As Hubble and Humason began to accumulate data on red shifts, they attempted spectrograms of fainter and fainter nebulae, pressing the telescope to the limits of its range. A spectrograph requires that the observer keep the light of the distant object on a narrow slit. For faint nebulae the exposure needed to gather enough light might run to eight hours or more. Sometimes a single observing session wasn't long enough to gather the light, so the spectra had to be made over several nights. To reduce the needed exposure times and extend the range of the telescope, they tried a new corrective lens on the one-hundred-inch telescope, built by W. B. Rayton at Bausch & Lomb. Working on the principles of a reversed microscope objective, the lens achieved the very fast focal ration of $f/0.6$, making it possible to record spectra for extremely faint, distant nebulae.

The observation procedures were demanding. For some objects the observer would have to twist and contort himself to be able to see through the guidescope. For eight or more hours, the observer would struggle to keep a barely visible image centered on the slit of the spectrograph. The temperatures were often cold enough to freeze breaths to mustaches and skin. The weight-driven clock for the right ascension drive of the one-hundred-inch telescope chimed whenever Humason pushed a correction button, and the balky controls sometimes had him leaning on the telescope or even climbing onto it to muscle the tube into the exact position he wanted. But Humason had the patience and persistence of a mule driver, and he got results—tiny glass plates fifteen millimeters square, with a spectrogram, only millimeters long, of the distant spiral galaxy.

When Hubble and Humason compared the lines on these spectrograms with the same lines on a calibration spectrum, they could measure how much the spectrum of the galaxy had shifted. The magnitude of the shift corresponded to the velocity at which the galaxy was receding. By 1928 Hubble and Humason had accumulated enough data to begin the search for a relationship between the magnitude of the red shifts of these distant galaxies and their distance. As Hubble plotted each new spectrum, he discerned a linear correlation between the receding velocity of the galaxies, as measured by the red shifts, and their distance—at least for the few galaxies whose distance he had been able to determine by the identification of the brightest resolved stars. The relationship was elegantly simple: The more distant the galaxy, the faster it was receding.

His earlier yardstick, the Cepheid variables he had discovered in nearby galaxies, required that the galaxy be close enough to resolve individual stars. Because red shifts were measured from the light of the galaxy as a whole, the reach of the telescope was suddenly extended a thousandfold. As equipment and technique improved, Hubble and Humason could record usable spectra of galaxies so faint that the galaxy could barely be resolved in the telescope. By combining the techniques— extrapolating from the distances to bright, nearby galaxies, which he could determine from the Cepheid variables, and using his linear relationship between the red shift, or speed of the receding galaxy, and its distance to measure the distance to galaxies that appeared only as a diffuse blob on photographic plates—Hubble had extended his yardstick to reach to the limits of the observable universe.

Hubble told Hale and his other colleagues about the work, but he hadn't read much of Einstein. Although Hale had corresponded with Einstein, he confessed his ignorance: "The complications of the theory of relativity are altogether too much for my comprehension. . . . I fear it will always remain beyond my grasp." Without a sound cosmological explanation for the data he had found, Hubble was reluctant to publish his findings too quickly.

Unknown to Hubble, a Belgian priest and mathematician named George Lemaître had already drawn a connection between Slipher's red shifts and Einstein's theory of relativity. Lemaître had made a tour of the United States, attending astronomy conferences. He learned of Slipher's red-shift data and, on his return to Brussels, in 1927, wrote a paper that connected the observed red shifts with the expanding universe predicted by general relativity. But Lemaître was an outsider. His paper was published in an obscure journal, and he got a brush-off from astrophysicists when he tried to present his views at conferences.

Cosmology hadn't caught up yet, but Hubble was certain that he was onto a very promising direction of research. He concentrated on what he could supply—measurements of red shifts that would reach farther into the depths of the universe. By 1931 Hubble and Humason had measured red shifts of 20,000 km/h, overwhelming evidence for the linear relationship between red shifts and distance. Humason was soon training the telescope on nebulae so distant that they could not be seen at the Cassegrain focus of the telescope. Humason's procedure for taking a spectrogram on these faint nebulae was to calibrate the slit of the spectrograph on known stars and then to move it by measured amounts (determined from long-exposure direct photographs) to the positions of the unseen nebulae. "The observations thus extend nearly to the extreme limit of existing equipment," Hubble wrote. "No very significant advances are expected until larger telescopes are constructed."

7

Old Boys

In 1926, in an article for the *Publications of the Astronomical Society of the Pacific*, Francis Pease wrote, "The question has often been asked, 'How large a telescope can be built today?'. . . My reply would be that anything up to a hundred feet in aperture can be built provided one wants to pay for it."

Some newspapers picked up the story, and a friend wrote to Hale, "Glory, Hallelujah, for the 300-inch telescope!" "Alas," Hale answered, "The report you saw was made out of whole cloth. Pease wrote a paper about the feasibility of building a very large telescope but there is no money in sight to pay for it!"

Money was always the first problem for a big telescope. Men like Lick, Yerkes, and Hooker had been eager to have their names attached to telescopes, especially when they were promised that the instrument bearing their name would be the biggest telescope in the world. But the cost of the telescope George Hale and his colleagues had in mind outstripped the resources or generosity of any one donor, especially after the advent of income taxes had put a noticeable dent in the largesse of the wealthy. The income tax raised enough revenue to finance World War I and the ensuing peace; it also accentuated the trend toward organized, as distinct from individual charitable giving. No one was willing to come forward with *after-tax* dollars for a telescope.

Nor was any institution or corporation willing to fill the gap left by the decline of individual benefactors. Andrew Carnegie's new pet project was the International Peace Palace in The Hague. During the war the federal government had supported weapons-related research by the navy and the hastily organized National Research Council—another George Hale project—but in peacetime the federal government was not in the business of funding scientific research at universities or other nonprofit facilities outside the government. Under the guise of its timekeeping mission, the navy ran the Naval Observatory in Washington, with some excellent equipment. Most congressmen were skeptical of any proposal to spend government money on scientific research with no direct practical application.

Hale refused to give up. At meetings of the National Research Council and the National Academy of Sciences, he listened and watched. He was also a frequent visitor to New York, the financial capital of the nation. Although he had never lived or worked in New York, Hale was a member of the University Club, a bastion of what a later generation would disparagingly label the old boy network. It was a good place to watch and wait.

Like the Bohemian Club in San Francisco, the Cosmos Club in Washington, or the Century Association on Forty-third Street, the University Club on Fifth Avenue at Fifty-fourth Street drew its membership from men of education and social standing, university graduates of respected, mostly Ivy League schools. The members were conservative though not extremist; if they had doubts about income taxes and foreign entanglements, they also accepted the notion that their privileged births, educations, positions, and fortunes conveyed responsibilities along with privileges. The rules of the club forbade business papers in the public rooms, but presidents of large corporations, distinguished academics, former and present government officials, foundation officers, politicians, and other shakers and doers routinely called on one another for advice and counsel and sought one another for informal and formal partnerships, knowing they could find the mutual trust of shared values in the retreat of their club.

One of the men Hale saw frequently at the University Club was Wickliffe Rose, the president of the General Education Board (GEB), a Rockefeller fund dedicated to improving education. With his moonish face, stiff collar, and slicked-down hair, Rose looked more like a small-town southern preacher than the president of one of the wealthiest foundations in the world. He had been an administrator of various Rockefeller-funded health organizations for years, waging war against hookworm, malaria, yellow fever, and other scourges. More experienced at administration and less impulsive than his better-known rival Abraham Flexner, in 1923 Rose was named president of the GEB. Before accepting the presidency he persuaded John D. Rockefeller to set aside an additional $28 million to establish an International Education Board (IEB), which would complement the domestic mission of the GEB. Rose became head of both.

Rose's education had been in the humanities, and he had been a professor of philosophy at Peabody College before coming to the Rockefeller funds. In his years as a headquarters commander in the wars against tropical diseases, his perspective changed, and he came to believe in the primacy of science over other forms of human activity, and of mathematics and the physical sciences over other sciences. "This is an age of science," he wrote. "All important fields of activity, from the breeding of bees to the administration of an empire, call for an understanding of the spirit and technique of modern science."

Rose's first official activity was a five-month trip through Europe, visiting nineteen countries and fifty universities, research organizations, and institutions to hunt out leaders in the sciences. Europe, he said, was "ground harrowed by the war, ready for new seed." In fields like physics, European laboratories and theoreticians were unraveling the secrets of the atom. These were the heady days when Niels Bohr "quantized" the electron in the classic orbits we see on symbols of the atom, Werner Heisenberg added matrix mechanics and the "uncertainty principle," Erwin Schrödinger developed an alternate theory treating the electron as a packet of waves, and Max Born demonstrated that Schrödinger's waves were packets of probability. As these sophisticated theories evolved, experimentalists in Berlin and at Cambridge University were probing the nucleus of the atom, discovering new sub-atomic particles and the incredible energy inherent in the atom.

The funds for this remarkably productive research were minuscule. When Rose visited, in the 1920s, the annual research budget of the famed Cavendish Laboratories at Cambridge totaled under one thousand pounds sterling. The possibility of Rockefeller money made Rose a very popular visitor. Wherever he traveled, to Copenhagen, Edinburgh, Göttingen, Stockholm, and Paris, he was received like a visiting monarch. On his trip Rose met Niels Bohr, J. Robert Oppenheimer, Enrico Fermi, Isidor Rabi, Samuel Goudsmit, T. Hogness, and Edward Condon—the leaders in theoretical and experimental physics—and made his first grant to Niels Bohr.

Rose unabashedly held views that later generations would label elitist. His favorite aphorism was: "Make the peaks higher." When someone on his staff pointed to the long tradition of grants the GEB had made to unknown agencies and institutions, Rose answered, "The high standards of a strong institution will spread throughout a nation and will even cross oceans."

His program was a radical departure for the GEB, which for a long time had directed its programs to rural schools in the south and improved education for blacks. To circumvent the inertia of the foundation staff, which was inclined to continue programs they had funded in the past, Rose reached beyond the foundation offices for advice and counsel, calling on an informal "club cabinet" drawn from the distinguished scientists he met in his travels, and from the men he saw regularly at the University Club, the Cosmos Club in Washington, and other familiar haunts. Foundation presidents make useful friends. Scientists who had known Rose before he became president of the GEB, and those who met him later, were eager to cultivate his friendship.

On questions of astronomy Rose turned most often for advice and counsel to George Hale, whose reputation, experience, and prestige were exactly what Rose sought in his advisers. On his visits to New York, Hale would often have lunch or dinner with Rose. Between visits Hale sent books and articles to Rose, and Rose wrote to get Hale's views on proposed GEB grants in support of observatories and astronomy programs. It was one of those mutually beneficial relationships that characterized the world of the University Club.

In the early spring of 1928, rumors circulated that Rose would soon retire from the presidencies of the GEB and IEB, and that a portion of the funds John D. Rockefeller had allocated for the IEB had not been committed. The unspent funds were not an endowment but a self-liquidating trust. When it was gone, if Rockefeller did not choose to renew the fund, the board would dissolve. The rumors among those who followed foundations were that the next president of the GEB would return the direction of the board's activities to fields like southern and black education in the United States, and that the IEB, which had been created at Rose's insistence, would be allowed to expire.

If the rumors were true, Rose's IEB was an orphan with money to burn. George Hale had been waiting for a long time to hear news like that.

Hale prepared his campaign carefully. In early 1928 he drafted an article for *Harper's Magazine* on "The Possibilities of Large Telescopes." *Harper's* was good seed ground. Foundation officials who see themselves as pacesetters, opening new frontiers of research, are profoundly influenced by the intellectual landscape around them. A serious but readable magazine was more likely to attract their attention than a scientific journal.

Hale had a knack for describing complex science in simple layperson's language. "Like buried treasures," he began, "the outposts of the universe have beckoned to the adventurous from immemorial times." Hale chronicled the continuing efforts to build larger telescopes, from the medieval Arab astronomers in twelfth-century Cairo to his own work on the Yerkes and Mount Wilson telescopes. "The latest explorers have worked beyond the boundaries of the Milky Way in the realm of spiral 'island universes,' the first of which lies a million light-years from the earth while the farthest is immeasurably remote. . . . While much progress has been made, the greatest possibilities still lie in the future."

Hale described the experiments Francis Pease had conducted on the one-hundred-inch telescope with an interferometer. The interferometer united the images from two smaller mirrors at the ends of a long beam into a single image. By comparing the images made with mirrors as much as twenty feet apart, Pease concluded that "an increase in aperture to 20 feet or more would be perfectly safe. For the first time, therefore, we can make such an increase without the uncertainties that have been unavoidable in the past." Hale's eloquent prose made the construction of a new telescope sound simple: "As there is every reason to believe that a suitable Pyrex or quartz disc could be successfully cast and annealed, and as the optical and engineering problems of figuring, mounting, and housing it present no serious difficulties, I believe that a two-hundred inch or even a three-hundred inch telescope could be built and used to the great advantage of astronomy."

It was a masterfully crafted article, clear enough for the untrained reader but with enough documentation and current information not to be dismissed as mere popularization by those who had followed recent developments. Hale elided over a few details that might have troubled a less optimistic reporter: No telescope disk had ever been successfully cast or annealed from Pyrex or quartz; no disk that large had ever been successfully figured; and there were no workable plans for a mounting that could precisely point an instrument that large or even support one in a stable configuration. Nor had anyone ever done the engineering for a structure that could house the instrument.

Hale's article also didn't admit that the challenge of casting, figuring, mounting, and housing the sixty- and one-hundred-inch instruments on Mount Wilson had far exceeded even the most pessimistic estimates. Hale did not mention the reservations of many astronomers, including Harlow Shapley, who questioned whether the seeing at any site was good enough to take advantage of the potential light-gathering ability of a huge telescope. Shapley's own interest was in seeing many midsize telescopes built, rather than a single large instrument. His views may have been motivated, at least in part, by jealousy of the growing fame of Mount Wilson, but his pessimism was widely shared.

Finally Hale did not venture a guess of the estimated cost for a telescope that large. It didn't take much extrapolation from the final cost of the one-hundred-inch telescope, which exceeded six hundred thousand dollars, to realize that it would require a larger grant than had ever been made by an individual or agency for any scientific instrument ever built.

Before the article appeared in print, Hale asked his editor at *Harper's* to forward a copy of the manuscript to Wickliffe Rose.

There is no record of when George Hale and Wickliffe Rose first discussed a large telescope. Their earliest talk was probably at the University Club. Neither seems to have kept a diary of these meetings, but Rose wouldn't have had to push hard to get Hale to talk about his favorite subject. They later met at the offices of the GEB at 61 Broadway. Rose, Hale reported to Walter Adams, was "so keen about a huge telescope that he [would] talk about nothing but the largest possible, and remark[ed] that there [was] no reason why fifteen millions should not be spent on such an instrument."

When Hale and Rose finally exchanged letters on the subject, their correspondence was in the stilted language of men laying paper trails to cover negotiations in progress. The salutations were formal—"Dear Dr. Rose:" and "Dear Doctor Hale:"—unusual for fellow club members, and in Hale's case correct only if you counted honorary doctorates. The language was formal enough to dispel accusations that these old boys had worked out too much in advance.

Hale's letter came right to the point:

> We now have definite observational evidence that at such a favorable site as Mount Wilson a large increase in aperature could be made with confidence. . . . May I ask whether the General Education Board would consider the possibility of making a grant to determine how large a telescope mirror it would be feasible and advisable to cast, with the view of providing later for the construction of a telescope considerably outranking the one-hundred-inch Hooker telescope?

Rose answered Hale within the week, in the imperial first person plural reserved for royalty and foundation presidents: "It is a matter that interests us. We shall be very glad to discuss it with you."

Rose had already scheduled a trip to Pasadena, to visit the California Institute of Technology, when he received Hale's letter. The distinguished faculty of the new institution comprised just the sort of men Rose favored with his fellowships. A visit to Mount Wilson and the Pasadena offices of the Mount Wilson Observatory was easily added to the trip.

H. J. Thorkelson, Rose's colleague at the GEB, had previously visited the observatory on a foundation scouting trip, in October 1926. He was given the standard VIP reception, met Hubble and van Maanen, ate dinner with the astronomers at the Monastery, saw Jupiter through the sixty-inch telescope, and was escorted into the dome of the one-hundred-inch telescope while it was busy on a multi-night exposure. After the tour Francis Pease showed Thorkelson his tentative drawings of a twenty-five-foot (three-hundred-inch) reflecting telescope, with a probable cost of some $8 million. Pease admitted that there were no real plans for construction or for securing the necessary funds. Thorkelson was impressed that much of the manufacturing work on the big instruments at Mount Wilson had been done by the staff at a saving of about "one-half the estimates submitted by manufacturing concerns."

Hale had another meeting with Rose in mid-March, before Rose's trip to California. The whole process seemed to be moving quickly, so Hale tried to get in touch with John C. Merriam, the president of the Carnegie Institution. Although the observatory functioned independently, with its own director, the Mount Wilson Observatory was a branch of the Carnegie Institution. Merriam was in Mexico and could not be reached before Rose and Thorkelson left for Pasadena and Mount Wilson.

The seeing was poor the night Rose and Thorkelson were on the mountain, but they were able to view the moon, Neptune, a double star, a nebula in Orion, and a star cluster through the one-hundred-inch telescope. The rest of the evening they talked with Walter Adams, Henry Norris Russell, and other astronomers who were at the observatory.

Pease again described his plans for a three-hundred-inch telescope. Rose asked what it would cost. Pease estimated the cost at $12 million, $4 million more than two years before. Pease explained that while a plate-glass disk could be impossible, "this disc could easily be made of Pyrex with very distinct advantages over glass," or of fused quartz, which would be better still.

Later that evening the group settled into a long discussion of what institution should run a very large telescope, if one were to be built. The astronomers favored an independent director and trustees who would administer an endowment and secure additional funds from time to time, with an advisory committee of astronomers—an administrative structure not very different from the organization of the Mount Wilson Observatory. They also argued strongly that one very large telescope would do more for science than additional telescopes the size of the one-hundred-inch telescope.

Rose disagreed with their suggestions for administration, arguing that a university was the right institution to run a big telescope. Walter Adams, who reported the discussions to Hale in New York, had the feeling Rose was talking about the California Institute of Technology. Adams was astonished: At the time the fledgling institution had no astronomy facilities and no astronomers or astrophysicists on its faculty.

On his return to New York, Rose met again with Hale, who had stayed at the University Club in New York during Rose's trip to California. Rose told Hale that he had been impressed by the facilities and research program at Mount Wilson, but that he had strong reservations about the administration of the institution. The observatory, he suggested, should be a separately endowed institution and not dependent on the Carnegie Institution of Washington for funds. Hale assumed that Rose had the Rockefeller Medical Institute, which he had administered for many years, in mind as an example.

As they talked that afternoon, Rose made no comments about the mechanical or optical plans for the telescope. From his trips to Europe and his hobnobbing with famous scientists, he had developed a style of his own, leaving the actual science to the scientists. His experience was as an administrator, and he considered himself an expert on issues of scientific administration. Hale was wise enough not to argue with Rose's institutional preferences, although he did point out that it would be difficult to build a large telescope without the advice and assistance of the staff of the Mount Wilson Observatory and their laboratories in Pasadena.

There is a peculiar, tentative choreography to the early discussions between a foundation and an applicant. Though both parties know they are discussing a potential grant, they dance around the notion of a formal application, wary lest they force a premature decision. Hale, from his experience with Yerkes, Hooker, Carnegie, and others, knew the quagmires in the way of any grant of funds. Donors, whether they speak for their own money or as officers of a foundation, have strong personal preferences that cannot be ignored. A slip of the tongue, a single reference to an unfavored concept, could kill the chances of the proposal. Yet Hale couldn't afford to be too cautious. The early rumors of Rose's retirement were now public, and the word among foundation watchers had it that when he was replaced, the boards would return to their previous funding priorities, which did not favor science. If a grant was to be secured, it had to be soon.

For hours that afternoon, as they sat in comfortable wing chairs at the University Club, Hale enthusiastically described the fused-quartz research that had been going on in the General Electric laboratory of the famed scientist and inventor Elihu Thomson, in West Lynn, Massachusetts, and the new Pyrex borosilicate glass that had been developed at the Corning Glass Works in upstate New York. Neither material had been tried in an astronomical mirror, but Hale assured Rose that both materials would be suitable for a very large telescope and that there were no technical problems in the use of either material for a mirror. The work would proceed in stages, Hale explained, working gradually up to a two-hundred-inch mirror, and perhaps to a mirror large enough for the three-hundred-inch telescope that Pease had designed. Hale was persuasive when he spoke about a pet project, and Rose was willing to be persuaded by a distinguished scientist with a reputation as expansive as Hale's.

Rose asked about a budget. They had only been discussing the possibility of a grant to explore the building of a large telescope mirror, but Hale had prepared for a budget question by getting figures from Gano Dunn, his friend and the president of the J. G. White Engineering Company, contractors for major projects worldwide. Building a series of mirrors up to two hundred inches in diameter, Hale said, would cost approximately half a million dollars. Building a two-hundred-inch telescope would cost in the neighborhood of $6 million. Endowing it with operating funds would require another $2 million.

"What about a three-hundred-inch telescope?" Rose asked.

Hale said that his estimates for a telescope that large were still vague, but the figure would be somewhere around $15 million.

"Well," Rose suddenly asked. "Do you want a two-hundred-inch or a three-hundred-inch?"

Hale was astonished. Until then they had only discussed the possibility of a grant to explore whether a large mirror *could* be built for a telescope. Rose's calm, unsurprised reaction to the budget figures was the first indication that his foundation might be ready to commit the funds to actually build a telescope. Neither of them had to say that a grant on the scale of Hale's proposed budget would be, by a substantial margin, the largest award ever made in support of a scientific project. Fellow club members could leave much unsaid.

Rose's question wasn't easy to answer. Hale knew the next telescope would be the last one he would shepherd into existence. The nervous breakdowns and the visits of tormenting demons were now coming so frequently that he had begun to husband the moments of clarity and peace, carefully dividing his energies between his own scientific research and the projects he held dear, like this one. He didn't have time to make mistakes.

The size of what the astronomers called the big telescope wasn't a virgin question. At Mount Wilson and in the library on Santa Barbara Street the astronomers had talked incessantly about the big telescope. Hale and his colleagues had spent countless hours going over Pease's sketch and model for a three-hundred-inch telescope. Still Hale was wary. It was one thing to write optimistically in *Harper's* for a general audience, or even in a preliminary feeler to a foundation. Now that the bait had been nibbled, how big a fish could they actually land?

Hale had spent years struggling with partially completed telescopes, fending off anxious donors who wanted to know why progress on the telescope that was to bear their name was stalled. Each leap in size, from the Yerkes forty-inch refractor to the sixty- and then the one-hundred-inch reflectors at Mount Wilson, had involved far more problems than anyone ever anticipated, and the new proposal entailed a leap in magnitude greater than any of the previous jumps. Much of the technology that had been extended and modified for each of the earlier telescopes couldn't be stretched any farther.

Hale didn't give Rose a direct answer that day. After the meeting he telephoned Walter Adams, the director of the Mount Wilson Observatory since Hale's retirement, and two other trusted advisers: J. J. Carty, the director of research at American Telephone & Telegraph, and Gano Dunn, the president of J. G. White. Neither Carty nor Dunn was an astronomer or had any special experience with large telescopes, but as the heads of the predecessor of the Bell Labs and of one of the largest engineering firms in the world, they did have experience with large-scale design and construction projects. All three men supported Hale's instinct to go for a two-hundred-inch telescope.

That night Hale had dinner with John Merriam. Hale repeated Rose's arguments favoring the California Institute as the body to administer a large telescope. Merriam made no objections. Hale stayed on at the University Club, waiting to hear back from Rose. He knew that Rose, too, was carrying on delicate negotiations with his colleagues and with selected members of the IEB.

On the morning of April 12, Rose and Hale met again at the club. Arthur Noyes, Hale's old professor from MIT and the chairman of the chemistry department at Caltech, joined them. Hale reported to Rose on his discussions with Adams, Carty, Dunn, and Millikan of Caltech. They had all supported the decision to go for a two-hundred-inch telescope and agreed that the proposed new observatory should be vested in the California Institute. In his diary Rose noted that even Carty, who was a member of the board of the Carnegie Institution of Washington, supported the decision that the telescope should go to Caltech, rather than to the Carnegie Institution, which ran the Mount Wilson telescopes. The process was suddenly moving quickly.

A few days later Rose sent a note to Hale: "I should appreciate it if you would write me an informal letter about the proposed telescope, in which you make all the essential points concerning the proposal in separate paragraphs as 1, 2, 3, 4, etc."

It was exactly the invitation Hale had awaited. Working in pencil, at the University Club, he drafted a proposal with the numbered paragraphs Rose requested. Most of his proposal was an expansion and rehash of the earlier letter and the *Harper's* article, but years of fundraising with men like Yerkes and Hooker, who liked to air their views, had made a good listener out of Hale and versed him in the art of praise. "As you have remarked," he wrote in the second-to-last paragraph, "the most important requirement in the construction and operation of the telescope is the close cooperation of the Mount Wilson Observatory and the California Institute."

The proposal included no details of the design of the telescope, because there was no design. Hale was selling an idea and a group of men. The budget in the last paragraph of his proposal was as vague as their talk at the University Club: "The best estimate of total cost I am now in a position to offer is six million dollars, to be expended over a considerable period (probably four to six or more years). . . . This estimate, which does not include endowment, is believed to cover the various contingencies necessarily to be reckoned with in a large undertaking of this kind."

When he finished the draft, Hale penciled in a few minor changes, then had it typed up and delivered downtown to Rose's office.

Hale had dealt with foundations often enough to know that the written proposal was a formality. The presidents of large foundations are quick to say, "I cannot commit the funds of the foundation to your proposal; only the board of trustees makes that decision." But foundation presidents do not solicit proposals for extensive and costly projects unless they plan to submit the proposal with their endorsement to the board. Unless the staff is out-of-step with its overseeing board, boards of trustees generally go along with the recommendations of the staff. The bigger the foundation, the more likely that the distinguished trustees are too busy to second guess the professional staff. And in this instance the grant was a terminal request to the IEB, which had been more or less Rose's private fund. With Rose's assurance that they could count on the support of the Rockefeller group, Hale left for Pasadena feeling confident about the proposal.

At the GEB offices, Thorkelson put together the supporting material for the proposal, requesting Mount Wilson photographs of spectra, spiral nebulae, star clusters, and exteriors and interiors of the Santa Barbara Street laboratories and the observatory, to accompany the formal presentation of the telescope proposal to the board. To wrap up the last loose ends, Rose scheduled a meeting at the Cosmos Club in Washington to discuss the project with Arthur Day, a member of the board of the Carnegie Institution and one of the inventors of Pyrex. At Hale's urging Millikan, Dunn, Adams, and others wrote letters to Rose endorsing the proposal.

For Wickliffe Rose there was only one step left before he presented the proposal to his board: a discussion with John Merriam, the president of the Carnegie Institution of Washington, to confirm its support for the telescope project. The call on Merriam was a formality. Rose assumed that the assurances of men he trusted (especially a University Club man like George Hale, whose name was almost synonymous with the Mount Wilson Observatory) were sufficient to guarantee the needed cooperation of the observatory and its labs on Santa Barbara Street.

Back in Pasadena Hale and his colleagues from Mount Wilson began discussing an appropriate site for the new telescope. The astronomers were deliriously happy in expectation.

8

The Politics of Money

Although they had not been direct competitors in business, there was little love lost between John D. Rockefeller and Andrew Carnegie. After reading Carnegie's essay "The Gospel of Wealth" with its pithy aphorism: "The man who dies rich dies disgraced," Rockefeller wrote to Carnegie: "I would that more men of wealth were doing as you are doing with your money." Before long the two men were competing to top one another's eleemosynary efforts. Carnegie announcements of new libraries and institutes fought for headlines with the grants various Rockefeller funds made for education, medicine, or science. In an era before baseball players and entertainers were famed for their salaries as well as their talent, Rockefeller and Carnegie were steady news.

Before income taxes the wealthy did as they pleased with their fortunes. Rockefeller's dining table used to be stacked with appeals he had gotten from Baptist missionaries, to whom he would personally send checks. Even after the introduction of income and estate taxes, the creation of foundations allowed the wealthy to maintain their incredible wealth intact, although with muckrakers exposing the sources of that wealth, controversy dogged some of the new foundations. Congress debated granting a charter of incorporation to the Rockefeller Foundation for three years, amid charges of "tainted money," "the kiss of Judas Iscariot," and "the Trojan horse," before turning it down. The foundation was incorporated in New York State.

By law, foundations had to maintain in their charters and in their public literature that they were *independent* decision-making bodies, and most observed the letter of the regulations scrupulously. When John D. Rockefeller established the IEB at Wickliffe Rose's request, he declined to serve as a trustee, on the grounds, as he put it, that "it is a technical field and I have no qualifications as an advisor." Yet even when he chose not to serve on the board, the presence of the founder was ubiquitous at the big foundations—not only in the paternal gaze from the oil portrait on the boardroom wall or the name chiseled in stone over the building's entrance and engraved on letterheads and business cards, but in the membership of the board and the officers of the foundation, who, as often as not, had been picked by the founder. When he set up his foundation, Rockefeller reserved the right to personally designate the recipients of up to $2 million from the annual bequests. He revoked that privilege in 1917 because it proved cumbersome, but even when he wasn't making grants himself, the staffs of his foundation knew where the money came from.

Men who owe status, respect, and an excellent job to the beneficence of a wealthy patron rarely forget the reciprocal responsibilities of patronage. His frail health had restricted him to a bland milk diet, but John D. Rockefeller was very much alive in 1928, issuing a constant stream of opinions, each with the force of a decree, on the many ventures and institutions that bore his name. Wickliffe Rose, the president of two Rockefeller-funded foundations, might be willing to trust George Hale on such questions as the technical possibilities of a very large telescope, and the exciting new technology that could be explored to build such a machine, but before a grant could be awarded, there were subtle questions of protocol, procedure, and politics that Rose had to consider. On those matters Rose was first and foremost a *Rockefeller* employee.

Hale's earliest conversations with Rose, and his first letter proposing the telescope, had suggested either the Carnegie Institution of Washington or the National Academy of Sciences as the logical organization to administer the proposed telescope. The Mount Wilson Observatory of the Carnegie Institution had more experience building and operating large telescopes than any other group in the world. The National Academy had no special facilities or experience with astronomy, but as Hale pointed out, they could offer a wide range of technical knowledge, and give the project a status "more nearly resembling that of the Marine Biological Laboratory at Wood's Hole." The National Academy as project sponsor also offered the promise of independence from existing facilities and institutions. But for Wickliffe Rose, both of Hale's suggestions were fraught with problems.

The Mount Wilson Observatory identified itself, on its stationery and on the entrance to its buildings, as part of the *Carnegie* Institution of Washington. Wickliffe Rose did not need special instructions to be wary of committing Rockefeller funds to a Carnegie institution.

Hale's other suggestion, the National Academy of Science, was also far from independent in the eyes of the president of a foundation funded by John D. Rockefeller. George Hale was an officer of the National Academy, and starting in 1914, he had repeatedly appealed to Andrew Carnegie for funds to build a suitable headquarters for the academy. Carnegie turned him down, but Hale was persistent, so much so that at one point he and Carnegie got into a heated argument about whether Hale had taken advantage of Carnegie's hospitality. The argument was patched over, and Hale's relentless pitches and a letter from Elihu Root finally persuaded Carnegie to donate the money to fund the new building in Washington in 1922. Carnegie's grant, and the beginning of work on the new building, were much publicized. To Rockefeller people, the National Academy of Sciences was another Carnegie institution.

John Merriam knew about Hale's proposal for the telescope long before Wickliffe Rose came to Washington to call on him. Hale had sent him a copy of the manuscript of the *Harper's* article at the same time he sent it to Rose. Merriam had written back that it was "extremely interesting," something they ought to talk about. Hale had also met with Merriam in New York to tell him about the proposed telescope. By then Hale and Rose had already agreed that the proposal would be submitted on behalf of Caltech. Since the Carnegie Institution was not the applicant, Hale had not made a formal request for the approval of the project by Merriam. That omission didn't sit well with John Merriam.

John Merriam had been second choice for his job as president of the Carnegie Institution, which included, at least on paper, oversight of the Mount Wilson Observatory. Hale had in fact been offered the presidency of the Carnegie Institution in 1919. He declined—the demons were already troubling him, and he wanted to stay close to Pasadena and the telescopes—and it was he who nominated Merriam, a University of California paleontologist, for the job. Although Merriam became Hale's titular boss, Hale's fame as a solar astronomer and his long association with the large telescopes—even long after he retired as director of the Mount Wilson Observatory—made George Hale's the name people associated with the great observatory.

Merriam seemed to many to have a permanent chip on his shoulder. He was stuffy, heavy-set, with long thinning hair swept carefully over the bald spots on his head. A goatee and tiny wire-rimmed glasses accentuated his stern expression. It was widely known that he had been second choice to George Hale for his job, and that he owed his job to Hale's recommendation. Like many men in positions of power, Merriam resented anyone to whom he was indebted.

When Rose came to Merriam's office and explained his reasons for directing the grant to Caltech, Merriam announced that he knew nothing about the proposal for a large telescope, that the trustees of the Carnegie Institution had not debated the matter, and that on his own he could not promise the cooperation that Rose was assuming from the Mount Wilson Observatory.

Rose was dumbfounded. He went back to his office and promptly wrote Hale:

After hearing from him [Merriam] a statement of the policy and program of the Carnegie Institution, it seemed clear to me that the program of the Institution should not be interfered with, and that if the
Board should appropriate funds for the proposed telescope it would
interfere with the normal development of that program and might be
a disservice. We are agreed that it would be inadvisable for us to
develop the idea any further.

The telescope proposal was dead.

As soon as he got the letter, Hale knew what had happened. The
"program of the Institution" that "should not be interfered with" was
John Merriam's ego and petty jealousy.

"As I understand Dr. Rose," Merriam wrote after Rose left his
office,

> he believes that Mt. Wilson Observatory is eminently fitted to do the
> work of planning, constructing and operating a two hundred inch
> telescope, but he believes the organization and policy of the Institu-
> tion inadequate to give proper guarantees for the utilization of such
> an instrument or the organization of its work in the future. . . . Much
> explaining would be necessary if a great telescope were given to the
> California Institute, without recognizing the fact that the Mt. Wilson
> Observatory is the leading observatory of the world today, that its
> program is well organized and effective, and its research is univer-
> sally recognized as one of the most effective programs in the world.

Men worried about appearances can be tough adversaries.

Fortunately George Hale's demons were temporarily at bay. He
called Walter Adams and Henry Robinson, the chairman of the
trustees of the California Institute. Together they telegraphed Merriam: "Have received report conversation between you and Rose which
Robinson, Adams and I feel must be distorted. Nevertheless am
informed it is likely effectively to prevent carrying out greatest project
in history of astronomy previously practically assured."

Hale then fired off a telegram to Rose, asking him to hold the
opinion of his associates "unprejudiced" until the "true attitude" of the
Carnegie Institution could be determined. Rose wrote in his diary that
he considered his letter of the same day—a confirmation that the proposal was dead—a sufficient answer to Hale's telegram.

Ignoring Rose's unambiguous letter, Hale mobilized his troops. He
got Robinson, Millikan, Adams, Dunn, and Carty to write Rose. He
telephoned Carty, who agreed to do everything in his power to change
Merriam's attitude and to convince Rose of the Carnegie Institution's
willingness to cooperate.

Once he had covered every base he could reach from Pasadena,

Hale took the next train east. Transcontinental trains weren't quite the novelty they had been in 1921—the funeral journey of President Harding from California to Washington in 1923 had introduced America to the miracle of the trains—but the East Coast was still a four-day journey from Pasadena in 1928.

On the train Hale tried to figure out Merriam's actions, jotting notes to himself on a yellow pad. If Merriam intended not to cooperate from the beginning, Hale asked himself, why hadn't he told Hale when Hale reported his initial talks with Rose and his desire to work out a plan of cooperation? The only answer was that Merriam was petty and insecure, more interested in his own status or how the world saw him and his institution than in the future of science.

As the train approached Chicago, Hale had to decide whether he should go to Washington to confront Merriam or to New York to try to limit the damage at the IEB. He spoke to Carty by phone from the station in Chicago and learned that Rose had called off a planned lunch meeting that had the proposal on the agenda, that the Rockefeller board had met without considering the project, and that as far as Rose was concerned, the entire project was dead. Hale opted for New York and damage control, caught the Twentieth Century in the nick of time, and when he arrived in New York, called Rose to request a meeting. Rose was cordial as he confirmed that the problem was Merriam's lack of cooperation and his refusal to commit the Carnegie Institution to the project.

On Hale's way out of the Rockefeller Foundation offices, Simon Flexner, a member of the Rockefeller board, told him that if he could bring Merriam around and guarantee the cooperation of the Mount Wilson Observatory, it might be possible to revive the proposal. The new associate for science at the Rockefeller Foundation, Max Mason, agreed to lobby the members of the executive committee to keep the proposal open. The problem, Hale knew, was that Merriam, who had a reputation for stubbornness, was adept at using the mechanisms of power, like his authority to call meetings or to control agendas, to get his way. Arguing with Merriam would do no good and would only stiffen his resolve.

Fortunately there was one man who could force Merriam's hand. The next day, May 5, Hale and Carty called on Elihu Root, the chairman of the board of the Carnegie Institution, at his apartment in New York. The old diplomat was eighty-three years old. He had served presidents and his country for more than half a century, untangled the most complex of diplomatic thickets, persuaded potentates and presidents, prevented and started wars, and counseled the mighty in government, academe, and industry. His bushy mustache and quick mind were recognized on every continent. His triumphs in diplomacy and statesmanship had earned him the Nobel Peace Prize, vast respect and fame, and the prestige of chairmanships and memberships on distinguished boards. Now, his days in a cutaway long behind him, he had retired from all but a few boards to care for his wife of fifty years, who was dying a slow and painful death from arthritis.

Hale found Root vigorous, cordial, and eager to help. He stood as erect as a "young West Point graduate," dressed, as always, in a salt-and-pepper suit, with an old-fashioned, flat-lapeled waistcoat, a soft blue shirt, a polka-dot tie, and the high-topped black shoes of his generation. Hale and Root had been friends for a long time, and Root, the son of a mathematics professor, was sympathetic to science. Root also had personal acquaintances or friendships with everyone involved in the dispute. He listened in his usual style, disconcerting to those who didn't know him. Without saying a word, all the while scribbling notes on little scraps of paper, Root seemed as though he wasn't paying attention at all.

When Root finally spoke, it was obvious that he understood the situation, that the strong personal rivalries of Carnegie and Rockefeller made it absurd to imagine that a *Rockefeller* foundation would make a gift on the scale of the proposed telescope grant to the *Carnegie* Institution. Root knew them both well, and characterized the old adversaries succinctly: "Carnegie . . . always on the warpath, armed with a tomahawk, and Rockefeller . . . smooth and compromising in most of his dealings." As chairman of the Carnegie Institution, a position he had held since 1915, there was no question where his own sympathies lay. With a sly grin Root said that it might just be an ideal partnership: The Carnegie Institution could supply the brains while the Rockefellers provided the funds.

Hale reminded him that the question of the hour was politics, specifically how to smooth Merriam's ruffled feathers and somehow get the grant application back on track. Root was reluctant to interfere in day-to-day operations of an institution he chaired, or to seem to overrule the professional staff. But if Merriam was stubborn when his authority was on the line, George Hale was persuasive when one of his projects was threatened. Hale brought out the draft of a letter to Wickliffe Rose, written for signature by Root and Merriam.

Root read Hale's draft silently. The language was direct, graciously enumerating the "appreciation" of the Carnegie Institution for the proposed gift to Caltech and the opportunities that would be offered to the members of the staff of the Mount Wilson Observatory to carry their research "beyond present possibilities" to the region "beyond the Milky Way" and promising the "most cordial cooperation" of the Carnegie Institution of Washington in the proposed project. When he finished reading, Root signed his name, and had the letter sent to Merriam in Washington for his signature. Root also gave Carty and Hale permission to contact the members of the board of the Carnegie Institution directly, bypassing Merriam, to lobby for support of the project.

In Washington, Merriam refused to sign the letter. He wouldn't say whether he was still hoping that somehow the big telescope would come to his own institution or simply insulted that the conception of the project had bypassed him. It didn't matter. John Merriam's pride had met its match in Elihu Root. When the report of Merriam's stonewalling came back to New York, the old diplomat summoned him to New York to discuss the matter. Merriam could not ignore a summons from Elihu Root. He took the next train to New York.

Hale watched the drama unfold from his command post at the University Club. He had recruited a delegation of Frederick Keppel, Carty, and Dunn—all board members of the Carnegie Institution—to meet Merriam at Penn Station and prepare him for the expected confrontation. There were no seats in the grand concourse of the station— Charles McKim thought that benches would detract from the Baths of Caracalla grandeur of his design—so the delegation lined up on the platform under the sunlit vaults of the concourse to await Merriam's train. With their hats in hand, the three men were exactly the sort of high-level delegation, befitting his position, that Merriam enjoyed.

When he got off the train, Merriam brushed the three men aside. He was going to meet with Wickliffe Rose, he announced. Carty urged him to consider the position and objectives of the Carnegie Institution over his personal feelings, warning that he could bear the sole responsibility for the failure of the two-hundred-inch project and the consequences to science that depended on a new telescope. Merriam refused to discuss the matter. "No one will have anything to regret," he said.

Carty telephoned Hale with the news. Even after years of dealing with Merriam, no one knew what to expect from the man.

At the GEB offices, Thorkelson, who had overseen the telescope proposal, was invited to join Rose and Merriam as Rose explained that the board was firm in its conviction: the only institution they would consider for a grant to own the telescope was the California Institute. To Rose's surprise Merriam didn't argue. At the time of their previous interview in Washington, he said, he had had only a brief conference with Hale, he had not received a formal request for cooperation from the California Institute of Technology, and he had no knowledge in detail of the plans for the telescope and its operation. Now that he had "further information," he and the executive committee of the Carnegie Institution had had an opportunity to consider the matter, he felt confident that the Carnegie Institution would offer its support, and he looked forward to the time when "one or more great telescopes and great coordinated programs for astronomical research may come into being."

Merriam was dissembling. He had no new information. The executive committee of the Carnegie Institution hadn't met, and the hurried discussions by telephone and telegram had been about his actions, not about the telescope. John Merriam was too proud to admit that he had backed down to avoid being taken to the woodshed by his boss.

Hale called Rose after he heard about Merriam's meeting. Rose assured Hale that with the formal cooperation of the Carnegie Institution, the grant application could now be reactivated.

Late that afternoon Merriam phoned Hale to suggest they have dinner together. The evening at the University Club was "as pleasant . . . as though nothing had happened," Hale wrote afterward. Merriam "explained" that his earlier position was based on the project as it stood when the subject was first broached to him. To mollify Merriam's bruised ego Hale prepared a formal request for cooperation, which he telegraphed to Millikan at Caltech. Millikan promptly got the approval of the board of trustees of the California Institute and wired the request to Merriam the next day. Everyone waited while Merriam went through the ritual of polling the board members of the Carnegie Institute, so *he* could announce their support for the proposal.

Even with all the formalities complete, Wickliffe Rose worried about the proposal he was about to bring before his board. Six million dollars was the largest request ever submitted to any Rockefeller foundation for a single project. The proposal for a very large telescope was at odds with the charter and the previous programs of the IEB. There was nothing international about the telescope project—except that a few consultants, none of them from Mount Wilson or Caltech, had suggested a site in the Southern Hemisphere for any large telescope— and the telescope grant, if approved, would deplete the funds of the IEB, leaving Rose's successor with nothing. Despite his confidence in the telescope proposal and in George Hale, Rose was afraid Mr. Rockefeller would consider the project inappropriate. When Rose told a colleague, Raymond Fosdick, of his concern, Fosdick said there was only one way to find out and brought the question directly to Rockefeller.

The old tycoon, modest in this as in few other subjects, said, "I have no competence in the field of astronomy. Six million dollars is a large sum of money, but I have complete confidence in Mr. Rose and the trustees, and if after careful investigation they decide that it is the wise thing to do, there certainly will never be any criticism from me." With John D. Rockefeller's blessing, the grant was assured. The proper procedure of a board meeting was now only a formality.

It would take months of negotiations between the lawyers at the Rockefeller Foundation and at the California Institute before a formal letter confirming the grant was ready. The letter was even less detailed than George Hale's proposal, though it did borrow from the language of the long tradition of bequests for telescopes. The two-hundred-inch telescope, according to the terms of the grant, was to be "as complete and perfect as possible."

Little did anyone realize what efforts those words would require.

9

Elation

George Hale celebrated his sixtieth birthday in June 1928. In a quiet moment at his beloved solar laboratory, he reflected on the task ahead: "I wish I were thirty years younger and able to jump into the task as I did at Mt. Wilson." He worried that he and others "may not outlast" the work ahead. "Building a large telescope, as I have found before, is not a rapid job."

Where would they begin? In presentations to Rose and other staff members at the IEB, and in authoritative-sounding letters to support the proposal, Hale, the scientists and engineers at Mount Wilson and Caltech, and friends of the project on the East Coast had all taken the position that the telescope was only one more step in the progression of telescopes. Gano Dunn wrote to Wickliffe Rose: "The proposed new five meter objective is not a Great Eastern but jumps ahead of existing construction in a lineal ratio that does not materially depart from the ratio of previous jumps which have been taken with complete success." It was a good argument to persuade a foundation that they were betting on a winner. But the new telescope wasn't just one more step in the progression of technology and science. Even at the start there were some who worried that the proposal reached too far, crossing the line from optimism to hubris.

America was still the land of optimism and superlatives. Thousands of Americans greeted themselves every morning with the ritual phrases of the French pharmacist Émile Coué—"Every day in every way I'm getting better and better." American technology seemed unstoppable. A new piece of furniture had appeared in American living rooms, a tall, handsome, wooden console—people called it "radio" instead of wireless by 1928—with brand names like Zenith, Philco, and Atwater Kent. Radio brought the immediacy of news, live entertainment, sports events, and reports of new technology to remote hamlets. Airplanes regularly flew mail and even passengers. Ambitious young men were hustling funds and machines for the challenge of flying the Atlantic nonstop. It was only a question of time before someone did it. The whole country seemed to follow Coué. Every day in every way ships and bridges were longer, airships were bigger, buildings and the stock markets were reaching higher. Only the most cynical of pessimists predicted it couldn't go on forever.

Yet amid the optimism and progress, science in America remained a cottage industry. If there seemed to be no limits to what technology could do, science worked on hand-to-mouth budgets. The previous big telescopes at Mount Wilson, the largest scientific projects in the country, had been built piecemeal, dribbling along as Hale raised the necessary funds. The designs were necessarily compromises, balancing budgets against technology. The size of the sixty-inch reflector was determined by the crucibles at the French glass foundries; the disk was the largest blank they could cast in a single pour. William Hale's gift for the disk would not cover the research or development of larger crucibles. The one-hundred-inch telescope had been a balancing act between the limits of plate-glass and riveted-structure technology and the constant struggle for funds from Hooker and the Carnegie Institution. The trials of casting and figuring the mirror—plate glass was the only material they could possibly afford in the budget—had been enough to precipitate the worst sieges of George Hale's demons. The mounting for the telescope was also a compromise; the closed-end English mount was the only way to guarantee a rigid support structure at least within the existing technology. The budget for the one-hundred-inch telescope did not leave room to explore new technology.

It was different this time. The funding for the two-hundred-inch telescope was committed, $6 million promised, under an arrangement that allowed the California Institute to draw down the funds as needed. But even with the money "in the bank," Hale and his colleagues faced a daunting task. No part of the new telescope could be ordered off the shelf. No company or institution had the experience to serve as the equivalent of a general contractor, because no one had ever built a machine to the scale and tolerances this one required. The closest parallels were enormous construction projects like the Panama Canal, but they were challenges of scale more than technology. The public loved stories about the plans for an Empire State Building, ships like the *Normandie,* the Boulder and Grand Coulee Dams, and the Golden Gate Bridge, with their superlatives about the tallest buildings and bridge towers, the deepest cofferdams, the longest ship, or the most tons of concrete poured, but each proposed project was only an increase of scale on an existing technology. For the new telescope there were few working technologies to extend.

Even the administration for the project put Hale and his colleagues on virgin ground. The earlier big telescopes had been funded through institutions—first the Yerkes Observatory of the University of Chicago, then the Mount Wilson Observatory—but in each case the institutions had been so closely associated with the name and fame of George Hale that the benefactors who had provided for the telescopes dealt directly with Hale and held him accountable when progress on the instrument that was to bear their name fell behind schedule. As complex as the projects were, they could be managed through the sheer energy of one man, and there was a convenience and efficiency in the simple administration.

Not even George Hale at the peak of his abilities had the energy to manage this project alone. The two-hundred-inch telescope was one of the first ventures into *big* science. No one had ever tried to research so many aspects of technology simultaneously, except perhaps the weapons labs the army and navy had established during World War I. The military, especially in wartime, enjoys a freedom that no civilian project can match: virtually unlimited budgets, the right of eminent domain to build what they want where they want, and the privilege of calling on every university, laboratory, corporation, and individual with the expectation that patriotism, if not payment, would elicit instant cooperation. The resources available to the military were a far cry from the hopes, dreams, and self-confidence of a handful of scientists gathered around the library table at Hale's solar laboratory on a side street in San Marino, California.

When Wickliffe Rose wrote, in late May 1928, to report that the IEB had given the observatory proposal "sympathetic consideration," his letter was vague. He did not say whether the IEB had approved the full amount of $6 million, or whether there were any conditions attached to the grant. It wasn't until Rose came out to Pasadena that he explained that the grant was only for construction of the telescope and the accessories needed for a complete observatory. The California Institute was to bear the expense of operation. When Rose asked Hale how much endowment the telescope would require, Hale said that Adams's latest annual budget for Mount Wilson was $250,000 and that he estimated the annual budget for the new observatory, with only a single telescope, at between $100,000 and $150,000. At 1928 interest rates, an endowment-grade investment to capitalize that annual return required between $2 and $3 million, which Hale would have to secure before they could be assured of the grant.

The next day, a meeting was scheduled at the solar laboratory to discuss the terms of grant with officials from Caltech. Before the meeting, Hale gave his estimate of the annual expenses of running the telescope to Henry Robinson, a wealthy Los Angeles businessman and the member of the board of trustees of the California Institute who had been most supportive of the telescope project. Robinson told him not to worry. At the meeting, Robinson said he would undertake to secure the necessary endowment. Robinson later told Hale confidentially that he would personally fund an annual endowment of $150,000, and ultimately capitalize the sum.

Rose, pleased by the prompt action of the trustees, assured Hale that they could begin work on the project immediately. Formal approval of the grant would not take place until autumn, but among fellow University Club members, a round of handshakes was enough to inaugurate the largest scientific project ever undertaken.

The first decision on the new project was to designate the solar laboratory as project headquarters. The lab was convenient for Hale. Even more important, it was at least symbolically distanced from both the Mount Wilson offices on Santa Barbara Street and the Caltech facilities on California Street. From the earliest discussions with Wickliffe Rose, there had been sore points in the arranged marriage of Caltech and the Mount Wilson Observatory. A neutral home for the project would minimize the complaints of too much influence from one set of in-laws or the other. Before long, as the pace of meetings and decisions picked up, the library of the solar laboratory became a command post, close enough to stay in touch but still removed from the inevitable pressures of expanding Caltech and the ongoing programs of the Mount Wilson Observatory.

Formally, the grant from the International Education Board was vested in an Observatory Council consisting of Hale, Millikan, Robinson, and Noyes, with John Anderson, an excellent optician and physicist from Mount Wilson, as executive officer. Although only Hale was an astronomer, Millikan and Noyes were the other members of Caltech's Big Three, the men who had been credited with attracting the faculty and funds to the small school. Henry Robinson was a member de facto as the chairman of the trustees of the California Institute. Caltech would pay one-half of Anderson's salary from the grant funds in return for his handling the day-to-day administration of the project.

An advisory committee was appointed to provide a forum for Hubble and the other astronomers. The Observatory Council and the Advisory Committee both met regularly, but the bulk of the work in these early days fell to Anderson, Hale, and the staff that they assembled. To the world George Hale was the man who built big telescopes. Long after it was announced that John Anderson was executive officer of the project, a disproportionate amount of mail and queries were directed to George Hale. Contractors, bidders, job seekers, reporters, publicity seekers with crackpot ideas, and the curious public all sought George Hale. To the extent his energies and the demons allowed, Hale assumed command from his headquarters in his solar laboratory, firing off endless memoranda, letters, and telegrams. His early influence on the biggest decisions for the telescope—design staff, siting, and technology—put his firm imprint on the project.

The first appointment was easy. Francis Pease of the Mount Wilson staff, who had designed much of the one-hundred-inch telescope and whose preliminary drawings and model of the three-hundred-inch telescope had been bantered around for years, was named Associate in Optics and Instrument Design. Pease had taken shop courses at the Armour Institute of Technology in Chicago as a young man, and that introduction to machines and mechanics, together with his training in astronomy and observational experience, gave him an ideal background as a telescope designer. He had been working on sketches and models of a big telescope for so long, and had shared his ideas with so many individuals, both staff at Mount Wilson and visitors, that his preliminary designs became the points of departure for the project. From the beginning Pease was pressured from both sides. The engineers considered him too much the astronomer, lacking the nuts-and-bolts problem-solving mentality of an engineer. To the working astronomers Pease was an astronomer manqué, who was now spending most of his time on machine design rather than research.

Hale was convinced that the new telescope was such a leap in scale that any design based on earlier telescopes and earlier technology wouldn't do the job. As a counterweight to Pease, he reached out for an inspired if unlikely choice for a second designer. Hale had been friendly for a long time with Albert Ingalls, of *Scientific American* magazine. One of the most popular features Ingalls had introduced to the magazine was a series of articles on amateur telescope making, written and illustrated by Russell W. Porter of Springfield, Vermont. Porter was a machine designer, an avid amateur astronomer and telescope maker, and a superb illustrator. His amateur telescope-building articles, later collected in a book, were so popular that Porter became a regular monthly columnist of the magazine.

Russell Porter had a background most employers would dismiss as too eccentric to trust. As a young man he abandoned his studies of civil engineering and architecture at MIT to sign up as an illustrator and surveyor on an arctic expedition. In the early years of the twentieth century, expeditions to the North and South Poles—then the great unknown for explorers—were much celebrated in the press. For all the publicity, there were often places available on the expeditions, because most men who had tried the arctic once couldn't bear to go back. Except Russell Porter. He found the stark landscape irresistible. He went back again and again, whenever an expedition could find room for him. He was attacked by a polar bear, had his gun freeze in his hands, got frostbite on four fingers when his mitten came off, and learned the threat of snowblindness to an artist. His duties on one expedition included observations with a Repsold Circle, an instrument designed to measure altitude and azimuth within four seconds of arc. The experience got him used to observing in conditions that made even the most rigorous observatory run seem luxurious: cold of –30° F, working barehanded because the instrument was too delicate to manipulate with gloved fingers. Often his hands were too cold to hold a drawing pencil, and once the kerosene in his lamp froze to a slush.

By the age of thirty-four, Porter had completed nine arctic trips and an attempt on Mount McKinley. He had become partially deaf from continued exposure to the cold.

When he finally abandoned arctic exploration, Porter tried his hand at architecture, then settled down to work for the Jones and Lampson Machine Company in his hometown of Springfield, Vermont. The president of the company, James Hartness, an ex-governor of Vermont, an old friend, and an amateur astronomer, recognized Porter's genius for simple, functional designs and got him interested in telescopes. Before long Porter's ability to render complex machines in pencil and charcoal drawings found a focus on telescopes, particularly instruments that could be constructed by an amateur. A series of innovative instruments of his design were built by members of the Springfield Telescope Makers. The group's observatory, Stellafane, which Porter also designed, became a mecca for amateur astronomers and telescope makers.

Hale was always on the lookout for people who knew something about telescopes, and by 1927 Porter and Hale were corresponding. Porter wrote an article for Hale's *Astrophysical Journal* on using photographs of knife-edge shadows for optical tests, and Hale sent him the plans for a spectroheliograph designed for amateur construction. Hale was impressed by Porter's designs and drawing skill. Early in 1928 he got Albert Ingalls to set up a dinner in New York for the three of them. Porter's hearing difficulties made him a poor listener; he compensated by talking throughout the meal, drawing sketches to illustrate one concept or another on any available piece of paper, including the menu and napkins. Hale was impressed.

Late in the summer of 1928, Hale wrote Ingalls to ask if he thought Porter would come to work on the two-hundred-inch telescope for a year for $3,500 to $4,000. Ingalls said it was hard to guess about a man like Porter, so Hale sent Anderson and Francis Pease to Vermont, to find Porter and feel him out.

Porter was off on an all-day picnic when the two men from California arrived at the Hartness plant in Springfield. James Hartness entertained the visiting dignitaries, giving them a hard sell on Porter's talents. Porter finally showed up in his old Ford and was sent in to talk to the two men. No one let him know who they were, and Porter as usual did more talking than listening. When Pease and Anderson managed to squeeze in enough words to explain that they had already missed several trains, Porter offered to drive them to the station, talking nonstop the whole way about the picnic, the virtues of Vermont, the glories of the arctic, the iniquities of the expedition leaders, and the limitations of being an artist. It took a half hour before Anderson was able to ask Porter if he knew anything about astronomy. "Not a great deal," Porter shouted. "It's a big subject. Some other time I'll be glad to tell you what little I do know, though."

Anderson and Pease were on their train, ready to depart, when they finally told Porter that they had come to offer him a job working on the two-hundred-inch telescope. For the first time that afternoon Porter was speechless.

Russell Porter had never worked professionally on a telescope, and never on an instrument larger than an amateur could build. He was fifty-six years old and settled in his ways. It took a lengthy exchange of telegrams before he agreed to come to Pasadena at the generous salary of six hundred dollars per month. It was to be a temporary stay, only until July 1929, so Hale urged him not to bring household goods but to plan on renting a house. Porter's job was deliberately described loosely: He was to help in designing the two-hundred-inch telescope, various auxiliary instruments, and the buildings to house them.

Like other employees of the project, Porter was put on the payroll of the California Institute. Porter, who had never gotten a degree in engineering or architecture, was an odd appointment for the new school, which had widely publicized the credentials of the distinguished scholars who had accepted early appointments. The engineers and physicists called him "the old guy," wondering who he was and why he was there. Soon the old guy was drawing architectural sketches for an astrophysics laboratory and machine shops for the campus of the California Institute. He had a unique style, using charcoal and careful smudges with his fingers to provide the "atmosphere" of big machines to his pencil sketches and renderings. Porter drawings were soon a familiar sight on the walls of the temporary buildings on California Street.

Even with Pease and Porter working full-time on the project, Anderson had his hands full. The heart of the telescope, the two-hundred-inch mirror, was his responsibility.

Stripped to its essentials, an astronomical telescope is a few grams of silver or aluminum, the reflecting surface that gathers and focuses the faint light of distant objects. Everything else—the precisely ground base of the mirror, the immense mounting, the electrical and mechanical controls, the instruments and cameras, the huge dome, the auxiliary optics—are only there to position and support the reflective surface of the mirror, to protect the instrument when it is not in use, or to enhance the image focused by the mirror. The crucial question for the two-hundred-inch-telescope project was whether a fine-enough mirror could be built.

The skeptics were convinced that it couldn't be done. The Bureau of Standards, *the* federal agency for science, had already pronounced that a telescope larger than the one-hundred-inch was a technical impossibility. The bureau's opinion carried some authority, because they had cast the first large glass disk ever poured in the United States, a sixty-nine-inch, three-thousand-pound blank for the Perkins Observatory at Ohio Wesleyan University. They had good evidence to support their claims.

"It is an open secret," the *New York Herald Tribune* wrote in 1928, "That the 100-inch reflector at Mount Wilson, now the world's largest telescope, has been something of a disappointment." There had been constant rumors of troubles with the one-hundred-inch telescope. Astronomers with grievances real or imagined against the Mount Wilson Observatory, including Harlow Shapley, who feared that money committed to the Pasadena observatories would not be available for the Harvard College Observatory, did their best to exploit the rumors. Shapley would tell anyone who would listen about the woes of the Mount Wilson telescope, including gossips like H. L. Mencken, who could be counted on to spread the bad news without checking it. What non-astronomer would question the judgment of the director of the Harvard College Observatory?

George Ritchey, fired from Mount Wilson at the end of the war, had gone to his lemon farm in Azuza, then to France, where, with the backing of a wealthy engineer, Assan Farid Dina, he was going to build a 104-inch telescope. The telescope project foundered, but in a series of published articles, Ritchey recounted the history of American telescope building, with himself as hero and Walter Adams and George Hale conspicuously missing. Privately he wrote scathing notes about the "wealth and power and egotism" of the officials at Mount Wilson, and rarely missed an opportunity to criticize the design of the one-hundred-inch telescope.

The mirror of the one-hundred-inch Hooker telescope *was* temperamental. Often, for much or all of an otherwise excellent night, the telescope would be unusable or marginal. The best explanation anyone could offer was that the great mass of plate glass was slow to adjust to changes in ambient temperature. The Mount Wilson opticians and astronomers fiddled with the mirror constantly. They tried packing the mirror cell with insulation, different cycles of opening and closing the dome to control the temperature inside, retuning the mirror supports, even removing the mirror during the day and storing it in a cork-lined cell until night. Some of the experiments helped, and the telescope was used effectively by Hubble, Humason, and others. But despite the constant attention, the one-hundred-inch didn't perform quite as they had expected. Hale and the Mount Wilson staff did their best to keep the bad news quiet, not only as a potential embarrassment but because it was powerful evidence for the skeptics who had already begun sniping at the two-hundred-inch project.

Everyone agreed from the start that the repeated frustrations in the efforts to cast the disk for the one-hundred-inch telescope were convincing evidence that it would be impossible to cast a larger disk of plate glass. Even if a two-hundred-inch mirror could be cast, the best estimates were that a mass of plate glass that huge might require several decades to anneal. The only answer to the troubles with the one-hundred-inch mirror, which seemed to be caused by the relatively high coefficient of expansion of plate glass, would be a new material. But what? The earliest reflecting telescopes, like Newton's instruments or the huge telescope Lord Rosse had made, had mirrors made of speculum, a bright metal alloy of copper and tin. Speculum couldn't be figured to the fine optical surface that modern standards required; it was even more sensitive to changes in temperature than plate glass; after a fresh polish the metal surface reflected only 70 percent of the light that hit the mirror; and every time the mirror was polished enough material was removed to effectively refigure the shape. No recent large telescope mirror had been made from any material other than plate glass.

Early in 1928 Hale had John Anderson draw up a memorandum on the qualities needed for a large astronomical mirror. The material used, Anderson wrote, had to be shapable into the form of a surface of revolution—paraboloidal, hyperboloidal, or plane—the exact geometry to be chosen contingent on the final optical design of the telescope. Because in use the reflecting surface would be exposed to the air, the heat exchange between the air and the mirror had to set up minimal mechanical forces that would tend to distort the mirror from its optical shape. Finally the disk had to be rigid enough to enable it to be supported in all positions without any appreciable change of shape.

The only materials that would meet the first requirement were glass or other "hard transparent substances," or certain metals or alloys, such as speculum, Magnalite, or chromium. Common metals, such as iron, nickel, aluminum, or silver were impossible to polish to an optical figure, because burnishing metal to a high polish would remove enough material to alter the optical shape. There had been some experiments that had achieved an acceptable optical surface on copper, gold, and tin, but never on pieces larger than two inches square, at exorbitant costs, and with the problem that the materials changed shape dramatically with changes in temperature.

Still the experiments went on. The Philips Lamp Works in Eindhoven, the Netherlands, had fabricated surfaces of glass fused to a chrome-iron backing. The problem with this approach was that if the coefficient of expansion of the backing were different from that of the glass, a change in temperature would introduce strains that would ultimately distort the surface.

That left glass or another transparent material. The surface of a glass mirror would ultimately be coated with a film of silver or another reflective material a few molecules thick, so the actual transparency of the material was not important, although it did have the advantage of allowing the opticians to check for internal strains in the disk. Ordinary plate glass was out; from the experience of the one-hundred-inch telescope, it was clear that there was no way it could be made to work on a mirror of even greater mass. The search was for an alternate glass or glasslike material.

George Hale knew exactly where to look.

Elihu Thomson was one of those inventor-scientists who found a natural home at the General Electric Company. Next to Thomas Edison, he was the greatest inventor in the company's history, with close to seven hundred patents for electric welding, transformers, centrifuge cream separators, three-phase AC windings, load regulators, magnetic switches, carbon-brush motors, electric-usage meters, electric refrigeration, and control circuits that made electric traction for trolleys possible. His company, Thomson-Houston, had merged with Edison's company to form General Electric, and his inventions earned a fortune for GE. In return the company provided generous compensation and built him a substantial research laboratory in West Lynn, near his Swampscott, Massachusetts, home. So much of what he tried ultimately proved successful for the company that GE gave Thomson virtually free rein to pursue his own directions in research.

Among his many interests, Thomson was an avid amateur astronomer. When he was thirteen, his mother had taken him to see a celestial display of meteors and comets; he later experimented with magnifiers that he sold to his friends, and before long he was inventing optical grinding and polishing procedures. Some historians give him credit for discovering that by rotating one flat glass disk over another, with abrasive between the surfaces, a spherical concave shape is produced in the lower disk—the essential procedure for figuring telescope mirrors. His private observatory at his home in Swampscott was as large and well equipped as the facilities at many universities.

Thomson's own telescope was a refractor he built from glass disks cast by the Paris firm of Mantois, but he understood the problems of mirrors for large telescopes. In 1899 he began experimenting with mirrors in the carriage house of his estate. He started with small concave mirrors, too crude to use in a working telescope. Thomson would focus the image of an artificial star (a point source of light) on two mirrors, one of glass and the other of fused quartz, then compare the focused images as he heated the backs of the mirrors with a flame. The heat would quickly distort a glass mirror, scattering the image. With the quartz mirror it took a considerable period of heating before the image was distorted. Quartz, Thomson concluded, could be an ideal mirror material: If a mirror could hold its figure under the heat of a torch, it would be all but immune to the effects of changes in the ambient temperature at an observatory.

George Hale, who kept his ear to the ground for new technologies that might be applied to astronomy, heard of Thomson's experiments and sent George Ritchey out to visit Thomson in 1904, offering a modest grant of three thousand dollars from the Carnegie Institution for additional experiments with fused quartz. At the time Hale was building the Snow solar telescope, one of the first instruments on Mount Wilson. For a solar telescope, in which the mirrors are exposed to the heat of the sun, the low coefficient of expansion of quartz promised a revolution in optical performance. Thomson experimented until the funds were exhausted and Ritchey had to return to the pressing work of grinding and polishing the mirror disk for the sixty-inch telescope. Thomson did not succeed in producing a usable mirror, and with a dozen other projects under way at the same time, some of them big moneymakers for General Electric, he was soon distracted.

Despite his failure to produce a working mirror, Thomson's experiments were too tempting to ignore. Fused quartz had such a low coefficient of expansion that Thomson calculated that a bar one meter long, raised from room temperature to 1000°C, or near the melting point of gold, would expand by approximately one-half of one millimeter. In the range of temperatures ordinarily encountered at an observatory, the expansion and contraction would be close to negligible. A fused-quartz mirror would be all but immune to the problems that plagued the one-hundred-inch telescope.

A fused-quartz mirror would also be far more efficient to grind and polish than plate glass. Figuring an optical mirror to its final shape is an exquisitely slow process, because the heat generated in polishing affects the optical shape of the surface. In the final stages of the figuring of a mirror, a brief stroke or two with jeweler's rouge on a tool or a fingertip can heat the mirror enough to produce distortions. The optician then has to wait hours, or even overnight, for the surface to cool enough to test the results of last effort before he can continue. Hours of waiting for minutes of polishing and testing extrapolates into years of work to put the final figure on the mirror of a large telescope. The alternative—grinding or polishing without frequent testing—would risk removing too much material. One slip, a single area polished too deep, could necessitate repolishing or even regrinding the entire disk, a task that could consume months or even years.

Thomson, who had a shrewd business sense as well as the creativity of a polymath inventor, knew the commercial market was too small, even with potential military and industrial applications, to warrant an enormous investment in quartz technology by General Electric. His early experiments seemed promising, but the research was expensive and time consuming. By the mid-1920s all work had stopped while they waited for a customer. At the annual meetings of the American Astronomical Society and the National Academy of Science, Thomson brought pieces of the fused quartz, which he would show to astronomers. Hale would corner Thomson at these meetings to find out the latest progress on fused quartz.

Hale and the Observatory Council made a show of considering a range of materials for the mirror, but fused silica promised to be so superior to any other material that for Hale the decision of what material to use for the new telescope was already made. This would certainly be George Hale's last telescope. For a long time it would be the primary research telescope of the world. There was no room for compromise. Fused silica was theoretically the best material for a mirror, and the obvious choice. The only problem was that no one had yet fabricated a functional telescope mirror for an instrument of any size from the material.

In March 1928, before he submitted the formal proposal for the telescope to Wickliffe Rose, George Hale met A. L. Ellis, Elihu Thomson's lab assistant, at the Commodore Hotel in New York. Ellis was carrying a beautiful piece of fused clear quartz. Hale asked him how much it would cost to fabricate a mirror for the two-hundred-inch telescope from the material. On the spot, on a scrap of brown paper, Ellis wrote up an estimate of the cost to fabricate a series of mirrors from the eleven-inch blank the laboratory had already produced, up to a two-hundred-inch quartz mirror. Ellis's rough figure for equipment, labor, and material—without profit or overhead—for a series of mirrors, from a twenty-two- to sixty-, one-hundred-, and a two-hundred-inch mirror that would be fabricated on Mount Wilson, came to $252,000. It seemed a modest sum in a budget of $6 million.

As soon as Rose gave the Observatory Council the go-ahead to begin work on the telescope, Hale had Henry Robinson, of the board of trustees at the California Institute, formally ask Gerard Swope, the president of General Electric, to undertake the fabrication of the mirror blanks. Swope wired back:

GENERAL ELECTRIC COMPANY WILL BE DELIGHTED TO DO THE WORK ON THE FUSED QUARTZ LENS UNDER THE PERSONAL DIRECTION OF PROFESSOR THOMSON WHO IS MUCH INTERESTED IN IT AT MANUFACTURING COST WITHOUT ANY OVERHEAD FOR COMMERCIAL OR ADMINISTRATIVE EXPENSES WHICH I ASSUME IS WHAT YOU HAD IN MIND.

Like almost everyone who had never looked carefully at a large reflecting telescope, Swope called it a *lens* instead of a mirror. No matter. If General Electric and the famed Professor Thomson could produce a two-hundred-inch-diameter disk of fused quartz, Hale and the astronomers would not only have the biggest telescope in the world, but the best.

Hale delayed a public announcement of the telescope as long as he could. He had always been wary of publicity, afraid that tentative, exploratory ideas and the meandering process of scientific research would be pummeled by a press too eager to "expose" science and scientists and to demand "results." Astronomers, he feared, could be tempted by the same pressures that had corrupted Sinclair Lewis's Martin Arrowsmith.

From his experiences with the Yerkes and Mount Wilson telescopes, Hale knew that once the grant for the new telescope was announced, the sheer scale and audacity of the project—more than the celebrated discoveries at Mount Wilson, or the famed debate in Washington—would open the process of building the telescope, and the observational astronomy program, to public and media scrutiny. The Observatory Council would be flooded with questions and offers. Developers with a mountain on their land would offer it as the site for the new telescope. States and counties would campaign to get the big telescope in their jurisdictions. Crackpots would come forward with their ideas of how to build a "giant eye." Shipping companies, foundries, machine shops, construction companies, and self-promoting entrepreneurs would offer their services as they sought some tenuous connection with the prestigious project.

California also had more than its share of fundamentalist and revivalist movements, led by evangelists who were quick to brand science and technology as the work of Satan. A few had already spoken out against the telescopes on Mount Wilson. A larger telescope, designed to reach even deeper into the mysteries of the universe, would be a prime target for their sermons. A campaign by fundamentalists would be an even greater threat than the union strikes and picket lines that increasingly disrupted businesses in Southern California, because the police couldn't be expected to show the same eagerness for scuffles with men and women of the cloth that they demonstrated against the unions.

Hale also feared that men like Harlow Shapley, at institutions fearful that their own plans would be slighted because of the funds committed to a big telescope, would be quick to join the public doubters. The debate in Washington, and the publicity that surrounded Hubble's work at Mount Wilson, had turned the attention of many universities to the possibilities of big telescopes. The academic world in 1928 was not exempt from jealousies, rivalries and backstabbing, and many an astronomy department chairman was willing to bad-mouth the new telescope if it would help channel funds or facilities to a local project. Whatever the motives of the critics, the vagueness of the plans for a big telescope would make it difficult to answer questions and challenges in the newspapers or on the radio. There were no working drawings, few calculations, no engineering studies. Hale and his colleagues needed time to experiment, to make mistakes, to explore possibilities that a strict budget analyst would no doubt rate as cost-ineffective.

In September a reporter for the *Los Angeles Times* got access to some financial records of the Mount Wilson Observatory and began asking Adams about the rumors of a new telescope. Adams and Hale debated using connections to squelch the news and concluded that the effort might backfire. Soon reporters were calling everyone they could find with rumors of the largest grant ever made in support of a scientific project, millions of dollars awarded to build the biggest scientific instrument in the world, a device that would require the largest single piece of glass ever cast and the largest bearings ever machined and gather more light and see farther into space and require more man-hours and demand finer tolerances and . . . It was impossible to put them off much longer.

Hale waited until the end of October. The big news in the papers were the presidential campaigns of Al Smith and Herbert Hoover, another victory by Babe Ruth and the Yankees in the World Series, and the *Graf Zeppelin*'s maiden flight across the Atlantic. The great airship drew record-breaking crowds to the airfield in New Jersey, eager to see the newest triumph of German technology. Congress awarded a special gold medal to Thomas Edison, in recognition of his scientific achievements. With a bevy of photographers as witnesses, Andrew Mellon presented the medal on behalf of a grateful nation. Science and technology seemed to dominate the news.

On October 28, 1928, a Sunday, Hale issued a carefully worded announcement, hoping to head off the hyperbole he feared would follow in the papers: "It is our strong desire to avoid all sensational or exaggerated statements," his release read. "With this type of telescope we do not expect to see very minute details on the moon or planets or to deal with the inhabitants or other hypothetical creatures. . . . Its object is not to detect skyscrapers on the airless moon or to search for indications of human beings on Mars."

Few paid much attention to his caution when the newspapers picked up the release for their Monday editions. The front pages outdid one another in point sizes and hyperbole, with headlines like BIG BERTHA OF SKY WILL AID SCIENCE. The *New York Times* was only slightly more restrained: GIANT TELESCOPE OF IMMENSE RANGE TO DWARF ALL OTHERS. With gracefully concealed pride, the *Los Angeles Times* touted the newest achievement of Southern California: "Standing on the threshold of a vast uncharted space to be penetrated by the 200″ telescope, the scientific world is frankly a-tiptoe with excitement. Men who ordinarily deal exclusively with uninspiring mathematical problems and cold, concrete facts find themselves engaging in imaginative flights. What, they are asking themselves, will the gigantic new telescope reveal?"

In response to the demand for more facts about the great machine, Walter Adams cautiously issued figures on the potential light-gathering power of the huge mirror. The newspapers, not sure what it meant, and insisting as often as not on calling it a *lens*, seized on his numbers. The telescope, they reported, would see stars "700,000 times as faint" as could be seen with the naked eye. It would be able to photograph "a candle flame at 40,000 miles." At least, Hale could console himself, they didn't report that the telescope would see Martians or moonmen holding the candles. A few papers, like the *London Daily Telegraph*, resisted the more extreme examples and explained quite simply that the new telescope would "penetrate to the limits of our universe."

The *New York Times* reported that although the size of the grant in support of the telescope had not been announced, it would surely be far in excess of the $600,000 that the one-hundred-inch Hooker telescope had cost. Whatever the cost, the editorials agreed, the telescope was a great leap ahead for mankind.

Public reaction everywhere was quick, the interest in the great instrument boundless. From all over the world, letters poured into Pasadena from well-wishers sending their salutes and blessings. Thomas J. Johnston, a patent lawyer in New York, praised "the most astounding feature of the proposition. That is, the harmonious concentration of the best trained minds in the world upon a project of pure science, with the certainty of fully adequate financing and technical manufacturing facilities. The history of science shows no such thing, or anything near it." Using an ethnic slur not uncommon in private correspondence in the late 1920s, Johnston reminded Hale of the struggle to get Congress to fund the old twenty-six-inch telescope at the Naval Observatory in Washington, and "the pitiable attempt in the Committee to jew down Alvan Clark the elder by $1000—What a contrast!"

A little girl sent a dollar to help the project. Hucksters, businessmen, amateur scientists, backyard engineers, promoters, quick-buck artists, dreamers, and schemers offered their ideas and themselves. Men of the cloth, seeing yet another threat in the hubris of science, sent curses and threats of eternal damnation to those who would unveil the secrets of the universe. Astronomers, hoping that they would someday get a chance to use the great instrument, sent their congratulations, sometimes barely concealing their envy.

Overnight, it seemed, the unbuilt instrument that would see to the limits of the universe became the symbol of an era of confidence, a metal-and-glass paean to science, knowledge, and progress. For the next twenty years the public and press would be relentless in their fascination with what the newspapers and radio reporters insisted on calling the giant "eye." Never before, and probably never again, would there be such widespread attention devoted to a scientific instrument with such a benign aim.

10

Beginnings

Rumors of the big telescope had circulated among astronomers for years. Visiting VIPs at Mount Wilson who had seen the sketches and models were suspicious of the repeated denials from the Mount Wilson staff. Scientists and administrators who had dealings with the various Rockefeller Foundations could tell from the hush-hush of the program officers that something big was in the works. Harlow Shapley, accustomed to being consulted by the foundation staff on astronomy proposals, hadn't been asked for his opinion on this one.

Shapley sent formal congratulations to Hale and Walter Adams, but they both knew he had been far from subtle in his opposition to the project. For years, he had been resentful of the money that the California Institute and the Mount Wilson Observatory had gathered under their control, and especially resentful of the publicity that the Pasadena institutions had gathered for the discoveries made at Mount Wilson. As the head of the astronomy department and the observatory at the oldest and most famous university in America, Shapley had the prestige to make his opinions heard.

H. L. Mencken, ever on the lookout for a chance to debunk the latest fascination of the "rubes of the booboisie," met Shapley for a lunch at the Harvard Faculty Club. At first Mencken thought Shapley, with his slicked-back hair and boisterous horselaugh, looked "inconspicuous and somewhat rustic." After they talked for a while the rusticity vanished. Mencken knew little about astronomy. His interest was news, and he welcomed Shapley's apparent candor about the much publicized two-hundred-inch telescope project.

"Practically everything it may be expected to accomplish could be accomplished by existing telescopes," Shapley told Mencken. Studying the millions of stars astronomers had already reached would occupy them for the better part of the century. The real reason for the new telescope, Shapley said, was publicity. In particular he singled out Robert Millikan, the head of the California Institute, quoting the famed physicist Ernest Rutherford, who had remarked that publicity seeking had finally become a learned science, with its own unit of measurement, the *kan*, a unit so large that publicity is normally measured in a workable fraction, the *Millikan*.

It was exactly the kind of cynical barb Mencken liked.

Shapley also found a way to make his reservations about the project heard in astronomy circles. The design and ownership of the new telescope were entrusted to the California Institute. There was little Shapley or anyone else could do to change that. But the question of where the telescope would be sited was not mentioned in the grant. It was an issue on which Shapley had strong views.

One of the grants Harvard had received from Wickliffe Rose's IEB had funded the move of the Harvard Southern Station from Peru, where they had a single year of good seeing during the initial testing of the telescope, and wretched conditions for thirty-nine years afterward, to Bloemfontein, South Africa. With the volumes of data they had accumulated over a long period at the Southern Station, Harvard dominated Southern Hemisphere astronomy. Henrietta Leavitt's original study of Cepheid variable stars was only one of the projects based on these plates. It was not surprising that Shapley began a campaign urging that the new telescope be sited at the Southern Station, or failing that, "Tibet, Kashmir, Peru, Chile, the Argentine, or Australia"—in other words, anyplace but Southern California.

Shapley directed his campaign at Trevor Arnett, the officer in charge of science programs at the Rockefeller Foundation. Writing with the authority of his position at the Harvard College Observatory, Shapley reported that "the opinion of astronomers in general" was that the big telescope project had been initiated by Hale and the California Institute without consulting other astronomers, and that "naturally they would only consider sites in relation to their own institution," despite the fact that there were superior sites in New Mexico or Western Texas that would make the telescope more accessible to astronomers from the East.

Shapley was joined in his efforts by other astronomers who feared that the new telescope, along with the sixty-inch and one-hundred-inch telescopes, would give a monopoly of deep-space research to Mount Wilson and what other astronomers saw as the Pasadena astronomy clique.

By the end of the summer the lobbying was so intense that Arnett went to Pasadena to show Anderson the voluminous correspondence he had received and his notes of meetings with visiting astronomers. Arnett assured Anderson that there was nothing personal against Hale in the questioning. Rather, he explained, there was widespread concern that Hale was skewing the search to make sure a site was selected close to the California Institute, instead of picking the best possible site for the telescope.

★   ★   ★

On the question of a site, as on most issues concerning the big telescope, George Hale had strong views. From the time he had organized
the Yerkes Observatory, Hale had been convinced that the old days of
an observer going up to a mountain with a sketch pad to record what
he had seen were long gone. The new problems of astrophysics
required that an observatory be close to well-equipped research
libraries and laboratories in related disciplines like physics and spectroscopy. The modern telescopes Hale was building needed not only
darkrooms and auxiliary equipment like blink stereocomparators, but
constant attention and experimentation with new sensors, emulsions,
photographic and spectrographic instruments, and auxiliary lenses.
Astronomers were constantly proposing new observation programs at
the limits of the telescope's resolution and light-gathering powers, with
complex instruments that could only be built, modified, and repaired in
dedicated optical and mechanical shops. The road from the offices and
shops on Santa Barbara Street in Pasadena to the telescopes and
Monastery on Mount Wilson was well worn.

When Hale first built the solar telescopes, and then the sixty- and
one-hundred-inch telescopes, Mount Wilson had been an ideal site. It
was close to the laboratories in Pasadena, and the peculiarities of the
local geography and weather created remarkable observing conditions
on the mountain. Mount Wilson was soon famous not only for the
instruments on the mountain but for the seeing. Many astronomers
thought that on a good night the atmosphere over Mount Wilson was
so still, the images of the stars so well defined, that it was perhaps the
best seeing in the world.

Over the years the seeing (atmospheric turbulence) at Mount Wilson had not deteriorated, but for dark-sky work—photographic or
spectrographic study of galaxies and other distant objects too faint to
record on a photographic plate when the moon is up—Mount Wilson
had begun to suffer from its proximity to Los Angeles. Other cities
grew up, becoming more dense. Los Angeles, already the fastest-growing city in the nation, spread. By 1928 the city and surrounding towns
reached right to the base of Mount Wilson. The San Gabriel Valley
below the observatory glittered with lights at night, more than even
the fogs could obscure. The Mount Wilson staff were sensitive to the
issue. When Rose and Thorkelson visited Mount Wilson, Hale warned
Adams to stay with them at all times; "if any questions about lights in
the Valley" came up, Adams was to "show how easily they [could] be
met."

For 90 percent of the observations that a new telescope would do,
including most spectroscopy, bolometric observations, and direct photography with moderate exposures, the conditions on Mount Wilson
were still superb. But for very long exposures on faint objects at the
limit of the telescope's reach—exactly the work for which the bigger

telescope was most important—even Hale acknowledged that "the illumination of the night sky [below Mount Wilson] may be sufficient to make trouble." To fulfill its mission of extending the limits of the observable universe, the new telescope would have to be at a site more remote than Mount Wilson, far enough from any city that even the unpredictable population growth and sprawling development of a Los Angeles wouldn't interfere with the future use of the facility.

For Hale the siting question was a tricky balance: How far from Pasadena would they have to go to achieve the dark skies and good seeing they needed? How far was too far from the Santa Barbara Street laboratories and optical facilities, and the new astrophysics laboratory that would be built on the campus of the California Institute? Ferdinand Ellerman and Milton Humason, observers with considerable experience on the big telescopes, and both veterans of the early days when Humason had led mule trains up Mount Wilson, cautioned against a site that was too inaccessible. Hale also liked to quote a Henry Norris Russell story about a mining engineer sent to investigate a claim offered at a suspiciously low price, who telegraphed back east: ORE THERE. UP TO SAMPLE. LOTS OF IT. WILL NEED PACK TRAIN OF BALD EAGLES TO GET IT OUT.

Long before the grant was confirmed, arguments about the virtues of one peak or another, and the potential seeing at various sites, were a staple of the conversations at Mount Wilson and Santa Barbara Street. By the end of the summer of 1928, Hale had instituted a regular program of calibrated measurements at the better potential sites. One of Russell Porter's earliest tasks was to design a small telescope to be used specifically for these seeing tests. He came up with a four-inch refractor of thirty-six-inch focal length, used with a 210-power compound microscope eyepiece. The resulting telescope magnified images 750 times, which would readily show up atmospheric turbulence. Porter's design was simple to build, with four steel legs that could be pushed into the ground to steady the telescope. The telescope was designed to observe Polaris, the pole star, which remains in the same position in the northern sky, at least over a relatively short period, so the telescopes could be made portable without motor drives that would require complex alignment procedures. The magnification was so high that Polaris would travel across the field of view of the little telescope in ten minutes, but that was long enough to record the seeing.

To complement Porter's telescope design, Anderson developed a technique of recording the size of the "tremor disc" of Polaris on a calibrated scale, so measurements by different observers, at different sites, could be compared. With practice a trained observer could get a reading in two or three minutes.

Ellerman, who had started with Hale at the old Kenwood Observatory in Chicago, developed a site evaluation program which recorded estimates of the seeing, the weather, the incidence of fog, and other factors that could affect the use of a large telescope. Humason was also recruited for site research, and Hale suggested equipping amateur observers with special eyepieces and instructions from Hubble or other experienced astronomers so they could aid the project by measuring and recording the seeing at different sites.

By fall, the list of potential sites for the telescope included Flagstaff, Arizona, various locations in the Mojave Desert, Winona, Bellemont, Barstow, and Hot Springs Mountain in San Diego County, Table Mountain behind Mount Wilson, Pine Flats, Union Flat, Holcomb Valley above Big Bear, Volcan Mountain near the Julian Flats, Rattlesnake Flat, Pleasant View Ridge, Lake Arrowhead, Mono Lake, and Catalina Island—all within a day's driving time of Pasadena. It seemed like a thorough program to evaluate the best site for the telescope.

Although he encouraged the site survey program, Hale already had strong opinions about the future site. He had never forgotten the inviting description of Palomar Mountain that W. J. Hussey had written on his 1903 survey for the Carnegie Institution. "Nothing prepares one for the surprise of Palomar," Hussey had written, ". . . a hanging garden above the arid lands." In 1903 Palomar had been too remote for an observatory. A quarter century later, what had been a disadvantage had become a benefit. Palomar was far enough from Los Angeles and San Diego to not be threatened by their light pollution. Before the measuring instruments for the broader site survey were ready, Hale suggested a program of observations at Palomar, including comparisons of the brightness of the night sky with the sky at Mount Wilson, which they would do by comparing photographs of the same star field taken from different sites with the same short-focus portrait lens. In those early tests, Palomar had none of the light pollution problems of Mount Wilson. Hussey had not measured the seeing, but he predicted that "the remarkable stillness, the steady temperature, and the evergreen covering of Mount Wilson could not be found on Palomar."

Hale was determined to prove him wrong.

Hale's strong ideas were a perfect target for the complaints from other astronomers. A few who had worked at Mount Wilson or Yerkes sent their comments and suggestions directly to Hale. Hale answered politely but firmly, explaining that to draw on the concentrated experience of the California Institute and Mount Wilson, the telescope had to be at a site no more remote than Arizona. Anderson told Arnett that the sole question in the site evaluation program was how many nights of good seeing would be available at each location. He conceded that southern Idaho, Nevada, western Colorado, or other sites that had been suggested might be clearer at times than the sites they were looking at in Southern California, and might even have *seeing* comparable to Mount Wilson, but they had storms, which are rare in Southern California. It wasn't a convincing argument, at least not one Arnett could use to silence the campaign directed at his office.

When he had his mind made up, George Hale could be stubborn and clever in equal parts. To blunt the controversy, he agreed to appoint a site committee for the telescope, with a membership that included outsiders: Charles Abbot of the Smithsonian Institution, who had arranged the great debate of 1920; Professor Charles Marvin, the chief of the U.S. Weather Bureau; Dr. W. S. Humphreys, a professor of meteorological physics at the Weather Bureau; and Robert Aitken, the assistant director of the Lick Observatory. Even Shapley couldn't find fault with the proposed outside members.

Arnett took the proposal to President James Rowland Angell of Yale University and Charles P. Howland of the Council on Foreign Relations, whose diplomatic skills presumably could broker between the warring parties in American astronomy. Both men urged that any site committee have a "free hand," by which they meant it should not be dominated by Hale.

Hale, anticipating the opposition, had already prepared an agenda for the committee, with criteria for a site:

- •The latitude had to be between thirty degrees and thirty-five degrees north, so the telescope could observe stars from the celestial pole to south of the celestial equator.

- •The altitude of the site had to be from six to eight thousand feet. The lower figure was the minimum to guarantee adequate sky transparency. A site higher than eight thousand feet would be subject to excessive snowfall, and heated air rising from bare rocks above the timberline.

- •The annual and daily temperature range had to be small, and the winds minimal, to produce the best seeing.

- •Freedom from cloudiness.

The only sites that met those considerations were in the southwestern United States or northern Mexico.

At that point Arnett asked Hale to come to New York to discuss the matter. In an era when it took four days to cross the country by train, an invitation from an official of the funding foundation was not to be treated lightly.

Hale, in no mood for a confrontation, stalled. The Los Angeles lights had become a problem, he admitted, but "the results thus far obtained point to Palomar as the most promising [site], as the 'seeing' is distinctly better than at Mount Wilson, while the sky is much darker and purer." His long years in astronomy, and especially in Southern

California, gave Hale an advantage in arguments: He drew examples from the results of double-star studies at Yerkes and at Lick, episodes from his own unsuccessful experiences at Pikes Peak, the threat of a single dust storm "such as I have frequently seen in Egypt," and the problems of extremely cold weather numbing the fingers of observers. "The problem . . . is much more complex than it may seem at first view."

To undercut Shapley's arguments for a Southern Hemisphere site, Hale cited Hubble's recent publications on the spiral nebulae, which had been celebrated in the popular press. Hubble had written that the nearest cluster of extragalactic nebulae, the only one with spirals large enough for study with a two-hundred-inch telescope, was in the *northern* sky.

By contrast, Hale explained, the Southern Hemisphere wasn't yet sufficiently explored by smaller instruments to be ready for the research efforts of the largest instrument in the world. In a naked jab at Harvard, he pointed out that the observations at southern stations were generally "conducted in a routine way by one or two assistants, cut off from contact with productive thinkers and of necessity pursuing their duties in a mechanical manner. It is not from such sources that prime advances in principles or methods of observation are likely to proceed. Not one but several investigators of the highest type, constantly stimulated by personal contact and by daily discussion with men of the same high calibre working in related fields, are absolutely necessary if we are to secure such advances as we have in view."

Finally Hale threw in an argument that could be expected to either win the day or ruffle the feathers of Wickliffe Rose's successors at the IEB: "A consideration regarded by Dr. Rose as paramount must be kept constantly in mind. This is the importance of establishing the two-hundred inch telescope within a few hours' ride of such a strong group of investigators as we have in Pasadena."

Arnett, with a foundation officer's hesitancy to interfere, assured Hale that he did not question the *merits* of a site in Southern California, but only the procedure followed in selecting the site. Perhaps, he suggested, they should hold a conference to discuss the site, with expenses paid by the IEB. With hindsight, it is easy to read between the lines of Arnett's letters. If the California group made a show of following an open search procedure, he seemed to be suggesting, they could have a free hand in picking a site. He wrote again and again, insisting that they either have a conference or that a special committee, completely independent from the Pasadena group, be appointed to study the site question.

Each time Arnett wrote, Hale stonewalled. As so often happened in tense moments, his demons came back to torment him, and the arguments of the whirligus were even more ferocious and more demanding than the brouhaha over the location of the telescope. Miss Gianetti made excuses for him when he retreated to his curtained room to seek the modicum of respite that darkness and silence could provide. Hale wasn't beyond taking advantage of the demons. If the doctors would permit him, Hale wired back, he would come to New York. For good measure he reminded Arnett that when they first discussed the grant Rose had agreed to a site within easy access of Mount Wilson and the California Institute.

It was the end of October before Hale allowed that he was well enough to make the trip to New York, accompanied by Walter Adams. They met at familiar stamping grounds, the University Club, with Arnett and Max Mason, who had just been named president of the Rockefeller Foundation.

By then the program of measuring the seeing at sites in Southern California and Arizona, and the plans for the Astrophysics Laboratory at the California Institute, were so far along that Arnett could do little more than explain that it was good politics to at least acknowledge the suggestions and comments of astronomers at other institutions. Harlow Shapley was only half conciliatory. It would take at least five years to make a two-hundred-inch disk, he wrote to Hale, citing private information he had received from General Electric's West Lynn laboratory. Hence he urged that Hale take plenty of time, two or three years, to test a site. It was the last Shapley would argue on the topic.

Hale had won the first battle over the telescope. But the pressure of the dispute and the trip to New York took their toll. The demons were relentless now. Within a month Hale made plans for an extended trip abroad. "The pressure I have been under ... a much larger amount of work than I can continuously carry, have delayed my correspondence and forced me to the conclusion that I must get completely away from work for two or three months." He booked passage for a Mediterranean cruise, sailing from New York on January 22, 1929, hoping "to take a complete rest of several months."

Before he sailed Hale wrote a memo to the Rockefeller Foundation staff, on a new letterhead of the Astrophysical Observatory of the California Institute of Technology, summarizing the progress to date on the project. He boldly listed the decisions that had been made: fused quartz for the mirror, under the personal direction of Elihu Thomson at General Electric; the J. G. White Engineering Company, headed by his friend Gano Dunn, for general supervision; Warner & Swasey, the veterans of the Lick and Yerkes telescopes, and recently of a number of large reflectors, for the mounting, with assistance from Francis Pease and General Carty. The project was in the hands of the University Club's best men.

The Mount Wilson opticians would grind and polish the mirror disk on the site of the telescope. Carty and Dunn were on the board of the Carnegie Institution, but the procedures he had outlined, Hale explained, were "designed to secure the best possible results without involving the Carnegie Institution of Washington in any expense, or calling for much of the time of its research men, or even of existing shop facilities during the construction period."

Hale made the project sound simple. He had assembled the most talented astronomers and telescope designers, gotten the largest grant ever, could call on the most distinguished leaders of industry and academe. He had the laboratories and staffs of both Caltech and the Mount Wilson Observatory available, and could draw on the collective experience of the men who had designed, built, and operated the largest telescopes in the world. Hale's enthusiasm was so contagious that everyone else thought his estimate that it would take four to five years to finish the telescope needlessly pessimistic.

Work in Pasadena began without him. Memorandums, correspondence, and minutes of meetings went back and forth from the solar laboratory, Santa Barbara Street, the California Institute, the Rockefeller Foundation offices, private contractors, and consultants, some solicited, some uninvited. Typewriters and carbon paper were already standard equipment in the business world by 1928. The handwritten correspondence of an earlier era had given way to a snowstorm of memorandums in multiple copies. Paperwork quickly began to fill file cabinets and old-fashioned correspondence boxes. The original Observatory Council and Advisory Committee begat subcommittees on site, optics, and myriad other topics, and before long the subcommittees were meeting regularly enough to have earned a permanence of their own. From an astronomer's dream the project had begun to take on all the bureaucratic trappings of a major enterprise.

Yet for all the flurry of business, there is a strange, otherworldly calm in these memoranda and minutes. In the newspapers the roller-coaster ride of the stock markets was front-page news throughout the summer and fall of 1928. In the minutes of meetings and the flood of correspondence from Hale and his colleagues about mountings and sites and disk designs and grinding machines and staff and budget and countless other details, there is nary a mention of the topic that seemed to dominate conversation everywhere else.

It wasn't that they were too far from the East Coast and the markets. California was no longer a hothouse hybrid. By 1928 the movies had made the once-strange images of Los Angeles an alternative norm of American culture, a mythical world that people knew from the silents even if they had never taken a trip out west. The reality of California was frighteningly close to the cinema images. The movie and aircraft industries fueled an exploding industrial and agricultural economy. Growth seemed to reach in every direction at once. Buildings, neighborhoods, even whole cities seemed to spring up overnight. From nowhere universities leapt into a national and world prominence that had taken East Coast institutions hundreds of years to achieve. Where once a rising young scientist would only have sought a position at one of the famed eastern universities, now there were institutions in California with top-notch faculties, with more money to offer for research and laboratories, and with the incomparable advantage of newness. Perhaps it was only illusion, but to a generation of bright young scholars, the California schools seemed more open to new ideas, less hidebound, than schools elsewhere. They seemed like places where a young scientist or engineer could make his mark unfettered by the constraints of precedent and tradition.

In 1928 men like J. Robert Oppenheimer and Ernest Lawrence, who had their choice of institutions, gravitated to California, Lawrence to the University of California at Berkeley, and Oppenheimer to a joint appointment at Berkeley and Caltech. Two years later the first meeting of the National Academy of Sciences ever held west of the Mississippi met in Berkeley. At the meeting Lawrence unveiled his first cyclotron, starting a chain of developments that would ultimately surpass even the two-hundred-inch telescope as big science.

Explosively, in less than a decade, Caltech established a reputation as a first-rate school for science and engineering. Powerful patronage in Pasadena and elsewhere in Southern California came forward with generous endowment and capital support for the new school, which multiplied the drawing power of the prestigious initial faculty appointments like Millikan and Noyes. The academic center of gravity of the United States was still on the East Coast, but in fields like aeronautics, the local aircraft industry in Santa Monica, Burbank, and San Diego made the Southern California location a magnet for top faculty. Engineers like Theodor von Karmann joined Caltech as senior professors or department heads. Their availability as consultants drew the attention and support of the aircraft industry, which in turn attracted bright younger scholars and students who could look forward to jobs at Douglas, Lockheed, or Convair after graduation from Caltech.

Even before the two-hundred-inch telescope project was announced, rumors of the telescope were enough to draw graduate students to Pasadena. A young Berkeley graduate named Olin Wilson thought about graduate study in astrophysics at Caltech before the school had a single faculty member in the field. Harvard, Princeton, and the University of Chicago were the famous departments for astronomy, but the chance of someday doing research on the big telescopes at Mount Wilson, or the unbuilt two-hundred-inch telescope, was an irresistible drawing card.

For engineers, too, the chance to work on a once-in-a-lifetime project was a powerful magnet. Engineers and opticians and mechanics sought jobs at the fledgling institution in the hope that somehow they could be part of the instantly famous project. Few of these engineers had any experience with telescopes, but engineers liked to see themselves as problem solvers, men of few words, ready to do battle with their slide rules and graph paper. Men who could do calculations of induced drag on airfoil sections or wind-stress calculations for tall buildings could do the same for an observatory dome, and men who had calculated the bearing loads for a battleship gun turret could also calculate the loads for a telescope mount.

There wasn't much of a campus yet at Caltech, and Pasadena, the newcomers were soon to discover, wasn't the California of the movies. By the late 1920s, motion pictures had become the leading industry in Southern California, with aircraft factories a close second. In Pasadena, the leading industry was clipping coupons from bonds. The assessed wealth in Pasadena in 1929 was $186 million, which would have been high for a city twice its size. Even as Los Angeles grew by leaps and bounds, Pasadena remained an isolated small city, nestled against the foothills of the San Gabriel Mountains.

Upton Sinclair lived in Pasadena during his losing bid for the governorship of California, but the presence of the old socialist warhorse didn't change the determinedly conservative politics of the city. Even the relatively staid campus of engineers and scientists at the California Institute seemed a radical intrusion to some. The Reverend Robert B. Schuller, successor to a long line of Southern California evangelists that included the infamous Aimee Semple McPherson, galvanized opposition to the ungodly threat of science at Caltech with revival meeting speeches on topics like "Evolution Unmasked" and "The Mark of the Beast." Pasadena voted three-to-one in favor of a statewide ballot measure ordering the King James version of the Bible placed in public schools. But for strong opposition by Northern California voters, the measure would have passed statewide.

Yet by 1928 the evangelists and even Aimee Semple McPherson's scandalous disappearance and reappearance were but brief interruptions to the steady news of the stock market that dominated headlines everywhere, including Pasadena. Despite periodic "corrections," the rise in the market seemed inexorable. There were fortunes to be made, the headlines cried. Old money, fearing the challenge of new, rose to the lure of the market. Even a bastion of conservatism like Pasadena wasn't exempt from the appeals that drew money from fixed-income investments to the rising market, where the mysterious yeasts seemed to renew themselves weekly. Men like Henry Robinson, of the board of trustees of the California Institute, who had personally promised to capitalize an endowment for the two-hundred-inch telescope, were soon heavily committed in the market.

The astronomers and engineers, in makeshift offices on California Street, at the Santa Barbara Street offices of the Mount Wilson Observatory, and in the library of George Hale's solar laboratory on Holladay Road, worked on, their future assured by the grant from the IEB. They were too busy for the stock market. They had a perfect machine to build.

11

Hope

Nothing went right in West Lynn.

It all seemed so promising when they started. Back in March 1928, Elihu Thomson's assistant, A. L. Ellis, had shown Hale a beautiful piece of hard, transparent fused quartz. Even under magnification it was free of striae that might cause strains or distortions in a mirror. The test results when the material was heated were spectacular. That summer Thomson showed off samples of the clear quartz to a visiting delegation from the AAS, including Harlow Shapley. Before long major observatories in the United States and Canada were asking about quartz blanks for astronomical mirrors. Given Thomson's reputation as a genius of engineering and production, the mirrors seemed assured.

Elihu Thomson was addressed as "Professor," although he was more a shirtsleeves scientist than an academic. He had been offered the presidency of MIT in 1897 and declined. In 1919 he accepted the position, but only for two years as an acting president, before he returned to his beloved laboratories. Like traction power, electric welding, or fused quartz, he saw MIT as problems that needed solutions. Two years was long enough for Professor Thomson to solve most problems.

He was at home at GE because the company had the same attitude. Although they conducted substantial research programs, the research was directed toward solutions, and specifically toward production. GE had converted the tinkering of geniuses like Thomas Edison and Elihu Thomson into lighting and electrical equipment, trolley traction drives, appliances, industrial diamonds, electric motors, monitoring and metering devices—products that ranged from consumer goods to massive industrial systems. GE's immense factories, like the River Works in Schenectady, supplied the world. Between the reputation of men like Elihu Thomson, and the sheer vastness of GE's manufacturing facilities, the company epitomized the technological optimism of the 1920s. The company, and the public, believed that GE could accomplish anything it tried.

In 1904 Thomson had already patented a process for molding quartz. He would pour fine quartz sand into a mold in a high-temperature vacuum furnace, seal the furnace, and slowly raise the temperature until at 500° the quartz sand would "explode," vaporizing any contaminants. At 1400° the quartz would begin to turn viscous. When the furnace reached 1700° (approximately the melting point of platinum) the quartz would fuse into the pattern of the mold. During the process powerful vacuum pumps sucked the air out of the furnace to remove gaseous contaminants and eliminate as many air bubbles as possible from the quartz.

But even with the most powerful vacuum pumps available at the West Lynn laboratories, the blanks emerged from the mold filled with tiny bubbles. The bubbles did not interfere with the thermal qualities of the quartz or weaken it, and for large telescope mirrors, the lightness of a bubble-filled blank might even be an advantage, reducing the load on the telescope mounting. But the surface of a bubble-filled blank could not be polished to the fine figure required for an optical mirror. True, a pock in the surface would only reduce the light-gathering capacity of the mirror by a tiny percentage of the total area. But there was no way to achieve a fine figure on a pocked surface, because the pocks would ruin the grinding and polishing equipment, and the resulting surface would scatter instead of focus light from faint objects. Thomson's challenge was to find a way to coat the fused silica that emerged from their molds with a layer of pure silica fine enough to take an optical surface.

The professor tended to lose patience with experiments once he understood the process, and after a bout with gout and asthma in 1926, he had withdrawn from day-to-day activities at the laboratory. He turned the laboratory experimentation on fused quartz over to Ellis.

If Ellis wore a jacket and tie and even a vest in the laboratory, this was more a reflection of the formality of business and science in the late 1920s than of personal style. On the project he wasn't afraid to roll up his sleeves and plunge into dirty work that front office scientists would shun. He was Thomson's kind of man.

Ellis started by standing short rods of clear quartz on end on the surface of the molded quartz, and reheating the disk in a furnace. The intense heat of the furnace melted the rods into a clear coating, but the resulting surface was marked with striae in the shape of the rods. When Ellis substituted a mosaic of sheets of pure quartz for the rods, the sheets fused in a patchwork of striae. Using ideas of his own and suggestions from Thomson, Ellis tried variation after variation of the shape of the rods and the pattern of the overlapping sheets. Nothing he tried produced a surface without striae. Ellis wasn't an optician, but he knew that any inconsistency in the surface material meant strains that would ultimately affect the stability and optical performance of a mirror.

He was close to exhausting the variations of sheets and rods on the surface of the disk, when an engineer named Niedergasse reminded him of an earlier experiment at the West Lynn laboratory. On a very different project, an early effort to create artificial sapphires and rubies, Elihu Thomson had introduced finely ground refractory substances like alumina into a high-temperature flame. Nothing else seemed to work with the quartz, so with Thomson's approval, Ellis tried the technique for the fused-quartz project.

He began by designing a burner that would blow pure crystal quartz powder, ground to the consistency of flour, into an oxyhydrogen flame at 3000°F. In a furnace the burner produced a sleet storm of quartz that fused to the base disk in fine layers, like ice coating trees after a winter storm. The process was painstakingly slow, demanding huge supplies of hydrogen to fire the torch, enormous quantities of superpure ground quartz, and infinite patience. By the end of the summer of 1928, Ellis finally sent an eleven-inch fused-quartz disk to Pasadena for testing.

The blank looked superb. At the optical laboratory on Santa Barbara Street, Anderson subjected the disk to violent heating "which no glass could possibly bear." After the tests Hale and Anderson proudly announced: "Its performance has been marvelous, and although this disk was not annealed at all, it has already returned to its original figure, leaving no doubt as to its internal quality." Enthusiasm was high, but Ellis knew the blank proved little. The surface of the test disk had been built up in a small gas-fired furnace with a single burner operated entirely by hand.

The troubles began when Ellis and his staff tried to make bigger disks. Ellis switched to a larger, electrically heated glazing furnace and trained a crew in the use of the high-temperature oxyhydrogen torch. The heat and the yellow glow of the molten quartz were so intense that even with heavy insulated suits and dark goggles, the men had to work at a distance from the furnace, using long rods to maneuver the burners. The opaque protective glasses and the awkward long rods made it difficult to judge how much material had been deposited in any one spot as they moved the burners over the disk. The first efforts with the new process, Ellis reported, looked "much like the Rocky Mountains." He fiddled with where the men stood and how they operated the spraying equipment, to no avail. He couldn't produce consistent disks. Ellis finally shut down the furnace and went back to the smaller gas-fired furnace for more experiments.

Spraying the surfaces of the disks by hand wouldn't work. What Ellis needed was an apparatus that would hold the spraying equipment a constant distance from the molded blank and move it over the surface of the blank uniformly and at a constant speed. He had the lab

shop fabricate a burner and piping to carry the fuel, oxidizer, and powdered quartz. The metallurgists experimented until they found a nickel
alloy that would work in the nozzle of the burner. To keep the brass of
the oxygen and hydrogen pipes, and the burner itself, from melting in
the intense heat of the furnace, Ellis added more pipes to his apparatus for cooling water, until the device began to take on the appearance
of a Rube Goldberg drawing.

After months of tinkering he had the workmen fire the gas furnace
to spray another disk. The process seemed to work, but the disk
emerged from the furnace with the surface pocked with bubbles. It
took weeks of sleuthing before Ellis traced the bubbles to iron particles and porcelain contamination in the supposedly pure quartz. More
detective work identified the culprit as the ball mill that had been used
to crush the raw quartz to a flourlike powder. The pages of the project
calendar began turning by months instead of weeks.

Ellis confidently reported that the problems were isolated and fix-
able, but by the end of 1928, he and his staff still hadn't produced a
mirror blank that could be ground to an optical surface. In Pasadena
the opticians were impatient for disks to grind. Pease and Porter
needed to know whether the mirrors could be cast before they went
ahead with design plans. Hale and the Observatory Council needed to
know whether they could build the telescope they had promised.

While he waited for the ball mill to be rebuilt, Ellis designed a new
mechanical support capable of moving three burners together inside
the furnace. The new apparatus was installed in the large furnace, and
everything was readied to attempt a second eleven-inch disk. The
machinery they were using for this disk, Ellis reported to Pasadena,
could be scaled up to build mirror disks large enough for use in the
telescope, at least as auxiliary mirrors.

When everything was finally in place to fire the electric furnace,
one of the three workmen trained in the use of the burners sprained
his back and was laid up for a week. As soon as the man recovered and
returned to the lab, he and his assistants got the flu. Two more weeks
were lost. The project seemed cursed.

Finally, at the beginning of the new year, Ellis fired the large furnace. The roar of the oxyhydrogen burners was deafening. Through
the peepholes, the inside of the furnace glowed an intense, hellish yellow. The process consumed tankcars of hydrogen and oxygen at a ferocious rate as the layers of quartz were slowly fused onto the base of
the disk. After seventy-two hours of continuous spraying, during
which he had gotten little sleep, Ellis realized that the coating was
being deposited unevenly, with lumps of fused quartz in some areas
and bald spots in others.

He tried to adjust the flow to the three individual spray burners,
but the only way he could reset the valves was temporarily to stop all
work, partially cool the furnace, relieve the vacuum inside, and finally

lift the cover of the furnace enough so the workmen could get to the equipment inside. After adjustments the furnace was resealed and fired, and the entire crew waited until the pumps drew down a vacuum before they could begin spraying again. Even the most cynical pessimists hadn't predicted the process would be this complex or slow.

The only aspect of the program that was ahead of schedule was the billing. Every piece of new equipment, each delay, and each breakdown increased the cost. When Anderson asked about the huge bills that Ellis was sending to Pasadena, Ellis provided detailed breakdowns, documenting every cent. The lab equipment, all experimental and specially fabricated for this job, was expensive. The process needed lots of men and huge quantities of fuel and supplies. The original agreement with GE, made in the summer of 1928, had been for one year of work. Six months had gone by, expenditures were already approaching the figure projected for the entire series of mirrors, including the two-hundred-inch mirror, and Ellis still hadn't shipped a mirror that could be used in a telescope.

Another month passed before Ellis finished a second eleven-inch disk. It was the first complete test run of the molding and surfacing process they hoped to use for the production of telescope mirrors. When the furnace cooled and the workmen lifted the cover, they found that the surface of the disk had cracked, probably from the repeated thermal shocks when the furnace had been partially cooled to service the burners. Despite the bad news, Ellis and Thomson hadn't lost confidence: "We have not encountered any serious obstacle," Ellis wrote to Anderson, "and are convinced that the production of the large mirror is merely an engineering problem."

The word "merely" troubled Anderson. Eager to see the promised progress, he visited the West Lynn laboratory. Ellis put on a good show but admitted that there were still enough problems with the spray process that he thought they should hold off on building a larger furnace and the associated spraying equipment, and assigning the additional men to the project, until all the problems were worked out with smaller mirrors.

In the midst of the bad news, Professor Thomson proposed a new idea. Instead of fabricating massive, solid mirror disks for the telescope, he suggested that they could rib the backs of the mirrors, like giant waffles. The ribbing would reduce the weight of the mirror while still maintaining the rigidity, and the pockets in the back would provide a means for supporting the mirror in the telescope. The arrangement, he assured Anderson, would be easy with quartz, though "practically out of the question for glass." Anderson liked the idea, and Thomson agreed that as soon as the spraying experiments were under control, the laboratory would produce a proposal for ribbed backs on the mirrors.

Thomson's ribs were an appealing idea. The whole project was filled with appealing ideas. Back in Pasadena they were beginning to worry when they would see a mirror for the telescope. The pessimists wondered *if* they would ever see one.

In Europe, Hale was supposed to be resting, away from the fray and the demons. But he was obsessed with the telescope. Against the orders of his physician, and from seven thousand miles away, he insisted on regular progress reports. When Anderson and Pease scheduled a trip to the East Coast, Hale sent a list of people they should see, obscure German publications on optics they should research in the New York Public Library, and questions they should pose to experts, consultants, possible subcontractors—anyone who might be useful to the project. He wanted them to ask the Zeiss people about the counterweight schemes they had used on some of their recent telescopes. He had questions for Gano Dunn and Elmer Sperry, of gyroscope fame, about bearings and mounting designs. H. H. Timken, the famed roller-bearing builder, was to be asked whether roller bearings could support the enormous weight of a two-hundred-inch telescope, and Hale wanted them to research the new low-heat-coefficient alloy Invar for possible use in the telescope tube.

Hale had appointed committees for every aspect of the telescope project. Anderson chaired the Committees on Site, Optical Design and Mirror Discs, Bolometric Apparatus, Design of 200-inch Telescope Mounting, Laboratory Design, and Design of Instrument and Optical Shops. The members included astronomers, engineers, and other scientists from Caltech, the Mount Wilson Observatory, and outside organizations like Warner & Swasey and GE. Dozens of astronomers, engineers, opticians, and others—including Harlow Shapley—were listed as official consultants to the project. Each committee theoretically had the authority to consider all options and to make recommendations. On paper the project was reaching everywhere, soliciting and combining opinions from the widest range of sources.

In fact many of the members of the committees served in name only, and many committees existed mostly on paper. The Committee on Optical Design and Mirror Discs was responsible for the big decisions about the mirror, the heart of the telescope. Their charter from George Hale requested that they

> begin immediately a theoretical and experimental study of the various possible forms of mirror discs (solid, superposed plates fused or cemented together, two plates *fused* to an intervening cellular structure, ribbed, etc.), and the efficiency of existing systems and new systems of supporting them in all positions they must take in the telescope.

It was an expansive charter, but the order for the mirrors had already gone out to GE a year before. Anderson briefly explored the idea of mirrors of metal coated with glass, and Sir Charles Parsons in England received a query about his plans for hollow disks, but neither investigation went beyond a few letters and a sample of the proposed disk material. The design criteria Hale sent to the Committee on Design of 200-inch Telescope Mounting were based on "the assumption that the weight of the mirror disc will be that of solid fused silica."

The committees met from time to time, but most of the decisions on the telescope emerged in notes and memoranda from George Hale. Despite his health problems, he kept his fingers in every pie. Hale asked for answers to his queries in writing, on notebook-sized graph paper that he could insert into the binders he accumulated in his study at the solar laboratory. On Anderson's own copy of the memo setting up committees, Hale had his secretary, Miss Gianetti, type special requests:

> Dr. Anderson:
> Please allow for the use of a 60-inch mirror on each side of the 200-inch tube, one for solar work (projecting far enough to permit the 200-inch tube to be completely shielded from sunlight) and one for stellar work. Both to be suitable for photographing in the ultra-violet.
> G.E.H.

The requests and suggestions of the man who had almost single-handedly shepherded the three largest telescopes in the world into existence, and had gotten the unparalleled grant for this one, could not be ignored. Sometimes Hale's questions were ahead of anyone else's on the project. Other times he was off on a tangent, and his questions wasted valuable time. He trusted the men he knew, members of his club, heads of institutions and corporations he had met personally, or who had been recommended to him by one of the old boys. Even if they pointed to the wrong man, to people who knew nothing about the telescope, and took up valuable time that was needed elsewhere, George Hale's suggestions couldn't be ignored.

When he returned from the extended trip to Europe, in the late spring of 1929 Hale checked into the sanatorium in Maine for more rest, woodcutting, and forced isolation. It was late spring before he finally went back to Pasadena and his solar laboratory. He joined the regular weekly meetings of the Observatory Committee, but his participation, like his occasional solar research on the spectrohelioscope at the laboratory, was increasingly frequently postponed or interrupted by his old bugaboos: the excruciating headaches, depression, and the demons. The reports from West Lynn, bringing more bad news about progress on the mirrors, aggravated the attacks.

The project seemed rudderless.

In Pasadena, Francis Pease tried to refine his sketches into working plans. So many decisions depended on the mirror that it was impossible to produce working drawings. Everyone pretended confidence—of course Thomson and Ellis would work out the problems in the fused-quartz technology and produce the mirror blank—but until they knew for sure that there would be a usable mirror, the basic engineering questions for a large telescope—How do you support a two-hundred-inch-diameter mirror so that it won't change shape as the telescope moves? Where can observers get access to the light gathered by the great mirror? How do you point the machine with the precision that astronomers require and keep it tracking faint celestial objects as they move with the sidereal rotation of the heavens? How do you move a machine that large with no perceptible vibration?—were on hold.

The design questions were complicated by the fact that the designers still hadn't agreed on the basic optics of the new telescope. Should it be a fast telescope, with a relatively large ratio between the diameter of the primary mirror and the distance of the primary focus from the mirror? For astronomers who were trying to photograph distant objects at the limits of detection of the telescope and photographic emulsions, the faster the telescope, the fainter the objects it would record with a short exposure. Those who had spent a whole night, or sometimes three whole nights, cramped and cold, with a bursting bladder, while they guided a telescope to keep the pinpoint of a faint star in the cross-hair of an eyepiece, knew the advantages of a fast telescope.

Spectroscopists usually favored longer focal lengths and used their devices at the Cassegrain or Coudé foci. But even for them a short focal length could have advantages by permitting a more compact telescope, which would ease the engineering requirements for the mountings and the dome to enclose the instrument and allow some flexibility in the siting of the alternate foci.

But the light-gathering power of a fast telescope comes at a price. Fast telescopes are more sensitive to light pollution. When Heber Curtis moved from the Lick Observatory to the Allegheny Observatory, in Pittsburgh, he discovered that he could only use slow, long-focal-length telescopes because of the light pollution. A fast telescope also requires a mirror of deep curvature, which is harder to grind and polish to shape. The deep curvature requires a thick mirror blank so that enough material will be left after the grinding to maintain the shape of the mirror. A thicker blank would also be more massive and take longer to adjust to changes in the ambient temperature in the observatory. For a mirror with a deposited surface, like the fused-quartz mirrors, a deep curvature would require either that the deposited layer be thick enough to accommodate the shape ground into the mirror, or that the molded quartz backing be ground into a rough spherical shape before they began spraying on the clear quartz layers. Either way it meant more complications for the already troubled work at GE.

Fast telescopes with traditional paraboloid mirrors also have small fields of sharp, coma-free focus. The f-ratio of the one-hundred-inch Hooker Telescope was $f/5$, which was typical for reflectors used for deep-space research. The uncorrected field of sharpness at the primary focus was less than an inch in diameter, because of an aberration introduced when the light from a parabolic mirror was focused on a plate. If they were to make the new telescope even faster, say $f/3.3$, the field of sharpness at the prime focus would be even smaller. A circle of film half an inch in diameter is a small area in which to concentrate the images of the heavens.

Hale asked Anderson, Adams, Pease, and Frederick Seares, Shapley's teacher at Missouri and now an astronomer on the staff at Mount Wilson, to explore the diameter of the "good field" at the prime focus of a two-hundred-inch telescope at focal ratios of 1:3.3, 1:4, and 1:5, and to calculate the sharp field that could be used if they tried curved plates instead of a normal flat glass photographic plate. "Considering the great investment in the telescope, and the value of short periods of the best seeing," he wrote in his memo, "It might easily pay to use such plates for several classes of work." Graduate students were recruited to do the calculations.

A curved plate, they discovered, would not increase the useful field of a fast telescope. In France, George Ritchey and Henri Chrétien had experimented with a new telescope design that bears their name. By using a hyperboloid shape in the secondary mirror, and a deep, fast primary mirror, the Ritchey-Chrétien design provides an image at the Cassegrain focus free from the coma, or distortions outside the central field of focus, that plague telescopes based on paraboloid mirrors. But the mirrors of the Ritchey-Chrétien design are difficult to figure, it requires curved photographic plates to realize the full promise of the design, and despite the promise on paper, no working telescope had ever been built to the design.

The alternative to increase the useful "good field" of a fast telescope was to use an auxiliary corrective lens to compensate for the aberrations. No one had ever designed a lens that could correct the field of a large $f/3.3$ telescope. Frank E. Ross, at Yerkes, thought he could come up with a corrector lens if the project could support him and his "computer," a woman named Margaret Johnston, who got fifty dollars for a half month of work. Ross planned to use the sixty-inch and one-hundred-inch telescopes as test beds for the corrector lens design. But his work was another experiment, with no promise of success. Every stage of design of the telescope, it seemed, called for research and engineering that had never been attempted before.

Shortly after the grant was awarded, Hale had written to his friends at Warner & Swasey, who had built the mounts for almost every large telescope since the first big Lick telescope, asking if their chief designer/engineer, E. P. Burrell, could come to Pasadena to assist in the design. Burrell had recently designed and supervised the construction of the seventy-two-inch Victoria telescope, the newest large telescope, second only to the Hooker.

Pease and Porter had already sketched different designs for a mounting for the telescope. Pease's was a refinement of the drawings and model he had been working on for almost ten years. It was a conservative approach, a blown-up version of the fork mount of the sixty-inch telescope, relying on massive girders and huge roller bearings for rigidity and smooth motion.

Porter's telescope design experience was with small amateur telescopes, many of radical and innovative design, like his garden telescope. He had never designed a large telescope. In his early sketches of possible mountings, he tried to combine the rigidity of the English-style mounting of the one-hundred-inch with the versatility of a fork mount. His designs evolved into a split-ring mounting, so different from any other telescope that had been built that the design was relegated to a curiosity.

Hale favored Pease's design, which drew heavily from features and solutions that had been worked out on the sixty-inch telescope on Mount Wilson and the seventy-two-inch Victoria telescope. The Pease design, which everyone had looked at and talked about, in one form or another, for more than eight years, seemed a safe, simple solution. "It simply remains to adapt the best of these, in the light of recent progress, to the needs of the 200-inch telescope," Hale wrote. "We now know beyond question that a tube and mirrors having a combined weight of 150 tons, involving a total weight for the moving parts of 500 tons, can be mounted equitorially and without troublesome flexure so as to afford access to the entire available sky."

Burrell drew up a design based on Pease's sketches and drawings. His drawings were passed on to Professors Paul Epstein and Romeo Martel of the California Institute, for calculations of the flexure in the mounting, and also to Hale's friend Gano Dunn and his colleague Samuel R. Jones of J. G. White Engineering. The old-boy network was in full swing.

While the engineers calculated, Warner & Swasey built a model of the Pease design for exhibition at the National Academy of Sciences. The model, with a huge fork mount carrying the entire weight of the telescope on oversize roller bearings, and with a filigree box-girder construction for the tube of the telescope, relied on the same Brooklyn Bridge school of overengineering that had produced the one-hundred-inch telescope. Like Pease's earlier model, which had been brought out for late-night discussions at Mount Wilson, the Smithsonian model was an exhibition piece, to meet public demands and queries. The actual design work on the telescope was in suspension, awaiting progress on the mirror.

While they waited Pease explored options for constructing the telescope. The mounting for the one-hundred-inch telescope had been built on the East Coast, but labor was 15 percent cheaper in California than on the East Coast, and 15 percent of $2 million—Pease's estimate of the fabrication cost—was $300,000. Los Angeles had more sunlight, cheaper power, cheaper gas, freedom from extreme temperatures, and freedom from strikes in the nonunion shops. G. W. Sherburne, a local machinist, estimated that they could erect their own local plant for the fabrication, with a salvage value of 50 percent. At Mount Wilson's own shops, overhead was less than 25 percent. By contrast, when they had paid the Fore River Shipyard to build the one-hundred-inch telescope mounting, the overhead had run from 35 to 110 percent, plus a 10 percent profit.

To explore another option, Pease organized a conference at the Llewellyn Iron Works in Los Angeles, a large foundry that assured him and Anderson that they could fabricate anything that could be built on the East Coast at considerable savings.

With the actual design of the telescope on hold, awaiting progress on the mirrors, Porter turned his efforts to designing an astrophysics laboratory for the California Institute campus. The building would provide laboratory space, offices, a library. Construction was scheduled to start in the spring.

Work was already under way on machine and instrument shops, which Porter had also designed, and Sherburne was persuaded to come in and take charge of the machine shop. Hubble was named chairman of the Astrophysical Observatory and Laboratory Advisory Committee. On the recommendation of the committee, the machine shop was equipped with forty-inch tools, large enough to do much of the fabrication for instruments, auxiliary telescopes, and some of the precision-drive equipment that would be required for the big telescope.

Porter, who had considerable experience with mirror grinding from his days of writing for amateur telescope makers in *Scientific American*, and more experience of cold than anyone else, argued that the original idea of grinding and polishing the mirror at the observatory site was less than ideal. The opticians at the Santa Barbara Street optical labs agreed. The laborious grinding and figuring of a large mirror was too delicate a job for a mountaintop. There wasn't room in the Santa Barbara Street laboratories for a two-hundred-inch mirror, and after the brouhaha over the application, the relationship between the Mount Wilson Observatory and the new project was still so tentative that an optical laboratory for the Caltech campus, large enough to house the mirror-grinding project, became the next item on Porter's drafting board. While the architectural details of the buildings were being fleshed out by a New York architectural firm, site work began on California Street in Pasadena—the first tangible evidence of the telescope project.

Visitors to the campus were told the purpose of the buildings, and some were even shown the Porter drawings of various designs that had begun to line the halls and offices of temporary buildings. But buildings and drawings were no substitute for a telescope.

12

Depression

The new year came without good news from the GE labs in West Lynn. Ellis and his staff fiddled for months before they had the furnace and auxiliary equipment ready to surface a twenty-two-inch disk, a substantial leap up from their previous efforts, and the last trial disk on their schedule before they began a five-foot auxiliary mirror that would actually be used in the telescope.

The equipment had to run twenty-four hours a day, spraying layer after layer to build up the clear quartz surface. The operation went well until the blank was half glazed. One of the three heating elements in the furnace burned out. Two elements could maintain the needed 1700° temperature, so work continued. Then another element burned out. Ellis ordered the furnace partially cooled, repaired, and refired. Before it was hot enough to restart the spraying equipment, the repaired elements failed again.

Ellis and Thomson concluded that the furnace had to be rebuilt. Ellis took advantage of the shutdown to do some planning for the big disks, extrapolating from their experience on the smaller one. The figures he came up with were shocking: A surface layer 2.5 inches thick on a two-hundred-inch-diameter disk would require seven million cubic feet of hydrogen fuel—enough to fill two *Graf Zeppelins*. To surface the disk, GE would need either an enormous hydrogen plant on the premises, or a gasometer one hundred feet in diameter and seven hundred feet tall. In an era long before the creation of the Occupational Safety and Health Administration (OSHA) or the explosion of the dirigible *Hindenburg* at Lakehurst, New Jersey, in 1937, no one gave much thought to the danger of storing that much hydrogen near an industrial plant.

The hydrogen consumption was so daunting that Ellis tried calculations for alternative fuels. A nearby plant produced dissociated (chemically separated) ammonia. A rail spur connected the plants. Ellis calculated that 3,700 tank car–loads, each of 2,500 cubic feet of ammonia, would be enough for the two-hundred-inch mirror. A shuttle train could carry the tanks, if the trains didn't break down and if they kept enough men on duty to load and unload the continuous stream of tank cars. With either fuel the lab would need a huge supply of oxygen for the furnace. An on-site plant could be built to produce it. The project was beginning to seem bigger than anyone had anticipated.

Ellis's mechanics finally got the electric furnace rebuilt. During testing the furnace broke down with nagging regularity, but Ellis persisted and produced two twenty-two-inch fused-quartz disks by mid-August 1929. The disks were suitable for testing, but Ellis warned Anderson in California that the quartz had emerged from the furnace with mysterious black specks embedded in the surface.

With the disks packed and shipped off to Pasadena, Ellis again shut down the furnaces, and had the room sealed off until he could figure out what caused the specks. Chemical analysis was inconclusive. He tried introducing traces of potential contaminants into the flame of the torch to see if he could produce comparable spots. After weeks of testing, iron turned up the likely culprit. When tests of the refractory bricks of the furnace, the piping and burner components, and the oxygen and hydrogen gases that had been fed to the burner could not detect quantities of iron greater than 0.0001 of 1 percent— too small to account for the specks—that left only the quartz itself as the source. The contaminations in the supposedly pure quartz were proving as troublesome and incurable as George Hale's demons.

One correlation that emerged from the testing was that the specks seemed to be most numerous when the quartz had been deposited onto the disk at relatively low temperatures. That meant an end to Ellis's idea of using dissociated ammonia as a fuel for the spraying. Ammonia would have been cheaper, safer, and easier to transport and store, but it would fire the burners at a lower temperature than hydrogen. It looked as if they still needed two *Graf Zeppelins* of hydrogen.

Charges for overtime and new equipment piled up. While they waited for the results of the optical tests on the disks that had been shipped to Pasadena, Ellis and Thomson planned for the fabrication of a sixty-inch disk that would actually be used for one of the secondary mirrors in the telescope. The telescope would ultimately require two or three sixty-inch secondary mirrors, on the optical paths for the Cassegrain and Coudé foci. The laboratory at West Lynn was big enough to house the furnace for the mirrors, but Thomson confidently authorized construction of a new building for the next stage of the spraying operation.

No one had ever seen a building quite like what Thomson designed: sheet iron sides and roof over a structural steel frame. And Anderson, in Pasadena, had never seen a bill like the one Thomson forwarded. With the connections for heat, light, water, steam, and gas, the electrical equipment to regulate the furnaces, and the construction of a furnace large enough for a sixty-inch mirror, the building cost more than $115,000, close to half of the entire original GE budget for producing all the mirror blanks for the telescope. When Anderson questioned the expense, Thomson pointed out the advantages of his design: The structure could easily be expanded to accommodate the spraying of the two-hundred-inch mirror, and the steel walls of the building would have a high salvage value when the operations were finished. Nothing was too good for a perfect machine.

At the Mount Wilson optical labs on Santa Barbara Street, an optician rough-ground one of the twenty-two-inch quartz disks to a spherical curvature, the first stage in the shaping of a telescope mirror. It didn't take long for his grinding disk to cut through the transparent quartz layer to the rough quartz underneath, in an area just off the center of the disk. Anderson reported to Ellis that the fused layer was only one-eighth inch thick.

"Impossible," Ellis answered, insisting that the transparent layer had to be consistent in thickness over the entire surface of the disk because the rough quartz had been ground flat before they began spraying on the transparent layer. A week later he admitted that while the edges of the disk had been ground flat, the center might have been "a little high." In any case, Ellis wrote, it was impossible to reglaze the test disk because the furnaces were already being reconfigured for the next mirror.

There was little hope of making a successful mirror out of the disk, but Anderson ordered the opticians to keep grinding. At least they could learn more about how the fused quartz behaved under the grinding tools. As the tool ground deeper into the quartz disk, layers of bubbles appeared. The deeper the tool worked, the more numerous the bubbles. The fused quartz was extremely hard, and the edges of the exposed bubbles were so sharp they tore the polishing tool. Each bubble had to be dug out with hand-grinding tools. It wasn't a good omen.

The progress on the disks was so unpromising that as a precaution Anderson quietly kept his fingers on sources of alternative materials for the mirror, as a "second line of defense." U.S. Steel, eager to promote its newest stainless-steel alloys, lobbied for a steel mirror. The Philips Lamp Works in Holland promoted its process of fusing a layer of glass to a metal base. Under contract from the Bureau of Standards, the English firm of Parsons, experienced opticians and telescope makers, were trying to build up mirrors from thin glass plates in a cellular mosaic. None of these alternative materials was accorded much of a chance by the opticians at Mount Wilson.

From his new base in France, George Ritchey was eager to get funding for a vertical three-hundred-inch reflector with a fixed mirror. A movable coelostat and secondary mirror would feed light to the primary mirror and back to a fixed focus, allowing the observers to remain at a fixed observation position. He wanted to build the telescope on the edge of the Grand Canyon, which he announced was the ideal site.

Ritchey designed cellular mirrors for his proposed telescope, built up of small sections that would fit together like a symmetrical jigsaw puzzle. He argued that a composite mirror could be lighter than a solid disk, the individual cells of glass could be thinner and hence less subject to thermal distortions, and the frame holding the cells in relation to one another would provide great rigidity to the mirror.

Adams, Anderson, Pease, and Seares all studied the Ritchey scheme and concluded that while a perfect cellular mirror might work, the demands on the framework that would have to hold the sections in alignment with one another were so great that the design couldn't be realized. Ritchey experimented with cellular mirrors, cementing the sections of glass together with specially fabricated presses. He made great claims for the process but was never able to eliminate the "quilt" pattern in the surface of the glass from the cemented cells. A coelostat telescope, like his designs, also has a problem of limited declination range, which limits the amount of sky the telescope can see even more than the English mount of the one-hundred-inch telescope.

Whatever the merits of Ritchey's ideas, no one in Pasadena wanted much to do with one of his designs, whether for mirror construction or the Ritchey-Chrétien optics with its promise of a wide field of sharp focus. Ritchey's last years at Mount Wilson had alienated almost everyone. His projects, Adams wrote, "seem to be a case in which Professor Ritchey's wish is father to the thought."*

There were just enough signs of progress from West Lynn that Anderson kept the alternatives on hold.

---

* Like so many ideas that emerged in the design of the two-hundred-inch telescope, Ritchey's idea took many years to realize. In the 1990s several telescopes are under construction with composite mirrors, built up of thin segments, and ultralightweight mirrors with precast honeycombed backs. These new technologies rely on computer-controlled devices to adjust the precise alignment and shape of the mirror many times each second.

Ritchey finally succeeded in building a working Ritchey-Chrétien telescope at the Naval Observatory in Washington. He proved as intractable there as at Mount Wilson. Once, while removing the forty-inch primary mirror, he dropped it from the lifting slings to the steel deck of the observatory. The mirror miraculously survived, but Ritchey was banned from the Naval Observatory. The Ritchey-Chrétien design, refined with a Cassegrain lens originally designed by Gascoigne in Australia, and fifth-order optics developed by Ira Bowen, is currently used for most large telescopes, including the Keck telescope in Hawaii and the Hubble Space Telescope. Modern mirror-grinding and -polishing techniques, using computer-controlled polishing machinery, have made the complex hyperboloid shapes easier to figure; modern instrumentation favors the Cassegrain focus for most research programs; and the relatively short tube adapts well to modern, lightweight mountings.

One year after the public announcement of the grant for the construction of the telescope, there were already signs of the work in locations around the country, not only Pasadena, but in architectural offices in New York, at the GE laboratories at West Lynn, and at sites where astronomers were researching the observational conditions. But despite the flurry of activity, the great "eye" that had captured headlines when it was announced had already faded from the news. Occasionally a flamboyant Southern California evangelist would gather temporary notoriety by blasting the telescope project as an enterprise of blasphemers, destined to bring the curse of perdition on the builders, and perhaps on mankind, for their arrogance. From time to time GE or another company with a contract for some part of the project would time a press release on its work to draw attention to its company in the newspapers. But the slow developments in Pasadena and elsewhere weren't really headline stuff—not when the nation was suddenly overwhelmed with much bigger news.

The stock market had been front-page news even before the telescope project started. The market had been edgy all through the latter half of 1928. In December 1928 a sharp "correction" was followed by a brief panic that had newspapers pulling out the 144-point type. The market rallied and pulled through, but when the Federal Reserve refused to extend speculative credit, the market collapsed again in February 1929. Call money soared to 20 percent as the New York banks poured money in at 15 percent to rescue the market.

Despite the frightening volatility, speculators still saw optimistic signs. Every slump, the analysts observed, was followed by a recovery, each time to ever greater heights. Those who had been successful with their investments, or who had borne their losses with aplomb, chided those who were still watching from the sidelines. Before long Americans who had never invested in stocks decided that they, too, had to have what the slangsters now called "a piece of the action." "It can't last," the voices of doom cautioned, but remarkably, the predictions of the optimists proved true. Between corrections, the market went up and up and up. Savings and borrowed money poured into the market. No one wanted to miss out on the apparently free bounty. By the summer of 1929, close to one million Americans held stock on margin.

The crash came on the first anniversary of the announcement of the telescope project, October 28, 1929. Overnight, paper fortunes collapsed. The falling market took good money with the bad, old money with the new: Once-solid accounts, committed to the market at the last minute in an effort to restore order, fell as hard as the margin accounts of speculators. The cautious few who could afford their losses painfully picked up the pieces. Those who had invested everything, often on margin, were ruined. Bankers weren't the only ones to jump off the fine bridge over Arroyo Seco in Pasadena. Before long the city of Pasadena would be spending twenty thousand dollars per year guarding the bridge.

There were brief signs of hope in the following months, as the market recovered a portion of the lost ground, but it was soon obvious to all but the most stubbornly optimistic that the crash had inflicted mortal wounds on the market and the American and world economies. Even the pessimists who had predicted the collapse hadn't imagined the consequences of an economic crash on a nation in which one-half of 1 percent of all Americans possessed one-third of the national wealth, and 80 percent of American families had no savings at all.

The real impact of the crash on the national and world economy wouldn't be felt for months or even years. But Black Tuesday shattered the last remnants of the optimism of the 1920s. A decade of mania ended overnight. In the months following the crash, there was no world-class prizefight, no great new athletic feat, no murder trial of national interest, no ticker tape parade, no spectacular flight to match Lindbergh's achievement. The Atlantic City beauty pageant was canceled. Shipwreck Kelly came down from the summit of the Paramount Building because no one was watching anymore. The stunts that had fueled public interest in the mad 1920s gave way to desperation. Two men drove a Model A Ford across the country in reverse, but no one paid much attention. Russia's Five-Year Plan and the reports of recent visitors to the Soviet Union were suddenly of more interest than the amusing stunts, as those who suffered sought relief in dreams of a different political and economic system.

Even those who had resisted the lure of quick money and margin speculation weren't immune to the infectious spread of the crash. Henry M. Robinson, the chairman of the Board of Trustees of the California Institute, who had personally guaranteed the endowment of the telescope, was ruined. A year before, a personal guarantee from him was beyond question; now his investments were in shambles. When the matter was brought up at a meeting of the Board of Trustees, A. H. Fleming, the president of the board, reported Hale's original estimate of $150,000 per year or an endowment of $3 million. After the meeting, a board member leaned over to the president to say, "Fleming, you're not worrying about this, are you? I'm going to take care of this matter myself."

The trustee who made the generous promise never came through with the funds. Progress on the telescope had become so halting by late 1929 that the council and the Board of Trustees stopped worrying where an endowment for the instrument would be found.

The grant to build the telescope was still secure, but California Street, where the new astrophysics building and machine shop were going up at the edge of the Caltech campus, was in the midst of its own depression.

The Mount Wilson optical laboratories were still testing pieces of fused quartz from GE. Anderson tried holding his hand in contact with the surface of a disk for a minute. With a glass disk, the heat of his hand would have raised the temperature of a portion of the disk measurably, distorting the optical surface. On the quartz disk Anderson and the opticians repeated the experiment three times and could detect no measurable distortion. Fused quartz still seemed an ideal material for a telescope mirror.

But after almost a year and a half of experimentation, GE hadn't produced a single satisfactory disk. Ellis sent samples of quartz bars off to the Bureau of Standards for testing. The native quartz was cheap but full of imperfections; Brazilian quartz was pure, but expensive. He came up with another idea: If he could find quartz supplies for the backing and the sprayed surface with nearly identical coefficients of expansion, Ellis suggested, they could use the cheap native quartz for the molded backing and the pure imported quartz for the sprayed surface. It would save money and was sure to produce a satisfactory disk.

Like most of the plans coming from West Lynn, this one sounded good. But with expenditures already over the original budget for the mirrors, the date when the two-hundred-inch mirror was to have been finished long gone, and not a single usable mirror finished, there were doubts in Pasadena that the West Lynn laboratory would ever produce a usable mirror disk.

Hale called Russell Porter over to the solar laboratory and showed him the letter of agreement for the grant from the IEB, pointing out a clause they all hoped would never be invoked: "PROVIDED, That if at any stage of the project it be decided that the construction of the telescope is not feasible, any remainder of the amount hereby pledged by the Board, according to the terms above prescribed, shall be and become null and void." Hale then sent Porter back to West Lynn to determine just how much truth lay under the promises and assurances Ellis and Thomson offered almost weekly.

Privately Hale questioned exactly what Gerard Swope had meant when he agreed that GE would do the work "at manufacturing cost." Swope's original telegram had said that there would be no charges for "commercial or administrative expenses," but it wasn't clear who was paying for the time and expense of filing patent specifications, in the name of General Electric, on every step of the process. Hale never put it in so many words, but there was a clear conflict between the commercial style of GE, which saw the project as prestigious publicity and an opportunity to develop processes with future commercial potential, and the scientists in Pasadena, who were concerned only with getting a working telescope before they spent their entire budget.

Efforts to pin GE down on costs got nowhere. Hale and Anderson raised the question with Thomson and Ellis. Robinson brought it up with Swope. Everyone at GE hewed to a consistent party line: Fused quartz was an experimental process; many bugs had to be worked out before it would work successfully; GE was committing valuable personnel and resources and making no profit; and any budgets were only estimates. As long as GE was making *some* progress, Hale was reluctant to press too hard on the budget. Given the record of the temperamental one-hundred-inch Hooker telescope, which still hadn't reached the theoretical resolution the Mount Wilson designers had predicted, Hale and his colleagues weren't ready to settle for a second-rate mirror for the Big Telescope, no matter what the difficulties.

In West Lynn, Ellis—despite the consistently optimistic reports he had sent to Pasadena—began to have doubts of his own. The whole process had grown too complex. The costs of building and fueling a furnace large enough to melt quartz sand for the base of a sixty-inch mirror were enormous. After the base was molded, they would face the experimental spraying procedure for the surface. The slightest mismatch in the temperature coefficients of the batches of quartz used for the base and the surface risked distortions or strains in the disk. For an engineer used to production processes, there were too many ifs.

The test reports on the trial blanks, and the nasty bubbles that had emerged during grinding, argued for thicker sprayed coatings than Ellis and Thomson had planned. The problem was especially acute with the sixty-inch mirrors. The finished telescope would require at least three of these secondary mirrors, with both concave and convex surfaces. To make the raw disks interchangeable, they would need clear quartz coatings thick enough so that the edge could be ground down for the convex mirrors and the center ground down for the concave ones.

As Ellis stared at his drawings he had an inspired thought. What if they didn't mold a base at all and built up the entire volume of the disk by spraying? Dispensing with the process to mold the base of the mirror—furnace, tooling, mold, fuel, and personnel—would save money and time. They would no longer have to face-grind the surface of a molded base before commencing the spraying, and there would be no chance of temperature coefficient mismatches between the different layers. Instead of two complex processes to set up, they would have only one. Once the spraying process was in operation, they could spray day and night, as much fused quartz as they needed. It was just a question of keeping the furnace going, and supplying more fuel and ground quartz to the spraying apparatus. They could do mirror after mirror, any size they needed.

The new plan made sense, but it was an incredible leap into an unproved technology. The largest volume of fused quartz they had sprayed was a layer approximately five-eighths of an inch thick on the surface of a twenty-two-inch disk. Despite every precaution the twenty-two-inch disks had not emerged as usable mirror blanks. The sixty-inch disks would need to be approximately twelve inches thick, an increase in volume of 225 to 1.

Before he could go ahead, Ellis needed a design for the disk. At the tolerances needed for a large telescope, even seemingly rigid materials like glass and fused quartz are so fluid that the telescope designer must cope with the potential changes in the shape of the mirror disk as the telescope is swung from the horizon to the zenith. The weight of the disk itself, and the forces acting on it from its own mounting, deform the mirror as its orientation changes. The acceptable tolerance is close to zero. Distortions too small to measure mechanically can still be detected optically, and even a minuscule change in the shape of the disk affects the quality of the images.

Telescopes like the sixty-inch and the one-hundred-inch had been built with massive, solid mirrors, thick enough to maintain their shape as the telescope moved. Solid disks might work for the sixty-inch auxiliary mirrors, but one fundamental concept of the whole mirror project—part of the plan from the time of Hale's earliest talks with Rose—was that each stage in the work would function as a test bed for the next. The auxiliary mirrors were the design models for the two-hundred-inch mirror. The rule of thumb was that the thickness of a solid disk should be one-sixth the diameter. Following that formula a solid disk two hundred inches in diameter would weigh more than forty tons. From his calculations of fuel consumption and quartz supplies, Ellis knew that a solid two-hundred-inch disk would be impossible to fabricate.

It would also be too massive to mount in a telescope. If the ratio of diameter to thickness is maintained, the mass of a disk increases with the cube of an increase in diameter. As the mass increases, even a temperature-stable material like fused quartz becomes vulnerable to changes in the ambient temperature. If the air outside the observatory dome was ten or fifteen degrees cooler than the air inside, not at all unusual in the early evening at a mountaintop observatory, when the dome was opened to use the telescope, the surfaces of the mirror would quickly begin cooling to the ambient temperature. The rest of the disk would cool at a slower rate, as the cold was gradually conveyed to the interior of the mirror. Depending on the mass and thickness of the mirror, for hours, perhaps for an entire observing session, there would be differences in the temperature of different portions of the mirror. With one part expanded and another contracted, the mirror would bend and flop, distorting the images it focused on the astronomer's photographic plates. The greater the mass of the disk the greater these distortions. Even if temperature effects didn't rule against a solid disk, the sheer weight of a mass of quartz that huge would make it unusable. A mounting to hold it would be prohibitive, and the disk would sag and deform from its own weight as the telescope slewed from position to position.

For more than a year Anderson, Hale, Thomson, and Ellis had been discussing alternatives to a solid mirror disk. They needed a disk sufficiently rigid to hold its shape, thin enough in each dimension that it would avoid the effects of differential cooling, and with provisions for attaching and supporting the mirror that would compensate for any changes in its shape as the telescope slewed from the vertical to the horizontal.

Hale, who despite the increasingly frequent attacks from his demons, had an idea for every problem, had already put his own suggestion for supporting the mirror in a memo to Anderson. Hale's idea was to use a liquid or air support under pressure that would vary with the inclination of the mirror. He urged that the idea be tested with a large disk of thin glass plate. An ingenious suggestion, it was never tested or explored because there was no pumping and valving equipment available sensitive enough to control the local pressure of a cell of water or air with the precision needed to shape the mirror as the mounting moved.*

As they went through different options, the best idea, it seemed, was the scheme Thomson had proposed to Anderson in the gloomy days a year before, of ribbing the back of the mirror like a waffle. If the ribbing were designed carefully, the ribbed back would mean that even with a two-hundred-inch mirror, no portion of the quartz disk would be more than four or five inches from the surface. The disk would respond to changes in temperature like a much smaller disk, and avoid the nightly problems of a thick slab. With the continuing problems of the mirror of the one-hundred-inch telescope in their minds, the ribbed scheme seemed a splendid solution.

Thomson carried the idea a step farther by suggesting that the pockets between the ribs could be used for an *active* mounting system to support the disk. If grooves and ribs were molded into the disk, they could be gently pushed and prodded with levers to compensate for the mirror's tendency to deform as it moved. Everyone began sketching his own version of the waffled back of the disk. Anderson tried dozens of variations, changing the thickness and layout of the ribs, and substi-

* Hale's concept was regularly employed after the 1960s by using a mercury torus, enclosed in a bag, under the mirror of new telescopes like the Du Pont one-hundred-inch telescope at Las Campanas. A later variation of the idea has been adapted in the 1990s for large telescopes built with meniscus mirrors, glass or quartz surfaces too thin to hold their shape without external support. The shape of the mirror is continuously adjusted, many times each second, with computer-controlled actuators. The actuators are mechanical rather than the air or water cushion Hale envisioned. Telescopes with this design are currently under construction in Arizona and Chile.

tuting or adding a pattern of round pockets sunk into the back of the disk along with or in place of the ribs. While Anderson worked on alternatives for the shape of the ribbed structure, Thomson sketched lever-operated actuators that would use gravity to create compensating pressures on various points of the back of the mirror as the telescope was swung from the horizon to the zenith. He used ball-and-socket joints, balancing arms with counterweights, and gimbels to create devices that could be fitted to each pocket in the back of the mirror. If the systems were designed correctly, the mirror would automatically assume the correct shape no matter which way the telescope was tilted or turned.

It was an ingenious idea. Thomson, an instinctive inventor who loved mechanical gadgetry, sent pencil sketches of the complex machines to Pasadena. No one had ever built such a device, and it seemed fantastic to assume that a mechanical device could sense and make movements on the order of one millionth of an inch to correct the changes in shape of the mirror, but it was also hard to find fault with the design. Thomson's reputation for genius made even fantastic ideas seem workable.

Pease, who had been working on the problems of the one-hundred-inch telescope, calculated that ball-bearings on the edge and bottom support systems for the mirror would reduce the friction of flexure and expansion between the glass and metal surfaces from ±0.1 inch to ±0.001 inch. His calculations, and the use of ball-bearings, added another element to the design. Suddenly, it seemed, they had a workable design for the mirror of the telescope. Like so many good ideas, it had emerged not so much from a deliberate research program as from very bright minds freely exchanging ideas, nourished by serendipity and the willingness to explore radical concepts. Ellis's idea of spraying the entire volume of the mirror meshed perfectly with the collective ideas of Ellis, Thomson, and Anderson for a ribbed back and a support system for the disk, and Pease's suggestion for ball-bearing supports to reduce the friction between the glass and the metal of the mounting.

Now, all they had to do was spray the quartz mirrors, grind and polish them to nearly perfect optical surfaces, build a telescope around them, and an observatory for the telescope. When things went well, everything seemed easy.

13

Orderly Progress

George McCauley was an orderly man.

Each workday morning he walked from his home on Fourth Street, up on the hill above the Corning Glass Works, down to his office or to one of the factory units where he had a project under way. He was a hard worker, with a reputation for concentration. Most days he would work without a break all morning. At noon he would walk home for his dinner and a twenty-minute nap, and then head back down to the factories in time for the one o'clock whistle and an afternoon of more hard work. In the evening, after supper, he would bring out a portable drafting table and a T-square, and do design work or calculations at the cleared dining room table where his children were doing their homework.

McCauley came to Corning via Northwestern University and the University of Wisconsin, where he had gotten a Ph.D. in physics. He had taught briefly at Northwestern, and worked at the Bureau of Standards during the war. His job at the Glass Works was research on special projects. He liked the company and the town. He had grown up on a farm in Missouri, and he enjoyed the quiet, rural beauty of the Chemung Valley. There were dairy farms and orchards in the surrounding countryside. It was a good place to raise a family. George was a warden of the Episcopal Church and a scoutmaster of the local troop.

Corning had originally been a railroad town. The company that became the Corning Glass Works was recruited from New York City by a commission appointed to seek new business at the end of the nineteenth century, when the railroads were threatening to pull out. Corning soon gave its name to the glass company, which grew large enough to dominate the town. For the most part it was a welcome domination. The Houghton family, majority owners of the Corning Glass Works and chief executive officers, were popular in Corning. They were fair bosses, and Corning built a reputation as a good company.

Corning had always been a research-oriented company. In the 1920s it developed a series of new glasses, based on borosilicates fabricated at very high temperatures. Under the trade name Pyrex, these new glasses revolutionized housewares. A Pyrex pie plate could go directly from the icebox to the oven, without danger of cracking or exploding. For housewives Pyrex ware was a timesaving revolution. Pyrex pie plates didn't crack because the borosilicate glass formulation had an extremely low coefficient of expansion—exactly the quality astronomers sought in telescope mirrors. It didn't take long before astronomers inquired about the new glass.

As far back as 1922, Corning had received orders from the Mount Wilson labs for small glass disks. In 1929 a query arrived from Yerkes Observatory. George McCauley pointed to the queries and the potential for customers at other observatories and got authorization to cast a twenty-seven-inch-diameter disk with a four-inch-diameter hole in the center for the sales department to use in approaches to observatories and at the annual convention of the AAS.

It seemed a simple enough project. McCauley had the Corning masons build a mold from insulating brick, using an iron strap around the bricks to hold them in place. The bottom of the mold was clay damper tile. To mold a central hole in the disk he had the technicians use more of the same C-22 refractory brick material, held in place with Hi-Tempite cement to withstand the heat of the molten glass. The mold was finished in May 1929, during a busy period in the old A Factory on the riverbank. McCauley had it set aside against the wall of the factory to await a slack period.

Through the summer of 1929 the factory was working long shifts. McCauley's mold became a convenient table for sundry objects, and by August, when he was ready to try casting a mirror, the central core of the mold had broken. A Corning mason did a quick repair, using wet clay and cement. As the mold was filled with molten glass, the moisture in the core launched a cloud of bubbles in the glass. McCauley told the technicians to add more glass to fill the voids. When the mold was full, he had it moved aside, out of the way of the usual traffic in the blowing room, so it could quietly anneal. Glass foundries are huge, noisy places. On the vast factory floor McCauley's experimental mirror mold was a small project. If this was all it took, McCauley thought, astronomical mirrors seemed an easy sideline for Corning.

The annealing went well until the temperature inside the mold reached 480°C. At that point, the heat triggered a sprinkler head on the ceiling, part of the factory's fire protection system, dousing the mold with water. After some laughter at McCauley's expense, technicians turned off the sprinkler, and the cooling process continued until the disk was cool enough to uncover. The disk emerged from the mold with traces of bubbles in one area from the hasty core repair, and the interrupted annealing was less than successful, leaving strains in the glass, but the disk had survived in one piece. McCauley pronounced the experiment a success and gave the go-ahead to the sales department to try to solicit orders for disks for telescope mirrors. As an added incentive for their presentations to observatories, he suggested that it would be easy to mold a curve into the face of the disks, which would cut down on the initial grinding when the disks were shaped into telescope mirrors.

McCauley's preparations came at the right time. George Hale's article in *Harper's* and the publicity surrounding the announcement of the grant for the two-hundred-inch telescope had started a wave of telescope building. The Universities of Michigan and Texas were shopping for telescopes in the eighty-four- to one-hundred-inch range. The Perkins Observatory was already building a sixty-nine-inch telescope. George Ward and Wilbur Foshay at Corning put together cost figures for different-size disks, based on McCauley's estimates.

Queries came in, but the Corning salesmen couldn't land a large order. They could argue that their product was superior, but their prices were high, at least compared to the quotes that the universities had received from glass foundries abroad. A flurry of interdepartmental memorandums went back and forth, questioning whether Corning should make an effort to quote cheaper prices for the big disks. The initial answer was no. On the basis of the special equipment and manhours the project would need, the small market for telescope mirrors didn't seem a profitable business in 1929.

A year later the answer changed. By 1930 the depression had touched almost every sector of the economy. In some industries, like steel in Pittsburgh, more than half the workers were laid off, and companies devised schemes like the "stagger plan" to let men share jobs (which left them ineligible for unemployment compensation). The Houghton family were reluctant to lay off any workers at Corning, no matter how severe the economic situation. The chance to keep a few more workers busy, to strengthen Corning's role in glass research, and to build more ties to university research departments was enough incentive to aggressively bid for the potential telescope business.

McCauley went to the Bureau of Standards, where he had worked before coming to Corning, to survey the procedures they had used to cast the mirror of the Perkins telescope, the first large glass disk ever made in the United States. The tank melting and ladling procedures Corning had already developed were as good as the Bureau of Standards procedures. But casting a disk was only half the battle. Annealing the disk, subjecting it to long, slow, controlled cooling to minimize strains in the glass, seemed to be the real challenge. The bureau experiences confirmed what McCauley had learned in his initial trials at Corning: The high temperatures and long duty cycle for annealing would destroy all but the most rugged, commercial-grade heating units.

McCauley spent more evenings at his oak dining table. To minimize the investment in what would clearly be a small sideline for Corning, he tried to use commercially available materials and facilities. The molds, furnaces, annealing kilns, cranes, and lifting slings could all be fabricated by Corning craftsmen, who might otherwise be idle. When McCauley demonstrated that the new business would require no large outlays for special equipment or custom-fabricated tools, and that it would provide work for underemployed Corning workers, the Corning bean counters recalculated their prices for disks.

In March 1931 the new cost estimates paid off. Corning received a firm order for an elliptical auxiliary disk for the Perkins telescope. McCauley was ready. But one order didn't make a telescope disk business.

At GE's West Lynn laboratory, Ellis got ready to spray a sixty-inch quartz disk that would be used for one of the auxiliary mirrors in the telescope. The huge sheet iron building, the custom furnace, built from special ceramics and graphite developed by the Carborundum Company and the National Carbon Company, and the huge, rigid copper bars and regulators to power the electrical furnace for this stage of the process had cost more than $115,000. Ellis justified the expenditure by noting that the building, the special ceramics for the furnace, and the electrical equipment would all be reused for the two-hundred-inch mirror.

Every time there was a choice in equipment or facilities, Ellis, with the approval of Thomson and the GE management, opted for the more versatile, and invariably more expensive, item, on the grounds that more versatile equipment could be adapted to larger mirrors and would have higher resale value. Instead of inexpensive insulated cables to bring electricity to the furnaces, he used expensive copper bus bars. The building was equipped with a seventy-five-ton gantry crane, far larger than needed to fabricate or move a two-hundred-inch disk. Whenever Hale or Anderson questioned the budgets, Ellis explained that each decision had been made in the interest of salvage value. No doubt his arguments were true, but it was also true that each choice favored continued productive capacity at GE over cost control and expediting the production of the mirrors for the telescope.

This time Ellis was confident. The latest disks to emerge from his furnace, though flawed, had been a great improvement over the earlier efforts, and the efficiency of the process had improved dramatically. At the beginning of 1929 the spray apparatus was depositing one cubic inch of quartz per hour while consuming two hundred cubic feet of gas. By the end of the year the nozzles were laying down two hundred cubic inches of quartz per hour and consuming only six hundred cubic feet of gas. A mirror could be built in one to 1.5 percent of the time the process originally required, which meant that the time during which the furnace and disk had to be maintained at the torturous temperatures was substantially reduced. Ellis had finally abandoned hydrogen as a fuel, because it would have required a gas plant larger than they were willing to build. Dissociated ammonia wasn't hot enough, so they had begun experiments on alternative fuels, based on research done at the Department of Agriculture, Du Pont Ammonia, Nitrogen Engineering in New York, the Fixed Nitrogen Laboratories in Washington, and their own experience in Schenectady and Lynn. The most promising fuel was cracked butane, which could be obtained in carload lots, evaporated at any desired rate, and promised a hotter flame than any fuel they had yet tried.

As Ellis and Thomson saw it, the entire development was progressing predictably. Ellis was "delighted" when the opticians in Pasadena found 80 percent of the surface of the last mirror shipped satisfactory. Ellis and Thomson considered the work that had gone before to have been experimentation, trials with different materials and procedures. The difficulties they had encountered were preproduction problems that could be expected in any experimental process. Although they hadn't yet produced a fully satisfactory mirror, that was almost to be expected, and wasn't really a problem in any case, since none of the smaller blanks were meant for use in the telescope. GE, and particularly Elihu Thomson, had been famed for turning experimental ideas into working products. The company's claim to fame, and its prosperity, had stemmed from the translation of the inventions and discoveries of Edison and Thomson into routine production processes. Ellis was following the company tradition.

What Ellis and Thomson never quite understood was that Hale, Anderson, the Observatory Committee in Pasadena, and the astronomers who were waiting for the telescope were not after a perfected production process. All they wanted was mirror blanks—three sixty-inch blanks for the auxiliary mirrors, a two-hundred-inch blank for the primary mirror, and a one-hundred-inch blank they could grind into a flat for testing the primary mirror. They needed only five disks, good enough to be figured into mirrors for a telescope. So far, after two years of experimentation, 80 percent usable was the best GE had achieved on any disk. The opticians couldn't make a telescope mirror out of an 80 percent usable disk.

Without the mirrors, or at least some assurance that a satisfactory mirror blank could be fabricated, Pease, Porter, and Anderson could not move ahead on a design for the telescope. And without a design, the rest of the work on the project—the search for a site, the mechanical and civil engineering for the observatory itself, the preliminary work on instrumentation and auxiliary lenses, the electrical engineering and calculations for a control system for the telescope—was on hold. No one objected if the work GE was doing on the fused-quartz mirrors ultimately led to techniques for regular production of astro-

nomical mirrors, if only they could produce the mirrors for this telescope.

The halting progress in West Lynn never stopped GE from issuing
a steady stream of press releases and articles for the trade and popular
press. While Anderson and Hale were having trouble coming up with
excuses for the lack of progress on a mirror, John W. Hammond of the
publicity department at GE was sending out articles with titles like
"Greatest Venture in Mirror Making Ever Attempted," "A Great Magnifying Glass to Help Read the Story of the Stars," and "Building a Looking-Glass to Mirror Unknown Stars" to any magazine that would print
them. Thomson, although he had left the day-to-day running of the
project to Ellis, never turned down opportunities to speak on the subject of the mirror and the telescope. One of his talks on the two-hundred-inch telescope, in December 1929, was broadcast on radio from
Philadelphia.

GE's appetite for publicity offended the reticence of the astronomers
in Pasadena, especially George Hale. Much that the GE publicity
department included in their articles was wrong. They wrote that the
smaller mirrors were needed as "finders" by which the heavenly bodies
are first located; in fact the auxiliary mirrors were needed to focus the
image of the primary mirror at different positions on the telescope, so
the same telescope could be used for both deep-space research and for
detailed spectrographic study of nearby stars. The GE publicists had
no qualms about announcing that the new telescope would open an
area of unexplored space thirty times greater than at present known,
or that the most remote of the charted stars were 150 million light-
years from the earth—both untrue statements. It wasn't only the inac-
curacy of the reports that bothered the astronomers. They were afraid
that the newspapers and radio stations would read between the lines
of the GE reports and speculate about the lack of progress on the project, and that the speculation would in turn feed the doubters like Harlow Shapley, H. L. Mencken, and others who enjoyed taking potshots
at the California astronomers.

As the effects of the depression spread, with daily reports of bank
failures, soaring unemployment, breadlines, and soup kitchens, Hale
and the others were also embarrassed and worried that the scale of
spending on the telescope project would prove difficult to justify. Six
million dollars was still a lot of money, and astronomers were already
the butt of jokes and cartoons about stargazing and heads in the
clouds. Men who had lost their jobs and were now selling apples and
pencils on street corners to feed their families might be less enthusiastic about a $6 million telescope than the ebullient newspaper readers
and radio audiences of 1928.

The differences between the astronomers in Pasadena and Ellis,
Thomson, and Swope at GE remained largely unwritten and unspoken. Theoretically, fused quartz would make the best possible mirror.

If GE could produce fused-quartz mirrors that would eliminate the problems that still troubled the one-hundred-inch telescope, the differences in style between the huge eastern corporation and its publicity department and the tiny West Coast university, the embarrassments over premature or inappropriate publicity—even the cost overruns wouldn't matter.

On the early experimental disks, even on the twenty-two-inch disks, Ellis could in a pinch get together enough usable quartz by putting a couple of men on mortars and pestles, and grading the material with hand sieves. For the telescope mirrors, Ellis ordered quartz by the carload, had it ground in a ball mill in West Lynn, then used graduated sieves to separate the powder by particle size. Extremely fine particles, which would foul up the jets of the spray equipment, had to be filtered out by an additional process. Traditional separation methods, like air-blast filtration, didn't work because the fine particles would become electrostatically charged and cling to larger ones. Ellis finally instituted a wet-filter process, in which the fine material would be suspended in a solution that was then drained away. The procedure was excruciatingly slow. It wasn't until late summer of 1930 that he was finally ready to begin spraying the first sixty-inch disk. The graded quartz waited in hundreds of large glass jars for the day when the spraying would start.

Two full years had elapsed since GE had agreed to fabricate the mirror blanks. The original budget of $252,000 had long been spent. Hale and Anderson repeatedly asked for a new budget. "It is of course impossible to make an estimate that will be hard and fast as final cost of producing the mirror, including the 200 inch," Ellis answered. His new ballpark figure was an additional $50,000 for the furnace and accessories, and $12,000 per month for materials and development work, for a period of eighteen months. The total of $266,000, added to the $308,000 they had already expended, made a grand total of $574,000 for three sixty-inch mirror blanks, a one-hundred-inch mirror that could be ground to a flat for testing, and the two-hundred-inch mirror. The price tag for the mirror blanks had doubled in two years.

If the actual progress on the blanks was slow, Ellis and Thomson were still ever ready with ideas. Ellis drew up his own ideas for a mounting for the telescope—a project on which Pease and Porter had been working for years—and for the kind of facility in which the disk should be ground, polished, and tested. To minimize vibrations and temperature variations, he wanted the room entirely underground. He had suggestions on how the two-hundred-inch disk should be moved to a optical shop, equipment for minimizing air movement in the optics shop, and a dozen other ideas that touched the project. When Porter visited the West Lynn laboratories, Ellis regaled him with those ideas, and with reports on their work on perfecting fire control on warships. The one thing Ellis didn't offer was a date when a usable disk would be ready.

In Pasadena, Pease's design work waited for progress on the mirrors. He turned his attention to the still-troublesome one-hundred-inch telescope on Mount Wilson. Hubble and Humason were using the telescope regularly, as were other astronomers, and on good nights it produced excellent results. But the falloff of the image quality in some areas of the sky was still disturbing, and the apparent inability of the mirror to settle down, after even modest changes in the temperature, remained troublesome.

Pease had worked out some ideas for edge mountings for the mirror of the two-hundred-inch telescope. He got permission to use the one-hundred-inch as a laboratory to test his ideas. What was happening, he concluded, was that the mirror of the one-hundred-inch telescope was changing shape, deforming from its own weight, as the telescope was tilted in various positions. The problem hadn't caused great concern on earlier telescopes because it was thought that the massive solid disks in telescopes like the sixty-inch and the one-hundred-inch would minimize the deformation. But the demands of a telescope increase with the size. A distortion of one-tenth of a wavelength of light—a distance measured in millionths of an inch—may be difficult to measure on a small telescope; it becomes readily apparent in the image quality on a larger telescope. Pease worked during the daytime and during light-sky portions of the month, when the moon was up and the sky was too light for deep-sky observations of remote galaxies.

Ellis's spray process for the sixty-inch mirror transformed the working area inside the steel building at West Lynn into a self-enclosed hell on earth. A bank of transformers along one wall hummed with the eight hundred kilowatts of power needed to heat the twelve-foot-diameter furnace. The steel walls reverberated with a low-frequency hum that made talk impossible. The temperature inside the furnace was over 1900°F, and heat waves caused the air in the building to undulate until the walls and equipment seemed to shimmer. The temperature at the nozzles of the burner was double that of the disk itself, close to 4000°F, so hot that the burner had to be shielded in an enclosure of fused quartz to withstand the temperature. A heavy outer pipe shielded the smaller pipes for quartz powder, hydrogen, oxygen, and cooling water to the burner. Technicians could only view the process through a thick, green glass window set into a steel viewing protector.

In the first trials the new burner apparatus worked exactly as Ellis had planned, laying down fused quartz at a much higher rate than even his optimistic predictions. The intense sleet storm of quartz that fell onto the mold surface appeared to fuse readily with the material laid down behind it, and the disk built up quickly. It wasn't until the trial disks cooled that Ellis discovered that the layers of quartz had failed to fuse. Instead of a single, fused disk, what emerged was a lamination of partially fused layers, a quartz *millefeuille* pastry.

Ellis's answer was more tests. Each cycle of heating and cooling the furnace took at least a week and an incredible quantity of fuel, so Ellis went back to the small furnace for testing. The only way he could get the quartz to fuse completely was to slow down the rate of spraying until it was too slow to be practical. More tests finally identified the problem: The new process for pulverizing and filtering the quartz filtered out exactly those fine particles that would have provided the fusion between the layers of sprayed quartz.

Ellis decided to start over again and prepare new batches of quartz instead of reprocessing the quartz that had already been prepared. He would use cheap native quartz for the body of the disk, instead of expensive Brazilian quartz. The native quartz had to be hand selected to avoid white streaks, which seemed to produce bubbles. Ellis assigned three men to the job of sorting an enormous rock pile of quartz into three smaller piles of quartz with no white streaks for the faces of disks, pieces with minimal white streaks for the base material, and a discard pile with excessive white streaks. Seventy-five percent of the quartz fell into the first two categories and was then ground in the ball mill.

Ellis tried the newly prepared quartz in another experimental disk. If the quartz had even minimal white streaks, disks came out looking "more like porcelain than quartz because of the great number of small bubbles." When he tried to fuse a layer of pure Brazilian quartz onto the bubble-laden base, the heat transferred to the base during the process expanded the bubbles, raising the mass "like yeast raises bread."

Guessing that the problem might be moisture in the quartz, Ellis tried drying samples for periods from one hour to one week, at a temperature of 450°C. The drying made no difference. No matter what he tried, he could not produce a satisfactory disk. He kept turning back to an earlier experimental disk, number thirteen, made by the older procedure, from the identical quartz, from the same quarry, even from the same part of the quarry as the material they were currently using. That blank had come out essentially bubble-free. When he tried a sample of quartz left over from the earlier disk, it fused beautifully. It was as if he had left out a magic ingredient that had made the earlier tests work.

Hands-on experimentation was the specialty of the West Lynn laboratory, the core of Thomson's reputation. Ellis had notes from the various stages of their experiments, but his methodology was trial and error, adding more or less of one ingredient or another, or substituting one grade of quartz for another. No one at GE fully understood the processes at work inside the furnaces. When Ellis offered a description of the problems, it wasn't the sort of rigorous explanation scientists like Anderson or Hale would expect:

> The pulverized quartz does not become melted while passing through the flame from the burner to the work, but is caught in the sticky surface of the work and melted there. Some of the quartz is vaporized from the surface and perhaps from some of the extremely fine material passing through the burner. This vapor condenses on the cooler parts of the surface and furnace forming a white spongy layer of varying thickness. To this deposit is added some of the larger particles of pulverized quartz by the action of the flame, or by pieces bounding from surfaces that have not yet reached the sticky state. The surface layer thus formed seems to be a very good heat insulator, and which must be melted by heat transferred through the layer of quartz being laid down by the burner as it passes over the surface.

Ellis concluded that even a trace of "white material" in the quartz would contribute excessive bubbles or contaminants of iron or chromium to the laid-down quartz layers. The bubble-filled material then became so effective as an insulator that it did not pass the heat to the layers beneath, preventing complete fusion of the layers.

Desperate to get the process working again, Ellis tried alternate materials, like flint shot, a pure silica sand from an enormous deposit in Ottawa, Illinois. He had used the sand successfully when he molded the bases of the earlier disks, but when he tried pulverizing a new batch for the sprayer, the quality and purity varied so much that a process to prepare production quantities would require "more development." He finally went back to the quartz pile in the yard outside the laboratory. For the next batch Ellis had a single man grade the quartz in an effort to assure uniformity of the grading standard. The man would study each piece of quartz against a black background, holding the samples underwater to minimize reflections that might hide the white streaks. A quick test of the new quartz appeared to fuse, and Ellis hoped that by December 1930 he would have enough quartz on hand to begin the long-awaited sixty-inch mirror blank.

At least, Ellis assured Hale and Anderson, the current problems were the last they would encounter. "Every man on the work is doing everything he can to produce a 60-inch mirror at the earliest possible date, for with this experience behind us the rest will be easy. There will, of course, be problems to be solved in making the 200-inch mirror, but it does seem to us that practically everything that can happen has already happened, and we have every hope that the coming year will be brighter."

He finally started spraying a sixty-inch mirror for the telescope on December 8, 1930. Miraculously, everything worked. On January 6, 1931, Ellis triumphantly telegraphed Hale: "We have laid one more ghost. The first sixty inch mirror blank has been reduced from annealing temperature approximately 1100 degrees C. to room temperature in eight days, an astoundingly short time compared with glass." Finally, it seemed, Ellis had licked the fused-quartz demons. Hammond in the GE publicity department prepared another article for the journal *The Glass Industry,* celebrating the achievement. A GE executive named McManus, at the main office in Schenectady, began negotiating with representatives of other observatories, including Harlow Shapley at Harvard, for future orders.

In Pasadena, Hale and his colleagues were delirious with excitement. Walter Adams and Theodore Dunham of the Mount Wilson staff traveled to West Lynn to see the sixty-inch when it came out of the annealing oven. They liked what they saw. Even Hale was encouraged. "If it were not for the 'fierce' cost," he wrote, "I should feel much encouraged. Their capacity for spending is appalling. . . . although they are two years behind their original time schedule some of us may live long enough to see a 200-inch disk, if the money holds out."

14

Change of Guard

Caltech was a tiny school with a big reputation. Only a decade after its founding, the superb faculty Hale, Noyes, and Millikan had recruited had brought the institution to the front ranks of the hard sciences. There were still only few dozen regular faculty members, but the list of speakers who trekked to the red-tile roofed stucco buildings under the visiting scholars program included Albert Michelson, Michael Pupin, and the cream of European physics: Niels Bohr, Max Born, Paul Dirac, Erwin Schrödinger, and Werner Heisenberg. Caltech had come a long way from its shaky beginnings as a polytechnic school.

In the spring of 1931 Einstein joined the faculty as a visiting professor. Einstein was a celebrity, his name celebrated in Cole Porter songs,

Your charm is not that of Circe with her swine
Your brain would never deflate the great Einstein.

and his theory in e. e. cummings's poems,

> *. . . lenses extend*
> *unwish through curving wherewhen till unwish*
> *returns on its unself.*

In Southern California, Will Rogers wrote, Einstein "ate with everybody, talked with everybody, posed for everybody that had any film left, attended every luncheon, every dinner, every movie opening, every marriage and two-thirds of the divorces. In fact, he made himself such a good fellow that nobody had the nerve to ask him what his theory was."

Except at Caltech. Richard Tolman, professor of physical chemistry and mathematical physics and dean of the graduate school, had been at Caltech for almost a decade, working on relativity, statistical mechanics, and cosmology. Ever since Hubble had discovered red shifts in the light from distant galaxies and had begun to calculate the speed of recession of those galaxies, Tolman had worked to integrate Hubble's findings into a workable cosmology of curved space. Hubble wasn't much of a relativity scholar, but he loved the publicity, especially the photographs, always in a tweed jacket with his trademark pipe. When Einstein went to Mount Wilson and posed for photographs at the one-hundred-inch telescope, Hubble gave his usual explanations about how the giant telescope was used to determine the structure of the universe. "Well, well," Mrs. Einstein said, "My husband does that on the back of an old envelope."

The one-hundred-inch telescope was working hard. Hubble and Humason had a near monopoly of dark time on the telescope to carry on their search for distant galaxies. They used every trick to reach farther and farther out with the one-hundred-inch telescope, including new emulsions for the spectrograms, and new auxiliary lenses. Hubble soon had images from enough nebulae to derive a classification scheme for galaxies, his famed tuning fork with elliptical galaxies on the handle and the spiral and barred-spiral types classified by their position along the two tines. The derivation of a morphology of distant galaxies, less than a decade after the very existence of "island universes" had been demonstrated, elevated cosmology into a hard science.

With each improvement in technique, Hubble and Humason measured bigger red shifts, which translated into larger velocities of recession and more distant objects. In February 1931, using a new Payton spectrograph objective, which had been developed as part of the two-hundred-inch project, they succeeded in recording the spectrum of a minute spiral nebula of the seventeenth magnitude, "far beyond the reach of all other instruments." To record the spectrum Humason had to keep the faint object centered on the slit of the spectrogram for seventeen hours, over three nights. "The nebulae are becoming so faint," he wrote, "that they are difficult to see." From the measured red shift, Hubble calculated that the galaxy was receding at close to twenty thousand kilometers per second. "Either the entire universe is flying apart," Hale wrote when he heard the report, "or a new and fundamental physical law must be elucidated to account for these extrordinary phenomena."

The consequences of Hubble's work were as exciting for cosmology as for observational astronomy. The universe was no longer the stable constant it had always seemed. Hubble and Humason had discovered thousands of "island universes," of infinite variety, streaming away from one another at inconceivable velocities. What could account for this seeming entropy? Tolman set to work on the problem and gradually concluded that the match between Hubble's data and Einstein's theory of gravity *without* the cosmological constant was too compelling to ignore.

For Tolman, Einstein's visit to the campus was a grand opportunity. Tolman, a witty man with a deadpan delivery, also served as campus toastmaster, a duty that reached fever pitch when Einstein arrived in Pasadena. In the evenings he would introduce Einstein at various functions. During the days Einstein attended colloquiums and private meetings with faculty and graduate students, including enough sessions with Tolman that he was finally persuaded that the cosmology of an expanding universe was probably correct. Einstein had actually wavered on the cosmological constant in a paper. When Sir Arthur Eddington chided Einstein about dropping the constant, Einstein said, "I did not think the paper very important myself, but de Sitter [Willem de Sitter, coauthor of the paper] was keen on it." De Sitter also didn't want the rap. "You will have seen the paper by Einstein and myself," he wrote to Eddington. "I do not myself consider the result of much importance, but Einstein seemed to think that it was."

Five months after he left Caltech, Einstein wrote Millikan from Berlin to report that "further thought regarding Hubbel's [sic] observations have proved that the phenomena adapts itself very well to the theory of relativity." The big telescopes in Southern California had already reached far enough to correct Einstein's original theory.*

Hale had envisioned a broad program of cooperation between the faculties at Caltech and the staff of the Mount Wilson Observatory. With Hubble, Walter Baade, who had originally come on a visiting fellowship from Hamburg, van Maanen, and a constant influx of bright younger fellows, Mount Wilson was the preëminent group of observational astronomers. Their offices on Santa Barbara Street were a short drive or a long walk from the Caltech campus. Despite the proximity, and Hale's hopes, the only open cooperation was the Astronomy and Physics Club, a joint effort of the Bridge Laboratory and the Mount Wilson Observatory, which met in a weekly colloquium for featured speakers.

The formal colloquiums were a Band-Aid over the scar tissue from the dispute that had almost lost the grant in May 1928. John Merriam, still chafing that the big telescope had gone to tiny Caltech instead of to the world-famed Mount Wilson Observatory, did all he could to limit the cooperation between the institutions. Caltech faculty were not granted the same observation privileges as Mount Wilson staff, and even those members of the Mount Wilson staff whose salaries were paid in part by Caltech because they were working on the two-hundred-inch telescope project—Anderson, Pease, and a young astronomer named Sinclair ("Smitty") Smith—were not allowed the full privileges of Caltech faculty.

---

* Recently cosmologists have begun exploring the possibility that a cosmological constant may be necessary after all.

The institutional distrust was mutual. After the experiences of 1928, and subsequent meetings that confirmed their impressions of Merriam as a stubborn, insecure, and petty administrator, Millikan and his colleagues on the Observatory Council concluded that as long as Merriam headed the Carnegie Institution, any meaningful program of cooperation between the two institutions was impossible. Hale usually wasn't one to bear a grudge, but this time even he agreed. Since Merriam wasn't scheduled to retire until 1938, and had made it clear that he was not leaving a day earlier, the council concluded that Caltech would need its own astrophysics faculty.

In Caltech tradition the new faculty of astrophysics started with a building, funded by the grant for the telescope. Porter designed the astrophysics laboratory. Hale wanted a solar telescope for the building, with a roof-mounted coelostat to follow the sun and a deep shaft to bounce the light in a long focal path to a spectroheliograph like the instrument he used at his solar lab. Everyone agreed that there should also be a small telescope for the astronomy students. Porter, who had worked with amateur astronomers in Vermont, had strong ideas about what features made a telescope convenient and useful. This was one area in which Porter was not disputed by others on the staff.

The two telescope domes on the roof came to characterize the new building on California Street, next to the Bridge Laboratory of Physics. From a fund-raising standpoint, separate buildings were a good idea. Donors eager to see their names in bronze would fund anything from an entire building to a lab or a broom closet. But separate buildings did little to encourage the cooperation between faculties and disciplines that Hale had dreamed of when he first proposed the telescope.

The first faculty member at Caltech who could be called an astrophysicist was Fritz Zwicky, a Bulgarian-born Swiss physicist who came to Pasadena on a Rockefeller fellowship. Zwicky was neither shy about his own abilities nor reluctant to voice his opinions. Most junior faculty trembled with fear in the presence of Robert Millikan. Zwicky, too young, naive, or bold to know what was expected, not only didn't turn to jelly but accused Millikan of never having had an idea of his own.

"All right, young man," Millikan answered. "How about you?"

Zwicky said, "I have a good idea every two years. You name the subject, I bring the idea!" Millikan on the spot ordered Zwicky to take on an astrophysics problem.

Forced to work with blackboard and chalk because Caltech didn't yet have a telescope, Zwicky turned his efforts to trying to explain the apparent radial velocities of the spiral galaxies. One explanation was Einstein's, now already famous even outside the narrow circles of astrophysics. Zwicky, undaunted by Einstein's fame or the widespread acceptance of his theory, took on the problem.

Zwicky wasn't successful in deriving an alternative to relativity, but in time he became an institution at Caltech and in the world of astrophysics for his often precocious and almost always eccentric ideas. It was Zwicky who first posited the existence of neutron stars, supernovas, dark matter, gravitational lensing, and clusters of galaxies—all long before anyone had evidence to prove or disprove his theories. The courses he taught were by reputation ferociously hard. In the required course in mechanics, Zwicky would assign terrifyingly difficult problems, then call on the least competent students to go to the board and work through the solution under his ruthless questioning. Years later, while walking across the campus, he admitted to one former student, "Some of those goddamn problems I could not do myself."

Even outside the classroom Zwicky was a terror. In a field usually characterized, at least in public, by gentlemanly behavior, Zwicky was consistently, and apparently deliberately, intemperate. His collaborations rarely lasted. His research on velocities of members of the Virgo cluster, which led to his questioning about the "missing matter" in the universe, came from data supplied by Sinclair Smith, an astronomer at Mount Wilson who had designed and built a special spectrograph for the purpose. Smith received no credit in Zwicky's subsequent work. Smith's spectrograph was based on the principle of the Schmidt telescope, a design Zwicky later claimed to have brought to America from Germany.

At seminars, he was as quick with an insult as with an answer. "Why are your ears so big?" he asked one faculty member. Understandably, he wasn't popular with his colleagues. With technicians, graduate students, and secretaries Zwicky liked to debate, in any of six languages, what kind of "bastards" his faculty colleagues were. His usual conclusion was that they were "spherical bastards" because no matter which way you looked at them they were still bastards. Other physicists learned to give him a wide berth.

Although Zwicky was the only faculty member assigned to astrophysics, others also worked on astrophysics problems. Ira Bowen, a bright young physicist with a specialty of spectroscopy, worked as Millikan's research assistant. When he got his Ph.D. in 1926, Millikan hired him for the Caltech physics faculty.* Bowen had begun working on spectra from the big telescopes at the Lick Observatory, where he had been a Morrison Fellow. In 1927 he identified lines in the spectra of gaseous nebulae, which had previously eluded explanation. The lines had been so puzzling that earlier astronomers had even invented a hypothetical element, not known on earth (nebulium), to account for the spectral lines. Bowen identified the lines as "forbidden" emissions

---

* Caltech seemed to have no fears of nepotism or inbreeding. By 1930 Bowen's colleagues on the physics faculty—Jesse DuMond, Charles Lauritsen, Victor Neher, and John Anderson—were all Caltech Ph.Ds.

of doubly ionized oxygen, ions of oxygen that were missing two electrons—a form unstable on earth but possible in the near vacuum of space. Walter Adams called Bowen's work "one of the most brilliant astronomical discoveries of recent years."

Hale was enthusiastic about Bowen's work. Spectroscopy was something he understood. Other new developments in astronomy were moving too fast for any but young men on top of the field. Einstein had written to George Hale in 1913 to explain how the sun would bend starlight passing nearby, but Hale found relativity impenetrable. In letters to Max Mason at the Rockefeller Foundation, Hale enthusiastically reported Zwicky's efforts to come up with an alternate solution. Mason, an applied physicist who had done his schooling and research before relativity was part of the theoretical physics curriculum, also was not comfortable with *space* and *time* as terms and thought they were "ridiculous" when applied to atomic structure, "perhaps it may turn out that they are as bad when applied to the universe as a whole."

Max Mason was a godsend to the telescope project. In 1928 a committee of trustees of the various Rockefeller-funded foundations concluded that the overlapping jurisdictions of the foundations were inefficient and unproductive and recommended that all programs from any of the foundations relating to "the advancement of human knowledge" be transferred to the Rockefeller Foundation. Rose, alone among the high officials of the foundations, opposed the consolidation, which went into effect with his retirement. The Rockefeller Foundation inherited the telescope project. Arnett was the program officer for physical sciences at the Rockefeller Foundation, but Mason was personally interested in the telescope project, and starting with the meeting when Hale was called to New York to discuss the site of the telescope, Mason gradually took over the role of foundation contact for the project.

Before coming to the Rockefeller Foundation, Mason had been a physicist at the University of Wisconsin in Madison. He had gotten his Ph.D. from Göttingen, the most famous of the German physics faculties, returning a difficult thesis problem ten days after it was assigned. During World War I, he had done major research on acoustic detection of submarines at the U.S. Navy's New London, Connecticut, research facility, adapting the Broca tube as the basis of the first workable passive submarine detector. From 1925 to 1928 he was president of the University of Chicago, was named director of natural science at the Rockefeller Foundation in 1928, and president of the foundation the next year.

Unlike Rose, Mason was a working scientist. He was comfortable with the language and style of scientists, eager and willing to listen to progress reports and to read preprints of forthcoming articles. He also had experience with applied research and the engineering problems that had to be solved to build research instruments. Mason quickly gained the trust of Hale and the other scientists working on the two-hundred-inch project. The feelings were reciprocated. As president of the Rockefeller Foundation, the foundation with the largest role in science, Mason was in a position to deflect the pressures directed at the Pasadena group from astronomers and others eager to take advantage of their acquaintance with foundation officials or trustees.

As the friendship between Hale and Mason developed, the two men began an exchange of clippings and notes. Grant recipients are usually reluctant to let on too much to foundations, but Hale gradually began to trust Max Mason with even the troublesome news, including the rumors of problems with the one-hundred-inch telescope and Hale's own doubts about the progress of the mirrors at GE.

By mid-February 1931, the surfacing of the sixty-inch fused-quartz disk was under way in West Lynn. The early progress reports were good. Hale reported to Max Mason that "our confidence in the ability of Dr. Thomson and Mr. Ellis to solve such problems was not misplaced, and they have now succeeded in making a 60-inch disk 8 inches thick. . . . There is every reason to believe that the same process will be successful on a very large scale, and that a 200-inch disk can be manufactured in this way."

The news was so promising that Hale proposed a revision to the project budget, shuffling various categories of the $6 million grant around to cover the increased cost of the mirror blanks, and formally proposing that instead of a temporary steel building on the site of the telescope for grinding and figuring the mirror, an optics laboratory could be built on the Caltech campus, adjacent to the machine shop. The advantages of a permanent optics lab had been demonstrated at Santa Barbara Street, where a staff of opticians was kept busy building new instruments and refining old instruments for the big telescopes on Mount Wilson. Having the optics lab next to a machine shop would facilitate the construction and maintenance of the grinding and polishing equipment.

Hale proceeded carefully, reluctant to ask the Rockefeller executive committee to approve funds for an expensive optics lab before there was some evidence that there would be a mirror to grind.

Porter was quietly working on plans for the optics lab and for the machines to grind the disks. On this as on so many other aspects of the project, the machines and designs that had worked for smaller mirrors couldn't be adapted or scaled up. The jump in size and weight from a one-hundred-inch disk to a two-hundred-inch disk meant a shift in both material, from wood to steel, and design, from the crank motion of the older grinding machines to a reciprocating-motion machine that would eliminate the high-torque crank arms and belt drives of the earlier machines. Porter's design cut the cost of the lab and equipment to half what Pease had once predicted, but it would still be an enormous structure. Hale estimated that at a minimum, the open internal space of a room to grind and polish the disk had to be two hundred feet long, sixty feet wide, and forty feet high, with walls strong enough to carry a fifty-ton traveling crane. The walls and roof would have to be "completely insulated from the heat of the sun" and the lab would need equipment for maintaining a constant temperature inside, as well as the machinery for grinding, polishing, and testing a mirror disk seventeen feet in diameter, weighing thirty tons. The machinery was so large that it would have to be built in place.

As the plans were refined, Porter's drawings went to the architect G. W. Iser. What emerged was an extraordinary building. The interior was a single room, which would be used for grinding and polishing the big mirror. The building would be windowless, the walls lined with cork, and with a special air-conditioning and ventilation system to hold the temperature and humidity steady and to maintain the air pressure inside the building higher than the pressure outside, to prevent the entry of dust. Additional rooms at the end of the building and in the basement provided laboratories for working on smaller disks and for experimentation with coatings for the mirrors.

In early March, Ellis finished surfacing the sixty-inch disk with clear quartz. For nine days the disk was sealed in the huge furnace, while the temperature was slowly reduced to anneal the surface of the disk. The spider frame on the big 135,000-ton lift hung overhead, waiting for the unveiling. There were no guidelines for annealing fused quartz, certainly not for a mass as large as the five-foot mirror, but Ellis had extrapolated from their experience with smaller trial disks and concluded that ten days should suffice.

A few hours before the scheduled annealing time was up, an impatient workman lifted the top of the oven to peek at the disk. Ellis was called over to take a look. He telegraphed the news to Pasadena: EXPERIMENTAL SIXTY INCH MIRROR COMPLETED FOURTEENTH. SURFACE QUALITY NOT GOOD. COOLED TOO RAPIDLY BELOW TWO HUNDRED DEGREES. STARTED CRACK.

The gloom in West Lynn and in Pasadena, coming on the tail end of the brief euphoria over the successful spraying of the base, was deadly. Harlow Shapley visited the River Works not long after the crack was discovered to discuss a disk for a Harvard telescope. He was "sufficiently disturbed by the expense and the uncertainties of this great experiment" to catalog his qualms for Walter Adams, his former colleague from his days at Mount Wilson.

Thomson and Ellis, Shapley reported, were "entirely too cocksure about the superiority of quartz." Shapley believed that the next effort at a sixty-inch disk would probably succeed, and that if the money held out GE could probably go ahead and produce the larger blanks. But he wasn't convinced that the quartz disks could ever be ground and figured to the precision required in a two-hundred-inch telescope:

The surface of a 60-inch, if I remember correctly, must be corrected to one-tenth of a wave length. The larger aperture and longer focus of the 200-inch requires correction to a fortieth of a wave length. Who can do that kind of work? With these quartz mirrors laid down with a sprayer working back and forth in one direction, how can we be at all confident that a cylinder will not appear? . . . Is it not doubtful if we know how to anneal quartz at the present time? Some inquiries of Mr. Ellis the other day elicited the surprising information that he did not know whether or not quartz has two critical points, as all glasses. The whole secret of glass annealing depends upon accurate knowledge of the two critical points. Ellis didn't seem to know that such existed even for glass.

Shapley tried to explain his concerns to Ellis—how the pattern the sprayers made might affect the way the quartz reacted to changes of temperature. A piece of rolled glass, he explained, responds to changes in temperature with cylindrical distortions* that reflect the rolling process; if two pieces of rolled glass are fused at right angles to the direction of rolling, the distortions are canceled out. He suggested that the burners for the quartz blanks be moved north-south for one layer, then east-west for the next to produce the same effect.

Ellis answered: "Quartz is not glass."

Ellis was in charge now. Thomson had lost touch with the fused-quartz project. At one point he had visited the lab and noticed that thousands of mosquitoes were attracted to the electric resistance furnace. He collected mosquito corpses to determine their sex, then consulted with an entomologist at Harvard to find that the mosquitoes were males, and that they had probably been attracted by the sixty-cycle pulses in the resistance circuits. At the time GE didn't capitalize on the discovery to produce electric mosquito traps, in part because by 1928 Thomson was already interested in yet another project, an effort to recover minerals from a meteor crater in the Southwest. Thomson was on a ship when he got the news that the sixty-inch disk had cracked. Other GE officials assured him by wire that the crack in the disk was "a blessing in disguise" because they had learned important lessons about how to handle the annealing cycle.

It took only a few weeks before the GE officials turned the latest disaster into positive publicity. Ellis summarized their experiments in a long internal memo that recalls some of the early battlefield reports from World War I. As Russian troops ran roughshod over millions of hectares of the Austro-Hungarian Empire, the Austrian high command triumphantly announced: "Lemberg is still in our hands." Ellis, conceding that they needed more work on the spraying process and that

---

* Cylindrical distortions are like the corrections put into eyeglass lenses to correct astigmatism. Rotate a cylindrically distorted lens, and the shape of the viewed objects changes with the rotation.

they didn't know what would happen to a mass as large as the proposed two-hundred-inch mirror if it were kept at 1100°C for weeks, triumphed that the furnace had not broken and that the fundamental process appeared to be sound.

On the basis of these last experiments, Ellis announced that they were ready for what he called "production" runs. The only impediment to taking orders for disks from Harvard and other observatories was the need to come up with a formula to amortize the initial expenses that had been paid by the Observatory Council. Assuring Hale and Anderson that he would soon be ready to try for a new sixty-inch disk, Ellis sought Swope's approval before he sent yet another set of budget figures to Pasadena.

In Pasadena they read different fortunes from the tea leaves. For the Observatory Council the problems with the disk had reached a crisis point. After three years and more than six hundred thousand dollars of expenditures, all they had received from GE were more revised (and inflated) budget figures, assurances that had begun to sound hollow from repetition, and bits of imperfect quartz for testing. Hale, Millikan, Henry Robinson of the Board of Trustees, Noyes, Adams, and Anderson scheduled a long meeting at the solar lab on Saturday morning when they wouldn't be disturbed. Ostensibly the agenda item was a review of the new budget figures from GE. The real question was whether they could afford to continue the effort to get mirror disks from GE. If not, what were the alternatives?

Before they turned to Ellis's new figures, Adams suggested that he had some test data that might be useful to the council. He had just completed initial tests of a new mirror-support system Pease had installed on the one-hundred-inch telescope at Mount Wilson, based on a design Pease had developed for the two-hundred-inch.

Instead of using defining arcs and springs to hold the mirror, the new supports mounted the disk on balls to reduce the friction* as the mirror moved. Adams reported that in the first tests with the new supports, the optical changes in the figure of the mirror seemed to be greatly reduced: instead of needing a long period to readjust, the mirror held its shape as it was slewed to different areas in the sky.

If the mounts worked as well in use as in the preliminary tests, he said, the one-hundred-inch telescope had a new lease on life. Hours or whole evenings that had been lost to waiting for the mirror to adjust to changes in temperature would be minimized or eliminated. The changes in figure of the mirror had limited the resolution of the tele-

---

* Hale called it friction. An engineer, describing the frictional inertia of a heavy component in contact with another, would probably use the term *stiction*.

scope in certain positions. Now, it seemed, Hubble, Humason, and other observers could extend their reach even farther.

Hale and his colleagues talked for most of the morning about Adams's report. If Pease's new supports made that much difference, it seemed that the decade of problems with the one-hundred-inch telescope had not been caused by the coefficient of expansion of the plate-glass mirror. And if plate glass would work for a one-hundred-inch mirror, maybe the two-hundred-inch mirror could be fabricated from a material less exotic than fused quartz. Plate glass still wouldn't work for a mirror that large, but one possibility was the new Pyrex borosilicate glass from Corning. Hale had used experimental Pyrex mirrors in some of the solar telescopes on Mount Wilson. He found the material optically stable and reliable, even in an application as demanding as a telescope element exposed to the full heat of the sun. He knew that Corning was already melting one hundred tons of Pyrex at a time in its furnaces.

After the meeting Hale optimistically wrote, "We believe that a Pyrex disk would serve admirably for the 200-inch telescope." He asked Anderson to quietly get quotes and a timetable on mirrors from Corning.

Shortly after the meeting Heber Curtis, who had gone from Lick Observatory to the Allegheny Observatory and then to the University of Michigan as chairman of the Astronomy Department, asked Hale for his opinion of the GE process for a large mirror blank for a new telescope for the University of Michigan. Citing the experiments with the one-hundred-inch telescope, Hale recommended that Curtis go to a Pyrex blank.

Emboldened by the possibility of an alternative source for the disks, the council drafted a telegram to Thomson at GE, questioning Ellis's newest estimates of $80,000 for another sixty-inch disk, and an additional $1 million for a two-hundred-inch disk. The language of the telegram was strong. Before he sent it Hale telephoned Max Mason to get his views. Mason urged Hale to wait before breaking off all work with GE.

Mason had reasons for his cautions to Hale. Like every other institution dependent on portfolio income, the Rockefeller Foundation had suffered from the drop in the stock market. The IEB, which had made the grant for the telescope, saw the value of its funds drop so precipitously that it could not make good on the balance due under the grant. With Wickliffe Rose gone, there was no continuing need or support for the fund that had been established at his insistence, so the remaining IEB funds were transferred to the GEB, which took over the financial responsibility for the grant. The lines between the various Rockefeller foundations had grown hazy enough that Mason could personally oversee the telescope project, with the checks being paid out of GEB accounts.

The drop in the stock market did not directly affect the funding for the telescope, but the telescope project could not avoid the consequences of the depression. The grant had been awarded in an era of unbounded optimism, when Americans trusted, even *worshiped*, science. The early twentieth century had produced electric lights, automobiles, airplanes, radio, telephones, refrigerators. The following decades, with the wondrous scientific work that was being reported on the radio and in the news weeklies, were supposed to bring even greater prosperity and ease to ever more Americans.

The depression rattled America's love affair with science and technology. Breadlines, soup kitchens, and tent cities stood in stark contrast to the buoyant promises of progress and prosperity. The economic critic Stuart Chase told the Women's City Club of New York that the advent of "talkies" had thrown ten thousand movie house technicians out of work. Men and women who only a few years before had rushed to read about or buy new inventions now turned their backs on machines and science, as Luddite arguments, once confined to a radical fringe, began to appeal to wider audiences.

Congress joined the reaction. The budgets for science research in the United States were small to start with. With the exception of the grant for the two-hundred-inch telescope, and Lawrence's successful fund-raising for his cyclotrons at Berkeley, most scientific research got by with the minuscule funds that universities could provide to their faculty. For the 1932 fiscal year Congress cut the appropriations for all science-related agencies drastically: The budget of the Bureau of Standards was cut by 26 percent. The few state-funded scientific agencies were cut even more. Private institutions devoted to science, like the Rockefeller Foundation or the Carnegie Institution, even if they were eager to keep up their funding efforts, felt the effects of the depression on their investments or in their fund-raising efforts. The annual income of the Carnegie Institution of Washington fell by $1 million, a substantial sum when the entire annual budget of the Mount Wilson Observatory was $250,000.

Signs of the depression were everywhere by 1931, even in prosperous towns like Pasadena. The movie business was still hard at work, churning out musicals that invited the world to sing or dance away its cares. The rest of Southern California's economy collapsed. The aircraft industry stood idle. The new refrigerated-produce industry, which was on the verge of revolutionizing California agriculture, went on hold. As prices for farm commodities fell, the agricultural exchanges tried dumping goods to preserve prices. In the citrus belt east of Pasadena, hundreds of tons of "surplus" oranges were thrown into dry riverbeds and sprayed with oil and tar so the unemployed and relief clients wouldn't be tempted to salvage from the decaying heaps.

When *Fortune* magazine sent a team of reporters to the Caltech campus, the reporters searched in vain for signs of the depression that had struck everywhere else. The reporters dwelled on the exciting discussions of cosmology and nuclear physics, Millikan's own research on cosmic rays, and the pathbreaking observational efforts on the one-hundred-inch telescope. What they didn't know was that the faculty had voluntarily agreed to a 10 percent pay cut, and that Millikan, Hale, and Noyes, the three great lights of the campus, had made up a portion of the school's deficits from their personal funds.

All construction on the campus had stopped, with the exception of the astrophysics laboratory and the optical lab, which were funded out of the grant for the two-hundred-inch telescope. The resentment from other departments was hard to ignore. Famous faculty members like Millikan and Noyes could always count on foundation grants to support their work, but the physics and chemistry labs were hard pressed for equipment and supplies to continue modest research programs by junior faculty and graduate students.

While other faculties scrounged for materials and space, astrophysics had a huge, well-equipped machine shop, its purpose to build parts for a telescope that couldn't even be designed until there was positive evidence that the great mirror could be successfully cast. An astrophysics laboratory was nearing completion on California Street, to design instruments and observation programs and to analyze the data from the unbuilt telescope. At ten different sites in Southern California and Arizona, telescopes had been set up to accumulate data on local weather and seeing, as part of the site survey to pick the best possible site for an unbuilt telescope.

The ever-volatile religious movements of Southern California, prospering in inverse correlation with the economy, found ready targets in Caltech, telescopes, and any other manifestation of science. The "work of the devil," the evangelists were quick to point out, exacted its tolls; the joblessness of the depression was punishment for decades of immorality, including the assaults of science on God's kingdom. In the early years of the institute, the gathering of scientists had been a ready target for the evangelists. By 1930 Millikan had come out for religion, or at least for God, which made an uneasy truce with the fundamentalists. The graffiti writers had a good time with that one: Under the signs saying JESUS SAVES, they would scratch, BUT MILLIKAN GETS CREDIT. The campus cynics didn't like Millikan's soft stance on religion, but it did mean that the fundamentalists found targets other than the telescope for their rallies and demonstrations.

Once at the center of headlines, the telescope had faded from public attention. The new focus of attention for the public who hadn't yet lost faith or interest in science was Albert Einstein, who announced that he was leaving Germany permanently.* American institutions competed

---

* Einstein saw the writing on the walls in Germany earlier than many other scientists. "If my theory of relativity is proven successful," he wrote, "Germany will claim me as a German and France will declare that I am a citizen of the world. Should my theory prove untrue, France will say I am a German, and Germany will declare that I am a Jew."

extravagantly to recruit him. Millikan's bids from Caltech were too late and too small. Einstein ended up at the Institute for Advanced Studies, in Princeton, a university without students. Photographs of the white-haired scientist who biked around Princeton and played his violin at dinner parties were soon familiar fare in the weeklies.

The individual members of the Observatory Council were mostly spared the effects of the depression. George Hale's investments had been conservative. Even after the stock market crash, his income was approximately four times what the highest paid faculty member at Caltech made. The other members of the Observatory Council—Millikan, Noyes, and Adams—were not dependent on their investments for income. The exception was Henry Robinson, an ex-officio member. Robinson had been a booster of the telescope project from the beginning, not only for Caltech, but for Southern California. He had lost a considerable sum in the collapse of the stock market, including the funds he had pledged as an endowment for the telescope. No one, even on the Observatory Council, seemed to notice. Progress on the telescope had been so slow that the endowment for operation seemed a distant concern in 1931.

Max Mason accompanied Noyes, who was heading east for other business, on a fact-finding mission to West Lynn in April 1931. Ellis and his staff put on a full show, with a presentation on the spraying process and a tour of the facilities. GE had accumulated an enormous array of equipment for the mirror disk program, from banks of transformers to power the electric furnace to the stockpiles of quartz, and the strange expandable steel building that housed the enterprise.

GE's show was a faux pas. If they were VIPs on the telescope project, Noyes and Mason were also working scientists who could see that there were fundamental problems with the quartz fabrication process. They brought up Shapley's suggestion of switching direction of the movements of the spraying head, first north-south, then east-west, to avoid introducing strains in the quartz. Hale had raised the same concerns in a telegram, asking that the tensile strengths of a piece of quartz from the flawed sixty-six-inch disk be tested "both parallel and at right angles to direction of burner motion." Ellis promised to carry out the tests. Mason also recommended that Ellis try pressing down on the surface of the heated quartz immediately after spraying the disk to minimize the ridges that formed on the surface.

Ellis had the shop fabricate a brick of recrystallized carborundum, the only material that would withstand the extreme heat inside the furnace, fastened to a metal bar wide enough to run across the top of the furnace chamber. Men in protective clothing, pressing down at each end of the bar, could exert a force of two hundred pounds on the brick to flatten the sprayed surface. Like most experiments in the deadly heat of the furnace, it didn't work. If the brick was as much as six inches from the flames of the spraying torch, the surface was already cooled enough to be rigid. If they got closer to the nozzles that were spraying the quartz, the brick took the full brunt of the flames. Chunks of the carborundum would chip off and embed themselves in the quartz surface. The resulting imperfections were worse than the ridges.

Despite the continuing problems, the GE publicity department was still sending off press kits, including a major release that found its way, almost verbatim, into the science section of the *New York Times* in mid-April. The article crowed that the successful fusing of the sixty-six-inch disk was undoubtedly the biggest step in the production process, with the next jump to a two-hundred-inch telescope only a magnitude of 22 to 1, after the leap of 225 to 1 in magnitude of the previous step. The GE press release and the article never mentioned that the disk was severely cracked and not usable.

After Mason and Noyes reported what they had seen and heard in West Lynn, confirming Hale's worst suspicions, Hale wired Thomson ordering that all work on the sprayed quartz mirrors cease on June 1.

Thomson, semiretired and involved in a dozen other projects, had lost interest in the project, but Ellis was reluctant to give up. As the June 1 deadline approached, Ellis reported that he had conducted tests on new experimental disks, using an interferometer to measure deflection of the disk under pressure on the two axes. The device was so sensitive that he could only run the tests between midnight and five in the morning on a weekend, when the other machinery in the factory was shut down and there was no traffic on local streets that might introduce a measurable vibration. Ellis's measurements, accurate to five millionths of an inch, confirmed that deformations of the disk on either axis were minimal.

"It is also evident . . . that a further development of the burner and spraying equipment is necessary to produce a surface layer," he wrote. He hoped, but couldn't promise, to have the additional work done by June 1, at which point they could spray another sixty-inch mirror, this one usable for the telescope and the final prelude to beginning work on the two-hundred-inch mirror disk. Trying to hang on to his lab when there was little work around, Ellis pleaded:

> We have been using a large part of our laboratory force directly on this mirror work day and night, without regard to union hours or personal inconvenience, and to the exclusion of other work, having taken on no new problems while the mirror work was in hand. As a result of this condition and the business depression, to comply with your telegram means that we will have to discharge the men on June 1. If this is done it will be very difficult to take up the work again months later should you then wish to go on with it. I am stating this case frankly, believing that you might have thought we could stop all work at a given time, place the men on other work while a review of the whole situation is being made, and pick up the work where we left off at some later date.

It was a sound argument, but too late. Sometime in the midst of the promises and frustrations, Hale had lost faith in GE, Thomson, and the whole fused-quartz project. It was no longer clear that the two-hundred-inch telescope *needed* fused quartz for the mirror, and the cost estimates, once reasonable, had escalated to the point where the budget of $6 million for the telescope no longer seemed adequate to keep up with the GE bills. After reading one report from Ellis, Hale asked Anderson whether—if Ellis did find a reliable method for selecting the quartz—the cost of "merely selecting enough quartz for a 200-inch mirror," would come within the budget.

Hale, Robinson, Millikan, and Noyes held another long meeting, this time to draft an ultimatum, over their four signatures, to Gerard Swope, the president of GE, reminding him that the original estimate was for the first sixty-inch disk to cost $75,000 and be ready in May 1929, and that after three years' work and $600,000, "We are not yet in position to know with any certainty whether a sixty inch disk can be successfully made and we have no definite information as to cost."

Swope telegraphed back immediately: "As far as we are concerned, this is an adventure into the unknown and outside our line of business. We have been giving our facilities and the best brains we have to this work to the neglect of other work and without any profit or hope of building up a business along this line."

Mason urged Hale to give GE the go-ahead for another disk, even at the estimated cost of $60,000 to $80,000, so they could see some return—if only a single disk—from the money that had been spent at GE. Hale thought more work at GE was throwing good money after bad, but he authorized the second try at a sixty-inch disk. "I don't see how we can go further unless some miracle occurs."

Ellis immediately started spraying. The base portion of the new disk went well. For the surfacing Ellis raised the disk on a platform of crushed quartz fragments mixed with binder clay to reduce the risk of cracking. With the familiar delays for breakdowns and repairs to leaking pipes and burners, Ellis completed the second sixty-inch disk before the end of June.

The disk emerged from the furnace uncracked. Thomson triumphantly wrote: "We fully believe that the completion of this 60" disk will point the way most assuredly to the building of the 200" disk of fused quartz." Later Ellis reported that the new disk was filled with "a good many bubbles" varying in size from a few thousandths of an inch up to a tenth of an inch in diameter; he was convinced that they would not interfere with the intended purpose of the disk. Ellis also suggested that they could saw the earlier sixty-inch disk in half. The cracked half could be "welded" together, and the two blanks could then each be surfaced. GE never stopped producing great ideas.

To the council's ultimatum, he had no satisfactory answer. "We cannot know that a 200-inch mirror can be made until it is an accomplished fact. It is therefore impracticable to give you a 'guaranteed price for a satisfactory 200-inch,' and if no other solution will be satisfactory to the Council we will have to shut up shop."

Hale wrote to a friend, "The way they swallow money without blinking is not good for one's nerves! We have just reached the end of this rope and must try another. . . . The financial depression has complicated the situation—we are fortunate in not completely losing the support of our backers. But perhaps you will partially realize why I have not known what minnit's gwine to be the nex', as Uncle Remus puts it."

Even if it had emerged from the furnace flawless, the last disk came too late. "It is evident that you have accomplished an important technical achievement," Hale wrote Ellis. "And that if further tests, including optical ones, prove the disk of suitable quality, it can be used for some purpose, such as solar work, where a glass mirror would not serve. . . . if we had not already spent the huge sum of $639,000, and if the estimates for the larger mirrors were not so far beyond our means, we should certainly wish to proceed at once with a larger disk." Under the circumstances, "We cannot be responsible for any further expense incurred."

15

New Light

The Corning Glass Works was a proper company. George McCauley had been ready to make mirror blanks since 1929, the company was eager for new work, and there were rumors that GE was having difficulties producing satisfactory disks for the Caltech telescope project—but no one from Corning formally approached Hale about the two-hundred-inch-telescope project while GE was working on the fused quartz.

Yet even good business manners couldn't keep the old boys of the University Club from talking. The Reverend Anson Phelps Stokes put in a word to his friends at the Rockefeller Foundation: "I believe that my friend, Mr. Houghton, our former Ambassador to England, is the head of the Corning Glass works." At a meeting of the National Association of Science in the spring of 1931, Robert Millikan got into a conversation with Arthur Day, a vice president of Corning and a respected scientist with the Geophysical Laboratory of the Carnegie Institution of Washington. Millikan told Day about the problems GE was having with the fabrication of a sixty-inch disk. "If we could get a good Pyrex disk from you for $20,000," Millikan said, "It would be a mere drop in the bucket in our expense account and would serve to guarantee us against the consequences of a second failure by General Electric." Day realized that Millikan had done his homework: Twenty thousand dollars was exactly the figure Corning had quoted the University of Michigan for a sixty-inch disk.

Day had followed the big telescope project for years. He was widely respected as an expert on glass and would have served on the mirror committee if he hadn't had an earlier falling-out with GE. Day believed that they had stolen his own process for making fused quartz. When he made his initial agreement with GE, Hale had asked Day's opinion of the possibility of a Pyrex disk for a large telescope. "The plan is entirely beyond any experience available from here or elsewhere," Day answered. "Courage is an essential asset to be reckoned with."

Now, after three years of unsuccessful experiments at GE, Hale wrote Day again, reporting the dismal progress at West Lynn and the recent good news about the mirror-support-system experiments on the one-hundred-inch telescope. Theodore Dunham, of the Mount Wilson Observatory, had gone over Pease's test figures and found the differences with the new mounts "little short of revolutionary." It was clear that a Pyrex disk would have a temperature coefficient well within the range needed for the two-hundred-inch telescope. Hale asked Day straight out: Could Corning produce a disk for the telescope?

Day answered that the Corning Glass Works officially and its laboratory staff individually had been interested from the beginning in the project to cast a two-hundred-inch mirror disk. "I think now, as I have always thought, that they could successfully make it out of glass of Pyrex type."

The problems at GE were open gossip at meetings of the National Academy of Sciences and the AAS. Day may also have heard about or guessed at the mostly unspoken differences in attitude between GE and the Observatory Council. The attitude at Corning, he assured Hale, would be different: "We should not overlook the fact that the making of one great disk in Corning would be exclusively a task for the Research and Development Department and not a manufacturing problem."

One thing held Corning back: "You have no wish or use for more than one disk and you could hardly wish to tax your resources unnecessarily with the development costs of two different varieties of these." Corning didn't mind secret flirtation, but it remained a proper front-door suitor, unwilling to get into extracontractual affairs. Day was making it clear that Corning wouldn't begin work on a disk until the Observatory Council broke off its relationship with GE.

In fall 1931, with his nerves in one of their increasingly rare periods of remission, George Hale took a train east to New York. "We mean to succeed," he told Arnett at the Rockefeller Foundation, "especially as the importance of the 200-inch has become far greater than originally appeared. This is the result of our recent work, as viewed by Einstein and [James] Jeans, who count on the 200-inch to give the key to their most vital problems."

Ellis made a final desperate try to keep the work at GE. He told Swope that GE's original quotes had been misunderstood because Porter, who visited West Lynn as Hale's representative, was hard of hearing. When Hale reported that the Observatory Council was considering at least two other methods of making mirrors, Ellis wrote, "I cannot believe they are seriously considering the grave uncertainties of any other methods, in view of the progress we have made and the money they have already sunk in this project." Despite his efforts, even Ellis seemed to know that the battle was lost. The GE publicity department in Schenectady was about to unleash a new salvo. Ellis urged that we "use our powder in celebrating the first 60-inch and leave the 200-inch until history has been made."

Ellis carried his campaign around Hale, going out to Little Compton, Rhode Island, where Max Mason was on vacation for Labor Day. He told Mason that Hale's demands were impossible. What would your best price for each disk be? Mason asked. Ellis said GE simply couldn't commit to a firm price. The next day, Hale and Mason met Ellis at the University Club in New York. Together the three went out to Gerard Swope's country house near Ossining. Swope had prepared for the meeting by discussing the matter with C. E. Eveleth, chief engineer of GE's Schenectady works. Swope's best estimate was that the cost of a quartz disk would be an additional $1 million, perhaps more. Given the experimental nature of the work, GE could not guarantee that it would be able to produce the disks at that price. Walter Adams later estimated that if they extrapolated the rate of expenditure at GE during the spring and summer of 1931, the total cost for the mirrors would be closer to $2.5 million, almost half the budget for the telescope. Everyone at the meeting knew it was a formal closure: a final polite lunch before the inevitable divorce.

Hale continued his round of courtesy calls by going up to West Lynn to talk personally with Elihu Thomson. The two men had known one another for more than a quarter century. They were titans in their fields, Thomson the inventor who could produce solutions for any problem, Hale the grand old man of solar physics and the builder of machines to see to the edge of the universe. For years, at the annual meetings of the AAS or the National Academy of Sciences, they had talked of pooling their skills, and most outside observers would have bet on the success of a collaboration. Neither man was accustomed to failure. Thomson had a reputation for doing what the scientists said couldn't be done. He had taken on much tougher challenges than this one, and had succeeded. If only Caltech would commit the funds, he no doubt told Hale, GE would produce the mirrors.

For the record, they later exchanged stiffly formal letters, congratulating and thanking each other for the contributions to science and technology that the experiments with fused quartz had provided.

Five days after he met with Swope, Hale held a meeting at the University Club in New York with Mason and three representatives from the Corning Corporation: O. A. Gage, the head of the Optical Glass Division; J. C. Hostetter, the director for special projects in the division; and George McCauley, the chief scientist. McCauley summarized Corning's experiments on telescope mirror disks. The variety of Pyrex that seemed the most promising had a coefficient of expansion three times that of quartz but one-third that of plate glass. The only potential problem McCauley foresaw was in annealing the glass—not that it couldn't be done, but that no one had ever tried to anneal large masses of glass.

Hale was impressed with the no-nonsense technical confidence of the Corning people, especially McCauley.

A week later, on October 14, the group met again, this time joined by Dr. Day and by General Carty of AT&T, still one of Hale's most trusted advisers. The Corning group was ready with samples of Pyrex glass, showing the kinds of surface that would emerge if different powdered materials were used to coat the refractory bricks of a mold before the glass was poured. Corning also presented a budget and a timetable— exactly what GE had never been able to deliver. Hostetter, McCauley's boss in the research division at Corning, laid out the budget. The minimum cost for three 60-inch disks, one 60-by-80-inch oval disk, a 120-inch disk, and the two-hundred-inch disk was $150,000. Unanticipated difficulties or repeated castings could push the figure as high as $300,000. The use of surplus equipment from the GE effort in West Lynn, or a disk design that would require less glass, such as a hole in the center or ribbed backs, would reduce the cost. He estimated that Corning could finish the mirrors in a little more than a year and a half from the time they got the go-ahead. If a second run were necessary for each mirror, the time might stretch to as long as thirty-one months.

The division of responsibility was as clear as the budget: Corning would expect Hale and his group to work out the optimum design for the disks, taking into account "best glass practice." Corning would take care of annealing and would investigate shipping possibilities.

The cool decisiveness of the Corning people impressed Hale and his colleagues. And McCauley liked Hale's idea that the each disk should be a model for the next. As a start he proposed that Corning would mold a 26-inch solid disk, then a 30-inch disk with a ribbed back like the designs that had been discussed at Caltech but never tried on a mirror. Once the 30-inch disk was ground and polished into a mirror, Corning would proceed to a series of Pyrex disks, a 60-inch, a 120-inch, and the two-hundred-inch—each based on the experiences with the previous ones. If the two-hundred-inch could be successfully cast, there would be further orders for additional 60-inch mirrors and other special-purpose mirrors. Hale agreed that if the project were to go ahead, the final price to the Observatory Council would be computed on a "cost plus" basis. The "plus" was a 10 percent profit that Corning would bill only after the disks were figured into mirrors and accepted by the Observatory Council.

Hale raised one other topic at the meeting. If the Observatory Council were to go ahead with an order for Pyrex disks, he said, they wanted no publicity of any kind. The failure of the quartz project was an embarrassment. More than $600,000, one-tenth of the entire telescope budget, had been expended on experiments that had not produced a single satisfactory mirror. At a time when eight million Americans were unemployed and the median income for an American family was $1,231 per year, $600,000 was a big expenditure to justify. Hale, with the reticence of a scientist, had resented the stream of press releases from GE. The research necessary to build the telescope, like the observations of an astronomer, were raw data, scientific work in progress, not fodder for the publicity mill.

GE's publicity had ultimately backfired. A few newspapers picked up on their final releases about the sixty-inch disk, and asked in editorials whether big businesses weren't "exhausting [themselves] in attempting to outbid one another for the privilege of building the 'world's greatest telescope' at a fabulous price." Three years before, GE had been proud to count itself among the builders of the telescope. Now the publicity department scurried to dissociate itself from the project.

McCauley went to work the day after the meeting in New York.

Trained as a physicist, he shared Hale's attitude about publicity. Although he had never been involved with a highly publicized project, McCauley had read enough journalists' reports to know that when reporters were eager for news, they often "insisted on forgetting the plain scientific facts and achievements given them while publishing a dramatical romance of their own invention which whatever scientific facts used garbled as to render them unscientific."

Fortunately Corning was a quiet company, secluded amid the rolling hills and narrow winding valleys of the southern tier of New York State, well off the beaten paths to the vacation areas around the Finger Lakes or Niagara Falls. Few tourists came to Corning in the early 1930s, and the town was content to go its quiet way. The company had been known for its research—Corning had supplied the glass envelope for Edison's incandescent filament to create the electric lamp, and Corning research had produced the Pyrex brand of heat-resisting glass—but the company had never been overeager to publicize itself. The Corning memorandums summarizing the meeting didn't even mention the publicity issue. An internal memorandum was quietly circulated to request that all reporting on the new project be oral only, with no written statements that might get out to snooping reporters.

McCauley had successfully cast a disk in his first try, years before, but he was a cautious man. He recruited Ralph Newman, a mold maker from the Corning Laboratory, and Wallace Woods, an experienced glassworker from the blowing room, for a series of experiments to explore the problems of molding of a large telescope mirror.

Compared to the ferociously complex process of spraying fused quartz, molding glass seems simple. A melting tank of refractory brick is filled with sand, soda, lime, and borax and heated with gas jets suspended from an arched roof until the mix melts into glass. Depending on the intended use, the mixture is fined—maintained at working temperature—for a period of days or even weeks to allow the bubbles that form within the mixture to rise to the surface. The glass in the tank is then ready to be molded or blown into shape. The process is straightforward and tried; for forms of glass used in undemanding applications, glassmaking hasn't changed much in thousands of years.

For optical glass, which would be sensitive to contaminants in the raw materials, from the bricks and cements used to build the tank, and from tools used to stir or transfer the glass, the process is a little more complicated. In glass for a critical application, like a telescope mirror, contaminants can weaken the internal structure of the glass or introduce strains that would ultimately distort the disk.

Even with ordinary plate glass, maintaining the purity and consistency of a large mass of molten glass while transferring it from a tank to a mold is a challenge. Borosilicate glasses, like Pyrex, are even more demanding, because at the extremely high temperatures required to melt borosilicates, tank and tool materials can give off gaseous contaminants or even begin to melt into the glass mixture.

Typically, a large production tank in the glass factory might produce twenty tons of glass over a twenty-four-hour period. The glass is dribbled out, depending on the articles produced, at anywhere from a few pounds at a time to a continuous stream of thirty to forty pounds per minute. At that rate of use the ingredients for new glass can be melted at one end of the tank as fast as the ready glass is removed from the working end at the other.

A large telescope disk would require twenty tons of glass all at once. No one had ever fabricated a piece of glass that large or come up with a procedure to relieve the stresses in the glass so it could be used for a high-precision instrument. The requirements for ribs in the backs of the disks, to reduce the thickness of the glass and the weight of the disk without compromising its rigidity, meant that McCauley needed a method of molding glass with complex structures. Corning had vast experience in molding and blowing complex shapes into small units of glass. But no one had experience with molds built to withstand the heat of borosilicate glasses for as long as it would take to fill a mold with twenty tons of Pyrex.

McCauley began with some basic assumptions: Given the schedule and budget of the project, they would not conduct costly experiments to develop new glasses and methods of melting or working glass. They would not build large machines for tasks that could be performed by manpower at lower cost. A surplus melting tank, idled by lack of demand, would be used to melt the glass, rather than a custom facility.

For the first experiments McCauley had Ralph Newman build test molds to a standard pattern, with two refractory bricks for the bottom, five bricks for the sides, and wire and angle irons to hold up the corners of the mold. Wally Woods, the experienced glass man, would then pour molten 702 Pyrex, the same formula that had worked in the earlier pours, from a test tank into each mold so they could study the interactions of glass and brick.

Newman started with the same refractory bricks that had worked successfully for a test disk years before, selecting good bricks, without cracks, for his molds. Each time Wally Woods poured hot glass into one of the new molds, a cloud of bubbles would erupt in the glass. Nothing obvious was wrong. McCauley dug up the records on the bricks and discovered that the leftovers from the shipment that had worked for molds in 1929 had been stored in a damp room under one of the glass factories. The bricks had wicked up moisture, which turned to steam when the molten glass was poured into the mold.

He had Newman switch to a new grade of brick. That cured the bubbles: Even when the new Armstrong C-25 bricks were soaked in water before the glass was poured, there were no steam bubbles in the molded glass. But success begat new problems: After a few hours in contact with the molten glass the surface of the brick emitted gas bubbles that left a rough surface on the glass. McCauley's solution was to brush the mold surfaces with a paste of silica flour and water before the glass was poured.

A satisfactory mold material was only the first step. At the meetings in New York, Hale had described the problems with the one-hundred-inch plate-glass disk on the Hooker telescope, especially the bubbles that marked the layers of glass from each of the three pours. Corning's production lines for Pyrex consumer and laboratory goods worked by filling molds directly from a big melting tank. The technique wouldn't work for a telescope mirror, because the mirrors had to be made of exceptionally pure glass, and the purest glass in any tank comes from the center of the batch, which is not in contact with the walls of the melting tank.

McCauley's solution was to try an old technology. Ladling, which had been abandoned for production glass work almost everywhere, had advantages for the work on disks. Ladles can take the glass from the center of the tank. They allow the glass to be inspected before it is poured, and because the glass begins to cure against the walls of the ladle, leaving a ladle heel that is later broken out, the glass that remains viscous enough to pour into the mold—generally between half and two-thirds of the contents of the ladle—has effectively only been in contact with glass, not with materials that could introduce contaminations. By introducing the glass to the mold in ladle-size installments, the mold does not face the sudden heat shock of a huge mass of hot molten glass. The alternative to ladles, arrangement of spouts to pour glass from pots into the mold, would have required experimental heated spouts, a turntable for the mold to provide even filling, and a procedure to cut off the spout from the mold and heal the scar. McCauley's instinct was exactly opposite to those of Ellis and Thomson and GE: He wanted the simplest, cheapest, and safest process.

Corning had done little ladling, and whatever experience they had, years before, was long before the development of Pyrex. To duplicate the process of filling a large mold, McCauley had Wally Woods try filling the molds with a small ladle, pouring half the contents of the ladle into the mold and the balance into a cullet can, where cooled glass was collected for later reuse. When Woods took two or more pours to fill the mold, strata would appear in the glass. McCauley suspected that the glass on the surface of the tank, exposed to the air, was devitrifying. When Woods rotated the ladle in the tank to fill it, the surface glass of the tank would end up in the ladle, and ultimately on the top of the molded glass, producing a layer cake with devitrified frosting. The devitrified Pyrex not only marred the appearance of the glass but introduced strains that would degrade the optical performance of a mirror. It wasn't good enough for a telescope.

The answer was to skim the surface glass off the ladle before pouring the balance of the glass into the mold. Woods explained that glass has to be skimmed with a paddle rather than a spoon. He fashioned the device he needed from a blow iron, and when he skimmed the surface off each ladle of glass before pouring it into the mold, the castings emerged with no strata. McCauley went home smiling that evening. He was ready to cast telescope disks.

Molding a large mass of glass is the first step of a complex process that physicists like McCauley were just beginning to understand. Large castings of glass are subject to strains that develop in the internal structure of the material as the mass of molten glass cools. The strains can be relieved, or sometimes eliminated, by annealing the glass—heating the mass of molten glass until it is free of strains, then gradually cooling the mass according to a precise schedule determined by the chemistry of the glass and the size of the mass being annealed. The larger the glass casting and the more critical the use, the more precise the annealing schedule must be. The process is demanding: If the cooling schedule is wrong, or if a failure of equipment drops the temperature too quickly, the casting will emerge with strains that show up as distortions, weaknesses, or instabilities. The mirror would be the largest piece of glass ever cast, and the optical demands put on the primary mirror of the two-hundred-inch telescope would be the toughest criteria ever applied to a large piece of glass. The annealing was crucial.

After the meetings in New York, McCauley asked two engineers, Howard Lillie at the Corning Laboratory and George Morey at the Geophysical Laboratory of the Carnegie Institution in Washington, to experimentally determine the annealing properties of the 702 Pyrex. The figures they produced for the annealing constant were inconsistent, but the required temperatures were high enough that McCauley knew he would need robust ovens, fine control of the temperature, and lots of power. It was theoretically possible to build a gas-fired annealing oven, but the fine temperature control he needed demanded an electrical oven.

Thanks to the many inventions of Edison and Thomson, and to General Electric's production of high-tension transformers and power distribution switches, electrical grids were bringing the miracle of electricity to vast rural areas of the United States in the 1930s. Factories switched over to electrical power even before homes. Electric motors allowed machinery to be portable, instead of requiring overhead belt drives that distributed the power from a central shaft driven by a steam engine or waterwheel. But even as electric power became ubiquitous, it was far less reliable than we have come to expect today. Generators and power lines failed regularly, lightning took out entire networks, and switching equipment and transformers fatigued or shorted out. For most usages, in an era before electronics, long-term reliability wasn't a critical issue. But for the heating elements and controls in an annealing oven that might have to run for a year or more, reliability was the main concern. It was the one area of the project in which McCauley assumed that GE's own work would prove useful.

In mid-November 1931, McCauley and his boss, J. C. Hostetter, went up to the GE laboratories in Lynn, armed with a letter of introduction from Gerard Swope. Ellis was friendly and accommodating. GE, Ellis assured them, could have built the disks of fused silica, and would have, if the new experiments with the one-hundred-inch telescope hadn't eliminated the need for quartz disks. He knew about the prices Corning had quoted on the series of disks and openly contrasted the numbers with GE's figure for a two-hundred-inch disk, somewhere between $750,000 and $1 million. When Hostetter pointed out that Corning had held discussions with the Observatory Council but didn't yet have a firm order for the disks, Ellis assured him that he had inside information on the project and that the order was as good as placed.

Elihu Thomson showed up at the lab, greeted the two men from Corning, and wished them well. He left after a few minutes, and Ellis talked freely about the professor's bitterness and disappointment that GE had been forced to abandon the project before producing a big disk. He proudly showed off the two sixty-inch disks the GE effort had produced—one good clear quartz, but split radially; the other filled with bubbles—and described his project to split the cracked disk and fuse the two halves together so it could be used for a solar telescope or other critical use.*

---

* When Ellis later tried to fuse the two halves of the disk, the power at the GE factory unexpectedly shut down. In little more than an hour, the temperature inside the furnace dropped from 1050°F to below 700°F. The alloy shield suspended over the mold to protect the disk from firebrick fragments contracted from the sudden temperature change and scaled off, contaminating the glass. That final disaster marked the end of GE's work on telescope disks. The sixty-inch disks sat for a long time at Lynn, ignored. In the 1950s Robert McMath of the McMath-Hulbert Observatory at Pontiac, Michigan, explored possible use of the disks for a solar telescope, but nothing came of the project. By then Thomson and Ellis were dead, and no one at GE seemed eager to revive a failed project.

Much of the equipment in the GE laboratory had been paid for by and was therefore the property of the Observatory Council, but after surveying the huge building and facilities GE had created, McCauley concluded that the only useful equipment for Corning was some large transformers, pyrometric devices for measuring high temperatures, and a large crane. Ellis said that he would be using some of the equipment to finish his work with the sixty-inch disks, so the only transfer arranged was for three large transformers that GE had never used. Those, and the information that GE used a special industrial grade of nichrome as heating elements for furnaces, were McCauley's inheritance from the years of work and hundreds of thousands of dollars GE had expended on the project.

McCauley wasn't a man to criticize openly someone else's work or plans, but he hadn't been impressed with what he saw in Lynn. The fiendishly complex apparatus and the yard full of quartz chunks, waiting to be sorted underwater by one man, seemed like parts of a process doomed to endless experimentation and cost overruns.

His own plan was conservative. Although they would be working with a glass mixture that had never been used for large castings, the procedure of filling a mold with ladles was tried and true. Glassmakers had practiced those skills, with glass formulations simpler than Pyrex, for centuries.

McCauley had considered and rejected two alternate techniques. One idea was to "sag" the glass into the mold, by placing a large block of pure glass over the mold and heating it until it flowed into the shape of the mold. It was an appealing idea because it eliminated the process of ladling, but for the ribbed disks that the Observatory Council wanted, McCauley would need a mold with cores on the bottom strong enough to support the weight of the entire block of glass. No one had experience building that sort of mold out of refractory brick, the only material that could withstand the heat of molten glass. The alternative was to sag a disk of glass large enough to cover the entire mold. That meant they would first have to cast the unribbed disk, which would entail the costs and risks of two moldings and the risk of handling heavy materials twice.

Still, the sagging idea was promising enough to merit at least an experiment. One of the Corning melting tanks was scheduled to be shut down for repairs. Normally they would shut off the burners, letting the melt inside cool rapidly. The resultant cracks and strains would then leave a mix of small chunks of glass. Instead McCauley had insulation installed on the sides of the tank before the shutdown, and he slowly reduced the heat in the tank. When the tank was cool, large blocks of cullet were left behind. None was big enough for a two-hundred-inch telescope disk, but several were large enough to suggest that sagging chunks of pure glass was a possible alternative to ladling.

Another physicist at Corning, Dr. Jesse Littleton, was famed for out-of-the-ordinary ideas. When he saw the plans for ribbed disks, he went back to his lab, muddled over a scratch pad, then suggested an alternative that he was convinced would be cheaper to fabricate, lighter, and substantially more rigid than a cast mirror. Instead of pouring glass into a shaped mold, Littleton proposed, they could start with Pyrex-brand custard cups, which Corning produced in immense quantities. The cups would be stacked inverted in layers in an open mold, which would then be filled with molten glass. The glass would flow around the cups and seal them into a single structure of regularly arranged thin glass walls enclosing a geometric pattern of air spaces. Littleton's calculations demonstrated that the resulting honeycomb structure would be light and rigid, and the fabrication procedure avoided the complexities of building molds in complex geometric designs.

Making a telescope mirror from standard Pyrex custard cups was too audacious not to try. McCauley brought in a supply of the cups and put the mold makers and then Woods to work. When Woods ladled the molten glass into the mold, it flowed into the gaps between the inverted custard cups, sealing off air cavities here and there in the pile. As more glass was added, the pockets of trapped air, expanding with the increasing temperature in the mold, formed odd-shaped bubbles, turning the neat geometric pattern into chaos. McCauley and Littleton, each with a Ph.D. in physics, had temporarily forgotten that hot air expands.

The custard-cup experiment settled the method of making disks. McCauley would stick with his original conservative plan. He had the Corning purchasing department order large ladles, capable of holding three hundred pounds of glass, and told the foremen in the Corning factories to be on the lookout for experienced ladlers among their glassmakers.

McCauley had been impressed with the heavy-duty nichrome heaters he had seen at Lynn and the big transformers that GE had transferred to Corning. Willing salesmen had plied him with specification sheets for electrical control equipment. No order was too small to ignore in the depression. He was especially impressed with the large theater dimmers Westinghouse manufactured. The controls were simple, ruggedly built, and fitted with calibrated dials that could be rearranged as a vernier scale, which would allow precise adjustment of the temperature in the annealing oven. For safety he had the GE transformers rebuilt so they would take 2,200 volts from the main feeder lines coming into the Corning factory and put out a nonlethal 35 volts to the heating elements instead of the stock 110 volts.

McCauley drew up sketches for an annealing oven and turned them over to George Ward, an engineer who had worked on furnace designs. Ward produced drawings and material lists for the Corning masons, millwrights, and· electricians. The first oven, big enough to handle a thirty-six-inch disk, would be used to anneal the twenty-six-inch solid disk and a thirty-inch ribbed disk, the first ones on the schedule for the Observatory Council.

By March 1932 everything was ready except the annealing oven. George Crown, the favorite Corning mason for lab work, had built molds for the first two disks, shaping the nineteen separate cores that would create the ribbed structure from the reliable C-25 insulating brick, and cementing them to the molds with the Hi-Tempite cement that had been used on earlier molds. Although the annealing oven wasn't ready, McCauley decided to go ahead by casting the disks, then cooling the blanks slowly under a layer of Sil-o-Cel powder insulation spread over the exterior of the hot molds. The cold disks could later be reheated and annealed when the ovens were ready.

Except for some bouts of stage fright among the ladlers, the first production casting, of the twenty-six-inch solid disk, went well. The first try at a ribbed disk was less successful. When Woods and his helpers ladled glass into the mold, small bubbles appeared in the spaces between the cores, and the cores slowly loosened and bobbed to the surface of the mold. Tests had showed that the cores should stay in place, but there was no denying the evidence. That evening McCauley went back to his drafting board at the oak dining table where Anne, Jim, and George Jr. did their homework.

For the next try McCauley had the mold makers use cemented dowels to hold the cores in the mold. The crew was reassembled at four A.M. on a Sunday, so there would be no curious audience of glassworkers to make the still-inexperienced ladlers nervous. They were alone except for a cleaning crew working in the steel girders and trusses of the roof, oblivious to the critical work on the floor beneath them. Dust and debris tumbled down into the open mold. Shouting and sign language finally routed the cleaning crew to another area of the building; a worker used a jet of compressed air to blow the debris out of the mold; and the ladlers started up again.

This time the cores held in place, but a cascade of bubbles rose from the surfaces of the mold, spoiling the disk. McCauley asked the mold makers and the men who had filled the tank to make sure that all materials in the mold and the batch of glass were exactly the same as they had used in the successful practice pours. An expert on gas analysis from the Corning labs punctured one of the larger bubbles in the disk and collected a sample. His gas spectrometer reported that the gas was from the combustion of a petroleum product.

For McCauley problem solving was a Sherlock Holmes game. He talked each member of the crew through the procedures they had used in the trial molds and the most recent efforts. Where had they gotten the bricks? Which cement had they used for the molds? Which lots had the material for the glass melt come from? The only change he found was that the early molds had been cleaned with a vacuum cleaner. The technicians couldn't get the nozzle of the vacuum cleaner between the cores that had been glued in place for the ribbed disk, so they had used compressed air to blow the mold clean.

Why would the switch from a vacuum cleaner to compressed air make a difference? The exhaust from a vacuum cleaner is expelled outside the machine. With compressed air a trace of oil from the compressor could have gotten into the airstream and ended up on the mold lining. When the molten glass hit the mold, at temperatures upward of 1000°C, a microscopic trace of oil would be enough to emit gases into the molten mixture, producing bubbles. For all future castings, McCauley ordered, the molds would be cleaned only with vacuum cleaners.

By April the first annealing oven was ready. The trial disks, which had been slowly cooling under Sil-o-Cel insulation, were consigned to the oven and held at 475°C for ten days, then cooled one degree Celsius per hour until the oven could be opened. At the unveiling McCauley found an opalescent twenty-six-inch disk, the strains in the glass obvious from the cloudiness, and a broken thirty-inch ribbed disk. The residual stresses at the rim of the smaller disk were eighteen times what he had anticipated. McCauley recalculated his annealing constants and concluded that the temperature they had used was 45° too low. The break in the ribbed disk, he concluded, was from a tiny fracture produced when the mold cores were removed from the disk. Reheating in the oven had aggravated the fracture, and the annealing temperature had been too low to refuse the glass. Corning was learning about handling the glass, but after six months of experiments, they still had no disk to offer the Observatory Council.

In May, McCauley tried again. This time he replaced Woods with two experienced glass ladlers from A Factory who had seen him struggling with the big ladles and thought they could do better. The ladlers cast three disks all at once, another solid disk, and two thirty-inch ribbed disks. One of the ribbed disks went straight into the annealing oven. Two weeks later it emerged, free of bubbles and with stresses within the expected range. The only flaw was that at one point the glass had not completely filled a portion of an outer circular rib. Knowing that the outer portion of the disk would be ground down to a slightly smaller size, McCauley sent it off to the Mount Wilson Optical Labs in Pasadena. The opticians found the material easy to grind and figure and pronounced it perfect in their tests.

After the long months of frustration with GE, Hale, Anderson, and Mason were delighted by the progress Corning had made. In February, Mason wrote that conditions were "ideal" for getting Corning to do the work: Production was slack at their plants, they were interested in as much work as the Observatory Council would provide, and Day and the other Corning people had a "fine scientific attitude."

After some preliminary grinding on the thirty-inch disk, Hale summoned another meeting in New York City, where he gave Corning authorization to proceed to the sixty-inch disk, and if it was successful to go ahead on the one-hundred-twenty- and two-hundred-inch disks. No one asked for a formal contract. Three meetings and an exchange of letters was enough to agree on terms. Corning gave their "best estimate based on experience in the manufacture of disks less than 30" in diameter," and agreed to bill monthly for "actual costs," which meant their direct expenses plus overhead. If the disks were successful, they would expect an additional 10 percent profit. The confirming letter from Corning left an escape clause: "Since there is some uncertainty regarding success in making these large disks, either party shall have the right to withdraw at any time."

The thirty-inch disk from Corning, the first successful ribbed disk, took a good figure in the optical shop—the opticians' way of saying that they were able to grind and polish it to a precise shape. More important, the ribbed back made the disk rigid, lightweight, and thin enough so that it did not suffer the slow adjustments to temperature changes of a solid disk. While there was no assurance that the Corning procedures would work on larger disks, with a successful ribbed disk in hand, the Observatory Council directed their efforts toward a ribbed two-hundred-inch disk for the telescope, with support mechanisms to maintain the disk's shape.

The ribbed back was essential. But what should the waffle look like? Hale had proposed the problem of achieving maximum rigidity for a given minimum weight of disk to the mathematical physicists Harry Bateman and Paul Epstein, and the engineers Von Karmann and Romeo Martel, all at Caltech. The four reported to Hale that the problem could not be solved. Arthur Day, surprised that Hale had given the problem only to men with no experience in the design of telescopes, wrote to Francis Pease and found that he already had a solution to the design problem for the mirror back—based on a series of wheel hubs, each of which would hold a support mechanism, with radial spokes between the hubs forming a rib structure.

Pease was devoting much of his time to the telescope now. A dozen years before, when the one-hundred-inch was new and sensational, Pease and Anderson had conducted a research program on the new telescope, based on an idea of Albert Michelson's, which used a huge interferometer erected on the upper end of the telescope tube and prisms near the focus to measure the actual disks of distant stars. By 1930, with Hale's encouragement, Pease had built a fifty-foot-long interferometer on Mount Wilson to provide even greater angular resolution. In 1932 he finished one last experiment, a refined measurement of the speed of light proposed by Michelson, and then put much of his own research interests aside to concentrate on the engineering of the two-hundred-inch telescope.

There still had been no public announcement of the work going on at Corning, but Dwight Macdonald, at *Fortune* magazine, had his antennae out, picked up some signals, and did an article on the progress of the telescope and the involvement of big business in the project. Max Mason, running interference for the Observatory Council, tried to get Macdonald's draft toned down. The phrase "which G.E. had done at cost," he wrote Macdonald, "is rather dragged in by the hair. Everyone mentioned did work at cost or for nothing at all." He also downplayed what Macdonald had called the "mystery" of the transfer from GE to Corning. The only mystery, he wrote, was the whole telescope, "because no one is willing to bet 100 to 1 that the 200-inch disk can be made."

With the cat out of the bag, Hale had no choice but to announce the work at Corning. The announcement was deliberately low key, focused on the switch from fused quartz to Pyrex rather than the change from GE to Corning. The press release caused little stir. By midsummer 1932 Americans had their own problems. Those who could afford it could live in a boarding house for a week for $3.50, could buy breakfast for $.10, or ride a trolleycar for $.05. Many couldn't afford even those prices, and lived in tin Hoovervilles or flophouses, where as many as three or five hundred double-decker beds were crammed into spaces that reeked of humanity, and where long lines queued up for meals of soup and bread. Andrew Mellon might call the depression a "hiccup" in the nation's economic life or announce that "America is going through a bad quarter of an hour," but families who scrimped with their meager savings or relief funds that averaged as little as $8.00 a week, trying to keep out of the poorhouses and Hoovervilles; and those who were forced to choose among heat, food, or clothing, had enough on their minds not to worry about the squabbles among the companies building a "giant eye."

The temporary respite from the press was a relief in Pasadena. The signs of progress from Corning meant that the design program could shift back into high gear. For years everyone had pushed their pet ideas about the mirror. Many questions were settled, but enough remained for Hale to ask Mason to head the design group for the mirror. It was unusual to have the head of the funding organization sit on a working committee of the project, but Max Mason was an unusual foundation president. As a physicist, and in his wartime research on acoustic detection at the predecessor of the Navy Sound Labs in New London, Connecticut, Mason had been more comfortable with solid applied problems, whether the wartime adaptation of Broca tubes to submarine detection or the later engineering problems of the telescope, than with the administrative and diplomatic duties of a foundation.

From Hale's perspective Mason was a superb choice to pull the mirror design together. He was respected as a physicist, which meant the scientists would listen to him; he had experience with applied engineering in his work on acoustic devices, which meant the engineers would listen to him; and he was enough of an outsider, working in New York, to be beyond the inevitable disputes of territory and personal preferences that arose between opticians and astronomers, Mount Wilson staff and Caltech staff, engineers and observers, telescope designers and glassmakers, and even among the scientists, between the spectroscopists and those who photographed distant objects.

The whole telescope depended on the mirror. When good news arrived from Corning, every other committee perked up. In 1929 the Site Committee had placed ten of the small telescopes Porter had designed at ten different locations in Southern California and Arizona. Each telescope was equipped with a high-power eyepiece that enabled a relatively untrained observer to measure the atmospheric oscillation of star images under a high magnification and to grade the images with the method Anderson had devised. The seeing tests were supplemented with weather records, using instruments borrowed from the U.S. weather bureau to record the extent of cloud cover and the number of days with sunshine. A year later the search was down to seven sites. The committee concluded that the summer rainy season in Arizona and the low winter temperatures there were unfavorable for a telescope site. The criterion that the site be within a few hours' ride of Pasadena reduced the list to three: Horse Flats, fifteen miles north of Mount Wilson; Table Mountain, twenty-five miles from San Bernardino; and Palomar Mountain.

By 1932 sunshine recorders and cameras were still in place at the other sites, but the preponderance of test figures were from Palomar. The initial readings for Palomar had been compiled by Mr. and Mrs. William Beech, who owned a ranch in the French Valley section with a cabin. Later a twelve-inch telescope was installed on the mountain, and observers from Caltech or Mount Wilson would go down to take more precise figures. Ferdinand Ellerman began a long series of observations to record the seeing through the seasons. He was troubled by the fog and noted that in optimum conditions the seeing at Mount Wilson was better than Palomar. But the average visibility—in terms of weather, light pollution, and atmospheric steadiness—was superior. Early in the search Hale had written that "results thus far point to Palomar as the most promising." Even as bouts with his private demons pulled him further from the project, the prophecies of the master builder of telescopes remained persuasive.

McCauley considered the thirty-inch disk only a partial success. The poured glass hadn't filled every crevice of the mold, leaving part of the ribbed structure unmolded. Again McCauley set off on one of his Sherlock Holmes sessions, exploring every possible cause before he concluded that the glass in part of the mold had begun to set before the mold was completely full. These disks used far more glass than Corning was used to pouring. To avoid worse complications with the larger disks, he decided that the disks would have to be poured into a heated mold. Direct heat, through the insulating refractory brick of the mold, was impractical. McCauley's answer was an "igloo," a bee-hive-shaped dome over the mold, lined with nichrome heaters. The igloo would be preheated and maintained at temperature throughout the casting process to keep the mold and the poured glass hot enough to prevent blank spots. A doorway in the igloo would provide access for the ladles of glass.

Working late at the dining room table, he transformed the ideas into sketches. The engineers and draftsmen turned the sketches to working drawings, and the masons, millwrights, and electricians built a new annealing oven, big enough for a sixty-inch disk, and an igloo for the mold, sandwiched into the space between two of the huge tanks used for ordinary Pyrex production work. The mirror project was substantial enough that Amory Houghton authorized a general policy of releasing workers from the various trades within the factory as needed. Corning reticence and Hale's injunctions against publicity were enough to restrict general announcements of the new project. Some factory workers realized what was going on, but Corning is a self-contained town. Word did not spread to the outside world. "We still enjoyed the quiet, undisturbing atmosphere," McCauley recalled, "perhaps not yet fully appreciated, of working together in the privacy of our own back yard without fear of being watched."

When the equipment, including an overhead traveling hoist to move the igloo cover and ultimately the disks, was in place, McCauley got his crew together to cast a sixty-inch disk. They rehearsed the whole procedure until the steps of opening the door in the heated igloo to receive the ladles of hot glass went smoothly. As ladles of glass were added, the level of glass rose in the mold, filling the crevices between the cores that formed the ribbed pattern of the back of the disk. The heated igloo kept the glass molten long enough to fill every corner of the mold. Another problem seemed solved.

Except that it wasn't that easy. The glass was barely over the tops of the molds when a rib form popped loose and bobbed to the surface of the molten glass. Others followed, ruining the neat geometric pattern of the back of the disk. McCauley stopped the pour.

A few days later the masons had built a new mold, to stricter specifications: stronger dowels for the cores, precise procedures for the cementing of the cores and the coating of the interior with silica flour. The masons were told to treat the entire mold as fragile when they vacuumed out debris. A try at casting the disk in the new mold went smoothly until the glass was almost at the required thickness. When the last ladle of molten glass was poured into the mold, a single core broke free and popped to the surface.

McCauley reacted quickly. Instead of calling off the pour, he ordered the ladlers to finish and then to use tongs to fish out the errant core before the disk was consigned to the annealing oven. After the disk was annealed, the back could be ground to produce the precise geometric shape of the missing core form.

When the surface of the molten glass leveled, the overhead crane picked up the annealing kiln. The crane order had been for a special "slow moving" model, but it still raced across the track, the kiln cover hanging at a rakish angle. A glassmaker jumped onto the kiln to balance it, then had to hang on as his weight sent the hoist rolling down its track, high over the floor. It took three husky millwrights finally to stop the swinging load and stabilize it. Plenty of hearts missed beats, but the Keystone Kops antics miraculously didn't harm the disk.

When the kiln cover was in place, the disk was consigned to two months of controlled cooling. It emerged in September 1932 intact, and with residual stresses as low as McCauley's calculations had predicted. The custom shop at Corning was able to grind the back to correct the rib pattern that had been marred by the core that had broken free during the molding, and the Observatory Council accepted the disk. There were no plans or facilities to grind and figure the disk in Pasadena—Anderson's plan was to grind and figure the auxiliary mirrors after the bulk of work had been completed on the two-hundred-inch primary mirror—so McCauley was told to store it at Corning while they went ahead on the next step, the 120-inch disk that could be used as both a practice run for casting the primary mirror of the telescope, and could also be ground as a flat mirror for use in testing the two-hundred-inch mirror.

In less than one year of work, Corning had gotten further than GE had in three. Optimism was so high that McCauley persuaded the Observatory Council that they could save money and time if they skipped a few steps and built the remaining equipment at Corning large enough for both the 120-inch and the two-hundred-inch disks, rather than going through the stages to build both a 120-inch and a two-hundred-inch annealing oven. With budget approval from Pasadena, McCauley got permission to have one of the large melting tanks in the Corning factory dismantled in November 1932. The site of the former tank was then covered with a steel floor, at the height of workbenches at the adjoining tanks. Two circular holes, each twenty feet in diameter, were cut in the floor on a north-south line, and rails were installed in the cave underneath the floor. A heated igloo, big enough for a two-hundred-inch disk, was suspended over one hole, with an annealing kiln over the other. Beneath the floor, a table large enough to hold the mold for a two-hundred-inch mirror was constructed on a 60-ton loco-motive screw hoist that could travel on the rails.

After watching the heavy kiln swing out of control over a disk, everyone had concluded that moving the disks on rails was safer than swinging heavy equipment over the freshly molded disk. The screw hoist was slow (two inches per minute) but reliable. The new arrange-ment also simplified the hookup of electrical connections to the igloo and the annealing oven. A separate room for the electrical transform-ers was constructed on the floor above and just west of the annealing kiln. To complete the arrangement, McCauley had a draftsman design trolleys for the casting oven to move it from its position over the mold, so that mold makers could have complete access to the mold before the casting began.

The arrangements looked good on paper. They took months of work by draftsmen, engineers, millwrights, carpenters, electricians, masons, and the outside suppliers who were called on for the struc-tural steel, insulating brick, electrical controls, and new ladles that would hold 750 pounds of molten glass each. The depression economy had hit much of the glass industry especially hard, so suppliers were quick to produce the needed materials, and Corning was able to assign dozens of men to work on the facilities McCauley needed. It looked as if they would move to the next step of the casting program ahead of schedule.

Then disaster struck.

Harrison Hood ran Corning's laboratory of glass scientists, who had the job of testing glass formulas under extreme conditions, and concocting new glasses to meet new challenges. A chemist in Hood's lab, Dr. M. E. Nordberg, had gotten hold of a chip of the glass that had been used for the successful sixty-inch mirror. One test in particular intrigued him.

Nordberg found that in 1923 a researcher named R. D. Smith had noted in a laboratory logbook that annealed 702-EJ glass was more soluble than unannealed ware. Nordberg decided to follow up on Smith's finding with a series of experiments. When he tested pieces of Pyrex that were heat treated, then rapidly cooled, the samples were fine in solubility tests in water or acids.

But when Nordberg put the chip from the annealed sixty-inch disk into water, it was far more absorbant than any of the test materials. The period of long annealing and slow cooling clearly had affected the glass in ways they hadn't anticipated. He tried more tests and con-cluded that the surface probably was too unstable for polishing with moistened or water-flushed rouge or for frequent mirroring with a solution of silver nitrate. And that disk had only been heat treated for two months in the annealing oven. What could they anticipate for the surface of a two-hundred-inch mirror that would be annealed for ten or twelve months?

McCauley went over the test results and could find no flaws with Nordberg's results. The 702-EJ glass he had been using was not suitable for telescope mirrors. The 60-inch mirror made from the material was a sponge. It was close to Christmas 1932; dozens of workers had dismantled a major melting tank and were constructing the facilities to mold the 120-inch and two-hundred-inch mirrors. He had just sent a revised and slightly accelerated schedule to Pasadena.

He couldn't bring himself to admit to the Observatory Council that the glass formula was no good and that Corning didn't have a glass with the proper characteristics for big telescope mirrors.

McCauley began the new year at a dead end.

16

Good News

The good news from Corning about the sixty-inch disk, and the successful tests of the trial Pyrex disk at the optical labs, inspired a flurry of work on the telescope design. Hale and his colleagues were confident that Corning could cast a Pyrex mirror disk, with a cored and ribbed back. The disk would be supported on ball-bearing supports, like those Pease had engineered for the one-hundred-inch telescope. The geometric pockets in the back of the disk would house a system of compensating supports to maintain the shape of the mirror as the telescope moved. Within those basic constraints, Porter and Pease were busy sketching designs for the telescope. The astronomers on the advisory committee, and others with strong opinions, were busy imposing requirements on the design.

The astronomers who focused their research on galaxies and other deep-space objects, like Hubble and Humason, wanted no limits on the orientation of the telescope. The one-hundred-inch, because of its mounting design, could not point to the circumpolar region around the North Star. Studies with the sixty-inch telescope showed this region of the sky to be particularly rich in galaxies. The deep-space astronomers also wanted the best possible facilities for spectroscopy and direct images at the prime focus of the telescope, which would have the greatest light-gathering ability for faint imaging.

The astronomers whose research centered on stars were interested in spectroscopy, which demanded an extremely stable temperature- and vibration-proof room at the long Coudé focus, where they could make detailed spectrograms of individual stars. In brainstorming sessions the astronomers came up with other ideas for the use of the telescope. The Cassegrain focus behind the main mirror would be ideal for some kinds of imaging and spectrograms, and someone suggested that instruments could be mounted at the Nasmyth foci on either side of the hollow declination axis. Each position offered advantages and disadvantages; in the interest of long-term versatility, no one was willing to abandon his favorite.

Normally it would take hours or even days to change the instruments in use on a big telescope to a different focus position. If the telescope had been used for spectrograms of the red shifts of distant galaxies at the Newtonian (prime) focus, the change would require moving new mirrors into position and realigning them before the telescope could be used at the Coudé or Cassegrain focus. In setting the research programs for the big telescopes, the time-allocation committees would try to string together programs that used the same focus and equipment: deep-space research during the dark of the month, when the moon was down; spectrographic studies on nearby stars when the moon was up.

But weather and the general seeing conditions didn't always pay attention to the plans of the astronomers and the allocation committee. The telescope and observers could be prepared for an evening of deep-space observing at the prime focus only to discover that the seeing wasn't good enough to make use of the light-gathering ability and speed of the telescope. The evening might be good enough for less demanding use at a different focus, but if it took six hours to change the instruments and focus of the telescope, the evening would be lost.

In the brainstorming sessions the astronomers asked if the telescope could be switched from one focus point to another in minutes rather than hours, so the balance of the night could be put to profitable use. What if the various auxiliary mirrors that would bounce the light path to the different foci could be mounted so they could be flipped into place with remote controls and still be rigid enough and well-enough aligned to preserve the optical quality of the telescope? The ideas sounded like a fantasy list. No one yet had an idea how to maintain the rigidity of the tube and mount of the basic telescope.

Depression pessimism actually encouraged the pie-in-the-sky ideas. If the two-hundred-inch telescope had once seemed like another step in a progression of telescopes, it was now beginning to seem more and more like the final step, the last big telescope for a very long time. This one had to do it all. No one used the term, but with so many demands to fulfill, the big telescope had to be a *perfect* machine.

George Hale was probably the only one who could have mediated the conflicting demands and requirements of the astronomers and engineers. When he was well he was lucid, quick, and judicious. His experience and prestige as the father of large telescopes were often enough to settle disputes. Unfortunately his nervous condition and its manifestations had grown more intense with the years. The periods when he was free of his demons were shorter and rarer, and the instances when Miss Gianetti would excuse him from meetings were so common that significant decisions were made without him.

John Anderson, as executive director of the project, was a member of each of the committees. As the work at Corning progressed, he was involved more and more in the day-to-day decisions on the planning for the mirror, and in the construction of the optical lab where the mirror would be ground. Anderson was an easygoing manager. He didn't fire off memos the way George Hale had, and he made fewer demands on the other staff members. The lack of a single strong guiding force may have been a blessing. Porter, Pease, and some of the young engineers on the Caltech faculty were free to play with ideas that might have been dismissed as too radical by a more rigorous manager.

One obvious question was how the observers would get to the various observation positions on the telescope. The Newtonian focus, used for deep space work on the sixty- and one-hundred-inch telescopes, had always been a problem. The eyepiece and instruments at the open end of the telescope tube were high off the ground and rotated with the telescope so that observers were constantly fighting fatigue and cramps as they scrambled from one position to another. The Cassegrain focus at the bottom of the mirror was less of a problem, but the sheer size of the two-hundred-inch telescope meant that the Cassegrain focus could be high off the ground when the telescope was pointed at an area of the sky close to the horizon. The Coudé focus, in a room beneath the telescope, was easy for observers, but the telescope had to provide for a light path that would reach the room no matter what the orientation of the telescope. If the telescope could dip low enough to see the polar region, it would need not one but two different paths of mirrors to bounce the light down into the fixed room.

The telescope was so large that even the secondary mirrors that would change the light path to the various instruments were heavy and unwieldy. Moving heavy mirrors in and out of the telescope tube each time the light path was changed would be cumbersome and dangerous. No one was sanguine about the idea of regularly suspending heavy equipment by crane over the priceless primary mirror of the telescope.

Gradually Anderson and Porter began looking at the scale of the telescope—the latest estimates were that the primary mirror would weigh close to twenty tons and the mounting more than five hundred—as an advantage. The size of the telescope tube and the stability of the immense mounting meant that the auxiliary mirrors for the Cassegrain and Coudé foci could be mounted in a permanent cell in the middle of the telescope tube, with gear-driven mechanisms to swing and lock them in place as needed. The mirrors and cell, approximately six feet in diameter, would rob the central portion of the two-hundred-inch disk of only one-ninth of its area, an acceptable sacrifice. And since the telescope could tolerate a six-foot cell, by extending the cell on the other side of the compartment that held the swinging mirrors, they had room for an observer to ride *inside* the telescope to use the prime focus.

Putting the observer inside the telescope had obvious advantages. With the observer at the prime focus, they would need no diagonal Newtonian mirror or the awkward platforms for an observer outside the tube.

The corrective lens and plate holder for the prime focus would be in one cell, separated physically and mounted independently of the cell that held the observer, so that vibrations from the observer's movements would not affect photographic plates during exposures. The observer's cell would ride circular rails and be adjustable at 22.5° intervals so the observer would always be upright even as the telescope turned. There would be room for a set of controls for the telescope and for telephone equipment to allow the observer to speak with the night assistant far below him on the observatory floor. Astronomers who saw the early plans were fascinated by the idea, although more than a few raised the obvious questions: How would they get up there? Those who remembered the cold nights and aching bladders of long winter exposures worried even more how they would get back down.

Porter answered the questions with sketches of an elevator that would ride up the edge of the shutter opening in the dome and an extending platform that would reach almost to the observer's cell. The observer would only have to step across ten inches from the platform to the prime focus cell.

For the Cassegrain focus behind the primary mirror, Porter came up with an equally ingenious observation platform. Because the end of the tube would rise and fall as the orientation of the telescope slewed in declination, Porter proposed a pivoting observation platform with a central portion on retracting cables. Electric motors would lower the center of the platform to the floor for observers or equipment. The pivoting platform would rotate so that the platform was always level with the floor, no matter what the orientation of the telescope. A few astronomers on the advisory committee raised questions about ascending to the telescope in what looked like a large children's swing, but the simplicity of the mechanism, without ladders to move and get in the way, was immediately appealing.

The details fell into place. Still, the biggest decisions—the design of the mounting that would hold and point the mirror, how and where to build it, and the final choice of a site for the telescope—remained open. Deciding those issues before they had a mirror was tempting the gods.

Corning's advertising motto was Corning Means Research in Glass. At the beginning of 1933, unbeknownst to the Observatory Council in Pasadena, the Corning engineers were going to have to make good on that slogan.

It had been McCauley's decision that they keep the disastrous test results on the sixty-inch disk from the Observatory Council. The initial thirty-inch test disk, which had been annealed only a short time, had passed the tests of the opticians in Pasadena. It was only the larger disks that suffered the increased water solubility of the glass, because of their long annealing. McCauley's plan was to use the period when they were building the large annealing oven and molds for the next steps to develop a new glass.

Harrison Hood, who had the nickname Sage of Glasslore at Corning, got the assignment of coming up with a new glass formulation. He had previously developed glasses that would survive long annealing, but none with the desirable low expansion of the 702-EJ glass McCauley had used for the trial disks.

Hood's initial formulations, tried in a small day tank in C factory, might have met the specifications, but they were so difficult to melt and work that the glassmakers rejected them. Each trial meant dismantling, emptying, and rebuilding the tank, and days or weeks to assemble the materials and get the melt up to the proper temperature, well over 1000°C for these borosilicate Pyrex glasses, before the glass could be ladled into test molds. Hood's next try used the same amount of silica as laboratory Pyrex, with more boric acid and less alkali. The mixture proved workable, and the glassmakers were able to pour a block, approximately thirty inches square and thirteen inches high. The block was covered with Sil-o-Cel insulating powder and allowed to cool slowly in an effort to duplicate the effects of long annealing.

The block survived the test unbroken and emerged with less opalescence than the same treatment would have produced in 702-EJ glass. Hood made a few more adjustments to his formula, ran it through another series of tests, and named the new formulation 715-CF Pyrex. On extensive testing the new glass held up, after the long cooling period, with no problem of water solubility. The formulation required higher working temperatures, in excess of 1500°C, but it also had a temperature coefficient 25 percent lower than the previous formulation and did not devitrify at the higher temperatures.

Arthur Day was already scheduled to meet with George Hale, so he was entrusted with the "good" news that Corning had developed a new formulation with reduced temperature coefficient that would make even better mirror disks. He mentioned not a word about the water solubility problems that had condemned the sixty-inch disk, and no one in Pasadena knew how close Corning had come to failure. Hale cheerfully welcomed the news about the new glass formulation and answered that there was no need to recast the sixty-inch disk because the latest design of the telescope would use smaller secondary mirrors that could be cast later. Sometimes a hiatus in communications can do wonders for a relationship.

McCauley's crews continued their work on the molds and annealing ovens and tried a test disk with the new formulation. The higher temperature in the melting tank required for the new Pyrex reduced devitrification on the surface of the glass in the ladles, so they no longer had to skim the ladles before pouring. At the same time McCauley discovered that the ¾-inch-thick ladles were heating excessively in the period between filling with glass and pouring into the mold. The high temperatures scaled iron oxide off the ladle rims, which in turn introduced traces of discoloration into the tank when the ladles were dipped in for another load of glass.

The obvious solution was heavier ladles. Corning had little experience with ladles, and queries indicated that even the big window-glassmakers had never used ladles thicker than ¾ inch. When even the window-glassmakers shifted to machines for plate glass, ladle technology had stopped. Fortunately, the purchasing office queries led to Bethlehem Steel in Buffalo, which still had the dies they had once used to press ladles for the window-glass industry. The depression had left their mill slack, and they were willing to bring out the old dies and press one-inch-thick ladles—all within fifteen days.

By early April 1933, McCauley was ready to cast a 120-inch disk. The masons and electricians built a new annealing oven, large enough for the two-hundred-inch disk. McCauley's conservative calculations came up with 516 for the number of oversize GE heating elements needed to maintain the temperature of the oven. To avoid the possibility of shorts, the heating elements were mounted in rows on the sides and bottom of the oven cover, with heavy-enough wiring and masonry to survive an annealing period, at high temperature, of up to one year. The thirty-five-volt circuitry reduced the insulation problems, but the increased current required large-diameter wiring from the panels of reactors and theater-dimming controls. Ten thermocouples were spaced around the upper, lower, and wall surfaces to continuously monitor the internal temperature. The controls were set up so an operator could control the internal temperature to within 1 degree in the range between 400° and 550°C.

The cores for the new mold were secured with cold-rolled steel rods to avoid an accident like the floating core that had marred the 60-inch disk. Just in case, McCauley had the masons prepare a spare set of cores. If the mold failed during the pour, they could quickly build a new mold for another try.

Corning masons converted one of the large melting tanks in A Factory to what the glass industry called a "day tank" for the 715-CF glass. Normally melting tanks had a bridge wall that separated freshly introduced materials in the back of the tank from the ready glass in the front. With the bridge wall removed, the tank could melt a single large batch of glass. Millwrights installed overhead tracks to support the heavy ladles, leading from three openings in the tank to three openings in the igloo over the mold, each 120 degrees apart around the mold. McCauley designed the tracks so the "tank and casting oven operated like the moon, with the same surfaces facing away from the workmen." In the heat and panic of filling a mold, he didn't want workmen confused by having to spin around and change directions while they were guiding ladles filled with 750 pounds of molten glass at 1500°C.

McCauley also had three special wheelbarrows fabricated for the ladle heels, with two wheels in front, to make it easier to avoid sideways tipping when the workmen were moving around the tank in the heat and confusion of a long pour.

At the end of April the tank was lit, and the filling began with 715-CF batch. Wheelbarrow after wheelbarrow went up the ramps to the loading shovel at the opening. Thirty tons of glass requires an enormous supply of sand, borax, soda, and lime. The level of molten glass inside the tank rose four inches per day. After a few days a workman discovered a break in the tank lining. McCauley ordered the tank cooled and rebuilt. There weren't enough of the special low-contamination refractory bricks on hand, so the masons cannibalized bricks from the floor of the tank to fix the walls and built a new floor from common bottom bricks like those used in the normal production tanks, with a four-inch layer of clay tamped over it. No one was sure a tamped bottom would work for a melting tank, but the alternative was to wait months for the Pot and Clay Department to prepare a batch of new bricks. After the months lost searching for a new glass formula, *delay* wasn't a welcome word.

Within the world of astronomy, observatory directors followed every move of the group at Caltech, hoping they could piggyback on the research efforts to obtain disks and technology for their own planned telescopes. Dr. O. A. Gage, head of the Aircraft and Instruments Division at Corning, took advantage of the talk to solicit more orders for telescope blanks for Corning. Telescopes would always be a small portion of Corning's business, but once a portion of the facilities had been set aside for the bulky casting and annealing ovens, a large melting tank had been dedicated to the special glass for the telescopes, and personnel—not only the ladlers but electricians, millwrights, carpenters, masons, tinsmiths, and glassmakers—had been assigned temporarily from other work to the telescope project. Aside from the juggling of the annealing ovens, more orders just meant an economy of scale for the operation.

Once he found out that Caltech was abandoning all work at GE, Harlow Shapley canceled his order for two 60-inch disks from GE and ordered them from Corning. The David Dunlap Observatory at Toronto asked for a 76-inch disk, which would ultimately become a 74-inch telescope blank, and the McDonald Observatory of the University of Texas ordered an 82-inch disk. Heber Curtis, Shapley's old adversary in the great debate, who had left Lick Observatory for the University of Michigan,* ordered a large disk of undetermined size.

---

* Curtis and Hale discussed the possibility of Curtis coming to Mount Wilson. Hale offered him access to the research facilities, but no job.

Each order for a primary mirror was accompanied by orders for one or two auxiliary disks.

In all it was enough work to keep the tanks and glassmakers busy for over a year. And since the available annealing tanks and casting igloos were limited to casting one large mirror at a time, and the 120- and two-hundred-inch disks for the Observatory Council would keep those busy—between casting and annealing time—for at least the next two years, McCauley had another set of molding and annealing equipment built south of the existing set, but close enough to the big 3A tank to use the same batches of glass. The new kiln was big enough for an 84-inch mirror. The smaller molds and kiln required less elaborate equipment for raising and moving than the big molds and kiln for the 120- and 200-inch mirrors.

By the third week of June the 3A tank was filled and the glass mixture had set (the glassmakers would say it had been allowed to "plain") for nine days until the glass was deemed ready. It was a Wednesday when the Pyrex was ready to pour. McCauley decided to use the 76-inch mirror for Toronto as a trial for the molding procedure, and to postpone the 120-inch disk for the Observatory Council until Saturday morning, when the regular blowing room personnel would not be in the factory to distract the ladlers and other crew.

The pour of the 76-inch disk went without a hitch, and it was consigned to the smaller annealing oven. The melting tank was topped up with enough batch mixture to replace the 2½ inches of glass that had been used from the tank. The crew to pour the 120-inch disk—the largest piece of glass ever poured—would be the same crew that had poured the 76-inch, augmented by the additional personnel needed for an operation that would keep three ladling positions busy. Two additional men operated switches on the overhead tracks that supported the heavy ladles. Two more men worked the additional ladles, and four men ran two more wheelbarrows to carry the ladle skins back to the tank. Because the journey from the tank to the mold was longer, three men with backpack spray tanks carried ring-shaped nozzles they could hold over the lips of the ladles while they sprayed water to keep the lip of the ladle cool. After each ladle was dumped, the sprayers would retreat to a recharge station to get their backpacks refilled with water and compressed air. From the first simple pours with one man handling a ladle, the operation had become a precisely choreographed ballet with a company of dozens of men.

On Wednesday evening, June 21, McCauley ordered the burners of the casting oven igloo lit in preparation to cast the disk the next Saturday, only to discover that the expansion of the metal anchor bolts that had been added to hold down the cores had dislocated the tops of the cores. The masons came in for repairs and finished on Saturday morning at 5:00 A.M., just as the pouring crew assembled. Except for a furnace staff to attend the fire under the tank, the factory was empty when McCauley gave the order to begin.

With no audience to distract the crew, the procedure went smoothly. Everyone worked according to the script. The area around a fired glass tank—with the roar of the furnace, the clanging of the metal doors, the rumbling of ladle carrier tracks on the overhead rails, and the rattle of the metal barrow wheels against the steel floor—was far too noisy for verbal cues from a stage manager. And the sheer level of activity—along with the danger of moving ladles, each filled with 750 pounds of molten glass, heavy equipment, and barrows of glass slag—was far too dangerous for visual cues.

As each ladle of Pyrex emerged from the tank, the switchers had to route it to the correct port on the mold. A sprayer stood ready, avoiding the two men on the ladle arm and the heat of the full ladle of molten Pyrex as he held the ring of his sprayer centered around the ladle's lip. The threesome walked under the track together, the ladle handlers maneuvering the ladle with its load of molten glass as the carrier rolled along the tracks to the opening over the mold. The sprayer pulled back just before the ladlers lifted the cup of the ladle into the opening in the igloo over the mold. When the molten glass was tipped into the mold, half would remain in the ladle, cooled enough by the journey to stick to the steel boilerplate walls of the ladle. As the ladle came back out of the mold, men were ready to empty the ladle skin into a waiting wheelbarrow, which was wheeled to the rear of the tank to be returned to the cauldron of molten glass.

Cycle after cycle, the pour went flawlessly. By 7:30 the mold was filled. The ladles and wheelbarrows were moved away as the crew gathered around the mold, staring at the disk of molten glass through the pour openings. When the last ladles of glass were added, the glass on the top of the mold was hotter than at the bottom. The cooled glass around the molds stood out, outlining the geometric pattern of the ribs and cores. While McCauley and the glassmakers watched, the glass at the surface began to cool, and the molds became radiant against the dark ribs of glass. The changing image was so beautiful that the crew held off, mesmerized by the spectacle, before they finally lifted the igloo off and maneuvered the mold down its tracks to the annealing oven.

By four o'clock in the afternoon the mold had been lowered from beneath the igloo and slid down its tracks to a position under the annealing oven and raised again. An insulating seal of Sil-o-Cel powder was added between the edge of the mold and the oven cover. McCauley had calculated and recalculated his annealing schedule. For 11 days the disk would be held at 520°C. Then it would cool by 1.6°C per day for 140 days, and finally be allowed to cool at its own rate, limited only by the insulation surrounding the disk. The crew went ahead pouring smaller disks during that period, but the 120-inch disk, the largest glass casting ever poured, was the test of McCauley's procedure and the glass formulation. If it worked McCauley was certain they could cast the two-hundred-inch disk. If not? McCauley didn't have contingency plans.

Theodore Dunham, of the Mount Wilson Labs, was at Corning to witness the pour, which he called a "magnificent spectacle." Hostetter was Dunham's guide in Corning, and he gave both Dunham and Hale the strong impression that much of the scientific and engineering work on the disk project was his doing. "Dr. Hostetter," Dunham wrote, "does not trust a crane for such delicate work and thinks his experience with this table might be useful in designing equipment for the optical lab." Hostetter announced that *he* was anxious to cast all the secondary mirrors at the same time with the two-hundred-inch in order to avoid the great expense of heating the furnace a second time with another melt of the special glass. Observers in Corning noticed how often Hostetter was calling the telescope project—which he oversaw as a project manager, but on which he had done no scientific or engineering work—*his* project.

While Hostetter claimed the credit, McCauley checked the annealing oven every day, including Saturdays and Sundays, for six months. As a senior engineer/scientist, he had no production responsibilities for the disk. He could have asked the production crew foremen to assign someone to monitor the equipment. But McCauley was an orderly man. He considered himself responsible for the telescope disk project, and he insisted on checking the oven himself.

George Hale fled to Europe again that fall, to get away from the pressures and the demons. On his way home he arranged his travel to stop in Corning while the disk was still in the annealer. Hostetter, who enjoyed the visits of the famous to Corning, joined McCauley in showing Hale the annealing oven in A Factory that held the disk, the casting equipment and molds, and the panel of electrical controls that regulated the heat. Hale was "delighted with everything I saw."

At Christmas 1933, six months almost to the day after the disk had been poured, McCauley lifted the cover of the annealing oven to peek at the disk. There were no cracks, no pieces of the kiln broken away and embedded in the disk, no displaced cores to mar the ribbed pattern. The glass was clear, with the characteristic yellow color of Pyrex. Under close examination the strains in the glass were minimal, close to what he had expected after annealing. The only flaws in the disk were small bubbles (what the glassmakers call "vacuum bubbles") near the tops of some of the round cores. McCauley knew the bubbles would not affect the disk.

"We were obliged to admit," McCauley wrote, "that our product, while wholly suitable for the service for which it was to be used, was not the sleek object, without blemish, for which we dreamed. We could only accept our disappointment and try for greater perfection in the 200 inch disc, the next chance to produce a flawless disk." McCauley's prose doesn't quite conceal the pride of achievement. To avoid bubbles on the tops of the cores in the next disk he decided to make the large cores hollow, so the surfaces in contact with the glass would cool rapidly.

In Pasadena the optical shop on the Caltech campus was under roof and work had already started on a grinding machine for the 120-inch disk, which was to be ground as an optically flat mirror to use in testing the two-hundred-inch disk. Hale was encouraged, but had been steeled by long experience to anticipate the unexpected. "Large telescopes," he wrote, "as I have learned before, are secular phenomena. But fortunately the Corning estimates of cost do not increase beyond their original figures." He had been surprised too many times not to keep his guard up.

Still, when he heard the good news from McCauley, he couldn't help a tone of confidence. "The point of doubt as to the possibility of getting a satisfactory 200" Pyrex disk," Hale reported to the Observatory Council, "had been passed."

17

"The Greatest Item of Interest... in Twenty-five Years"

At the annual meeting of the American Ceramic Society in Cincinnati, in February 1934, Arthur L. Day delivered the Edward Orton Jr. Memorial Lecture. Day, the director of the Carnegie Institution Geophysical Laboratory, began with the requisite pseudo-Shakespearean paraphrase—"All the world is a ceramic product"—as he traced the history of ceramics from the formation of natural glass in the earth's crust to the current frontiers of ceramics research and to the most difficult of optical challenges, the mirror disk for the planned two-hundred-inch telescope. The ideas that had been pursued for the telescope disk, he reported to the audience, included everything from sawing off slabs of the obsidian cliffs at Yellowstone Park to fabricating disks of fused silica. There were problems with both extremes, which was why Pyrex-brand glass had become the compromise choice. Everyone at the meeting knew that Pyrex meant the Corning Glass Works.

Reporters hurried to ask Day questions. He confirmed that Corning was casting mirror disks for the big telescope, including the two-hundred-inch disk for the primary mirror: "This disk in all its details is a whale! Every detail of the process is on a scale so much larger than anything heretofore attempted that the setup is already somewhat appalling to contemplate." When would the telescope be ready? a reporter asked. "So far as astronomy is concerned," Day answered, "the existence of a 200″ disk will remain a dream until such a disk emerges from the annealing furnace at ordinary temperature in one piece and free from strain. After that I am at your service for any account of the disk or the manufacturing operation you may wish to publish."

Corning—the company and the town—had not been accustomed to the attention of the outside world. There had been occasional hoopla when a new product, like the first Pyrex utensils, was announced, but marketing publicity was very different from the persistent questions of newsmen. When a reporter called Corning for details on Day's comments, Leon V. Quigley, the newly appointed director of publicity for Corning Glass, dutifully explained the status of the project, the successful casting of the 120-inch mirror, and the preparations for the casting of the two-hundred-inch mirror. When the time came, he explained, Corning glassmakers would ladle the 20 tons of glass into the mold to create a two-hundred-inch-diameter, 26-inch-thick disk of Pyrex for the telescope. The reporter's story became one more newspaper item, lost among the reports of Roosevelt's frustrated efforts to deal with the deepening depression, Hitler's first stabs into the maelstrom of European politics, and the morbid daily details of the Lindbergh kidnapping.

The story of the mirror might have remained buried in the news if an NBC researcher hadn't picked up the item and put it into a script that was sent over to Lowell Thomas's office in the new Empire State Building.

In 1934 America listened to the radio. Whole families gathered around a console in the parlor, or perhaps a tabletop unit in the kitchen, in the hours after supper. The immediacy of radio meant that for the first time, an entire nation could be focused on a single program, event, or news item. Rich and poor, black and white, men and women, recent immigrants and Mayflower descendants—all heard the same broadcasts.

The most popular program on the air, from its beginnings in 1929, was the nightly *Amos 'n' Andy*, broadcast for fifteen minutes on NBC at 7:00 P.M., just after Lowell Thomas's news broadcast. People were so eager not to miss a word of the stereotyped antics of the Kingfish, Brother Crawford, and Madam Queen that most tuned in early enough to hear Lowell Thomas and his news broadcast. It was said that the resonant tones of Thomas's trademark sign-off, "So long until tomorrow," and greeting, "Good evening, everybody" made his the most recognizable voice on the planet.

Thomas had made his fame traipsing across the desert in pursuit of T. E. Lawrence, Lawrence of Arabia. A decade and a half later, books, lecture tours, and publicized adventures in faraway places had made the young, handsome, mustachioed Thomas famous. Damon Runyon claimed that Lowell Thomas was successful because he gave the impression of saying, "Now here is the news with some human slants on it and you can interpret it to suit yourself." The columnist Cy Caldwell offered a different explanation in a proposed epitaph:

Here lies the bird
Who was heard
By millions of people—
Who were waiting to hear
"Amos 'n' Andy."

Whichever the reason, for many Americans in the midst of the depression, the news was what Lowell Thomas reported. The superlatives in the story—*twenty* tons of glass, poured to create the *largest* glass casting ever made, for the *costliest* scientific instrument ever designed, the *biggest* telescope in the world, an instrument that would see *farther* into the cosmos—were the sort of tale Thomas delighted in reporting. He was never averse to hyperbole, and a list of superlatives was just what he needed to make a story that some would find "ordinary" into the kind of news that people remembered. Thomas knew a depressed America craved good news. He recognized a story about America's greatness in the NBC script.

In his broadcast Thomas described the plans to cast the great mirror in the quiet upstate town of Corning, New York, then added a few superlatives of his own. This event, he reported, the creation of this mirror, this step forward for science and technology, was "the greatest item of interest to the civilized world in twenty-five years, not excluding the World War."

The next morning Leon Quigley, the telephone receptionists at the Glass Works, and the Western Union office in Corning couldn't keep up with the requests for information about the project. News services, newspaper reporters, and radio broadcasters wanted tickets to view the casting. Amateur and professional filmmakers besieged Quigley with requests to document the pour. Scientists and industrialists, eager to be present, rang up their old friends from school or clubs to request tickets for themselves, families, and friends. When Corning's own paper, the *Evening Leader*, asked if Corning residents would also be allowed to attend the great event, Amory Houghton yielded to the circus atmosphere and announced that Corning employees and their families would receive tickets and that the Glass Works would make provisions for public viewing.

In Pasadena the Observatory Council tried to discourage the crowds that Corning seemed to be welcoming. Arthur Day, in Washington, was glad that Corning, *his* company, had finally gotten the attention of the press. Day even insisted on written confirmation from George Hale and Max Mason that they wanted only a few tickets for the great event.

Until Thomas's broadcast the preparations for the two-hundred-inch disk had gone on quietly. Even the success of the 120-inch mirror wasn't announced to the press. The optics labs in Pasadena weren't ready for the 120-inch mirror, so it had been crated in 8-x-8-inch timbers and left standing on its edge in a corner of Building 31 of the Corning Glass Works while efforts turned to the preparations for the two-hundred-inch mirror.

The 120-inch mirror had come out so well that McCauley decided not to tempt the fates with changes. The only modification in the casting procedure for the two-hundred-inch disk would be the use of hollow mold cores, which would cool rapidly with the disk, to form the pockets and ribs in the back of the disk. It would mean no more of the spectacular light-and-shadow show they had seen with the 120-inch disk, but also the end of the bubbles that had formed around the cores.

Building up the new hollow forms was laborious. Thirty-eight different forms were needed to create the complex geometry of ribs and pockets for the mirror supports. Altogether, 114 core forms had to be built up, out of 4,800 pieces of brick. McCauley derived formulas to calculate the cutting angles to build circular forms from pieces of rectangular brick. Draftsmen and engineers then worked out the exact dimensions of each piece for the mold makers, and the Corning masons, now practiced at the construction of these strange molds, cut the bricks on a modified table saw, finished them with a lathe, used a portable grinder to fair the edges of the assembled molds, and painted the inside with silica-flour coating. The task was complicated by the ultimate shape of the disk. Although it would be molded with a flat surface, the face of the mirror would ultimately be ground to a radius of curvature of 111 feet, which meant that the surface would be dished approximately 4 inches from edge to center. To maintain the proper thickness of glass over the forms, the cores had to decrease in height from 20 inches at the rim to 16 inches near the center.

To make sure there were no repeats of the floating core disasters with the 60-inch disk, McCauley increased the size of the steel rods that held the cores down. There had been only mild oxidation of the heads of the steel rods that held the cores for the 120-inch disk, so the larger rods, which would be cooled with forced air inside the hollow cores, seemed more than adequate for the job. Water cooling was rejected because a tiny leak in the mold would have produced a spectacular explosion of steam. Air cooling also permitted the use of electric thermocouples to monitor the temperature inside the cores. A full 9 inches of solid refractory brick on the top of the molds protected the ends of the anchor rods from the heat of the molten glass. The masons didn't construct a second set of mold cores this time; it was hard to justify the cost after the extras for the 120-inch mirror had been consigned to the brick wasteyard.

The 3A melting tank in A Factory had to be rebuilt for the larger pour. McCauley's calculations showed that the level in the tank would drop approximately 15 inches as the glass was removed, which would make it difficult for the ladlers to dip down to the surface of the glass late in the process. Adding new glass during the pour was not a solution; the glass was being used much faster than normal production glass, and the special 715-CF batch was slow to melt and fine. McCauley had the production department rebuild the top 12 inches of tank wall under the ladling doors out of removable refractory bricks, fastened to the tank with a steel strap. As the level of glass dropped, a workman with a torch could cut away the steel strap and knock out the row of refractory bricks (which had been beveled to be easily removable), letting the ladlers reach deeper into the tank.

These preparations, and the the sudden attention of the press and public, came just when orders for Pyrex ware and other Corning production-line products had picked up from their depression lows. While Quigley juggled the requests for tickets, McCauley had to negotiate with the production department for space for the telescope mirror project. Pours had to be scheduled when the casting would least disrupt the regular production of blown glassware, Corning's bread and butter. Having reporters and guests present, and Amory Houghton's decision to admit the public, complicated the logistics. Carpenters were put to work prefabricating an elevated walkway for the south wall of the A Factory blowing room, with doors arranged so a steady procession of employees and townspeople could see a portion of the operation. A platform with corral rails was erected on the floor of the blowing room, where the press, industry and science bigwigs, and invited guests could stand close enough to hear the roar of the furnaces and to see the glow of molten glass during the pour.

The fires for the tank were started in early February, and two tentative dates were blocked out, February 18 and 25. When McCauley discovered iron contamination in the melting tank from a ladling door frame, the tentative dates were canceled, and March 25, a Sunday, was set as the new date. Formal invitations and tickets went out, printed on colored paper to indicate the seating and standing assignments for the anticipated crowds. Principals like Francis Pease and Walter Adams were coming from California to see the pour as representatives of the Observatory Council. Corning officials who had previously worked hard to keep the project quiet invited friends to the great day. McCauley invited his former professors at the University of Wisconsin and friends from Cleveland to come with their families. The whole world was coming to Corning.

The Baron Steuben Hotel quickly filled as the newspaper and magazine reporters, photographers, and filmmakers crowded into Corning. Quigley had the assignment of finding rooms for the press and invited guests, arranging entertainment for the bigwigs, answering questions for the reporters, and providing what were not yet called "photo opportunities" for the wire service and newspaper photographers. He lined up Corning executives for interviews with the magazines and newspapers that wanted more than the press releases. Film crews, discovering that there wasn't enough light or room in the area around the casting ovens, came up with alternate topics. One crew did an entire *talkie* segment on how the 120-inch disk would be used to test the bigger mirror while it was ground and polished. Footage of the mirror, which had been brought out for public viewing, was supplemented by a talking-head blackboard session with Dr. Gage of the Optical Glass Division.

One reporter from a Buffalo newspaper arrived in Corning on Saturday, only to be told by the other reporters that the "pouring" had already begun. Convinced that he had missed the big event he was there to cover, the reporter labored to concoct an alibi, until he realized that he was the butt of a joke: The "pouring" that had begun was of a different sort. Another segment of the glass industry, bottle makers like Libby-Owens-Ford and Owens Illinois, were enjoying a sudden business revival since the repeal of Prohibition in December 1933. There was a good deal of "pouring" that night in Corning, as Glass Works managers were recruited to entertain the distinguished visitors in hotels, restaurants, and private homes.

All day Saturday the carpenters worked to assemble the platforms and walkways around the molding area, while photographers and artists took photos and made sketches of the tank, ladles, mold, kiln, controls, and the locomotive hoist that would move the disk from the casting igloo to the annealing oven. Color-coded arrow signs were put up to direct visitors with different-colored tickets. In the evening the casting-oven burners were lit to preheat the igloo to working temperature.

McCauley came in for a last check. The 3A tank was brimming with fined 715-CF batch; the equipment had been checked, tested, and rechecked. He had asked the two ladlers who had done the 120-inch disk if they wanted assistance, since this disk would be so much larger. They said they preferred going it alone. McCauley concluded that their pride was important and said they could pour the whole disk.

The Sunday date for the pour had been chosen to minimize the disruption to the production lines in A Factory. The day dawned bright and crisp, a good omen. But no omens were enough to deter the fundamentalist ministers, who read between the lines of the press releases and concluded that the choice of the Sabbath for the pour had an obvious, sinister meaning. A telescope built on the Sabbath, they preached, was clearly the work of the devil; those who would build it were Satan's agents. In pulpits where the usual targets of Communism, fornication, racial mixing, or the New Deal had begun to seem weary, satanic technology drew believers eager to prove that fundamentalism hadn't died with the Scopes trial.

But the fundamentalists, for all their clamor, were a minority. For most of the country, and especially Corning, New York, Sunday, March 25, 1934, was a long anticipated celebration of science, progress, and the Corning Glass Works. Father Lynch, the rector of the Episcopal Church where McCauley served as a warden, agreed to hold a special early-morning service on Sunday for McCauley, his family, and others who wanted to pray for the success of the effort. The reporter from *Time* magazine, hearing about the service, christened the scientist in charge of the project "Pious McCauley." McCauley got a chuckle out of that: Unknown to the reporter McCauley's father and grandfather were both named Pius.

The crew assembled at 7:00 A.M. for a final hour of rehearsal with the ladles and other equipment. The procedures were all familiar, but the visitors' galleries on the floor and overhead added a new element, and McCauley was eager to have no surprises. It was 8:00 A.M. when Charles Wilson, the chief ladler, signaled McCauley that they were ready.

The visitors' platform and gallery were already full. In the front row of the VIP platform, the orchestra section for the performance, Mrs. McCauley and their two sons, James and George Jr., shared the railing with Walter Adams; Francis Pease; Sir William Bragg, an exchange professor at Corning from England; Lyman Briggs of the Bureau of Standards; Max Mason; G. M. Chant, of the Dominion Observatory in Canada; Arthur L. Day; and A. E. Marshall, president of the Society of Chemical Engineers. Reporters and out-of-town guests with blue or red tickets shared a balcony that overhung the operation. The first of the thousands of general public who were admitted in fifteen-minute shifts stood on the gallery along the south wall; a serpentine line outside, growing longer by the minute, was routed past the building where the 120-inch disk was on display—a divertissement to keep the public entertained while they awaited their brief period inside the factory. Officers of the Corning Glass Works, including Vice President Hanford Curtiss, mingled among the guests.

The reporters saw McCauley nod. The center door on the melting tank opened, and the audience got their first glimpse of the dazzling bluish white mass of molten glass at 1525°C. Wilson, at the head of the center ladle, wore an asbestos apron and held a protective shield in place over his face by a mouthpiece clenched in his teeth. The men behind him wore goggles and gloves. There were no OSHA regulations: All the workers wore ordinary street shoes.

Wilson guided the huge ladle into the tank, signaling the men behind him to pivot and dip the long handle, which was suspended by a fulcrum from the overhead track, then to swing it up and back with its load of 750 pounds of white-hot, glowing molten glass. As the cup of the ladle emerged, the ladle cooler—wearing a jacket and tie and a tweed cap along with his protective gloves—placed the ring of his cooler around the lip and opened a valve to spray a ring of water on the lip of the red-hot ladle. The water sizzled off the hot rim, spraying front-row visitors as the ladle swung around the track toward the waiting mold. The two switchers made sure that each ladle went to the correct opening on the igloo, and a gateman stood ready to swing the gate of the igloo open.

At the mold Wilson, holding the ladle at the fulcrum point, directed it into the igloo and signaled the men behind him to turn the long handle by the crosspiece at the end, pouring the molten glass into the heated mold. As the ladle came out, a cheer went up from the crowd. Someone in the gallery started singing, "I'm looking at the world through rose-colored glasses." Successive groups in the gallery picked up the melody until it became the theme song of the day.

About a third of the glass in the ladle ended up in the mold. The rest, which cooled into a heel in the ladle on the trip from tank to mold, was broken into a waiting two-wheeled wheelbarrow. The ladle went to a cooling water bath next to the tank, while the wheelbarrow carried the cullet to the rear of the melting tank, where it was dumped into a box on a crane that returned the material to the tank. The wheelbarrow was then cooled in a water bath.

As soon as one ladle had left the tank, another started, filling at a different door of the three. Only eleven men made up the actual ladling crew, as the two experienced ladlers alternated among the three ladles and the three filling ports on the mold, filling and pouring, clearing and cooling the ladle, hour after hour, ladle after ladle. They would pour one hundred ladles of glass before they were finished. Two dozen other men worked behind the scene, tending the furnace fires, moving wheelbarrows, switching tracks.

McCauley walked around the platform, watching, taking an occasional sample of glass for testing. Spectators who recognized him extended a hand. Some applauded. He answered with a nod. Like most of the guests, he wore his fedora all day.

The procedure fell into a rhythm. Visitors who had watched a ladle poured moved on, letting others up to the front row. The public on the high gallery moved by, letting the next group in for a viewing. Copies of the *Evening Leader* were passed around by both the public and the reporters. The stories fed the crowd the superlatives that Lowell Thomas had promised: the *biggest* piece of glass in the world was being poured, the *biggest* scientists in the world were there, part of the *biggest* crowd ever to assemble at the Corning Glass Works, the *biggest* collection of newspaper- and cameramen ever assembled in that part of the world, the *biggest* "soup" spoons ever built were being used for the process, and a *big* time was being had by all. By late morning the man taking count of yellow gallery tickets had reached three thousand. The line snaked all the way around the huge factory building.

During the morning George Maltby, Hostetter's assistant, heard one of the Corning engineers say to McCauley that it wouldn't work, that the thermal calculations were all wrong. McCauley shook it off. The pour was going perfectly. The samples of glass he had tested showed no contamination. The only mishap was that Corning Vice President Curtiss, who had tried to scale one of the plank fences that had been erected as safety barriers, had fallen and broken a rib. The pouring went on without interruption until noon, when McCauley called for a lunch break. After the workers had at their lunch pails, the operation resumed. The ladling seemed so routine that some visitors who had fought hard for tickets and then for positions close to the front of the platform chose not to return from lunch at the Baron Steuben Hotel or local restaurants.

Dr. Littleton, who was doing the lab tests of the glass samples for McCauley, told him in the middle of the afternoon that there was a phone call. McCauley ducked into an office. It turned out to be a reporter from a London paper. "When did ladling begin?" the reporter asked.

McCauley, eager to get back to the factory floor, answered tersely: "Eight o'clock this morning."

"When will it be finished?"

"Early evening."

"Why are you making the disc on a Sunday?"

"To avoid interference with the normal production operations during the week."

McCauley remembered that phone call for the rest of his life. Before he got off the phone, all hell broke loose.

Littleton was waiting for McCauley at the door to A Factory. "Mac," he said. "A core has broken loose from its place in the mold." Littleton was quick to add that it was only one core, that the others were still in place. McCauley knew the cores were all fastened the same way. Every precaution had been taken with each core. If one had failed, he told Littleton, the other 113 cores couldn't be far behind.

From the days of the first failed disk, the ladlers had orders to halt for instructions if a core floated. When the ladling suddenly stopped and McCauley was seen hurrying to the mold and peering through an opened gate, rumors started in the gallery that the igloo over the mold had caved in. As McCauley emerged, his smile of the morning gone, the rumor grew. The word in the gallery was that many of the twenty thousand refractory bricks that had been used to build the "doghouse" were falling into the mold and that the disk was a complete loss.

What McCauley saw through the pour opening in the igloo was a single tapered round core bobbing on the surface of the molten glass. His mind whirled with ifs: *If* only the crowd were miraculously swallowed up. *If* only "we" hadn't been so cocky with success to think it safe to invite an audience. *If* only one core had floated up on the 120-inch disk to demonstrate the weakness of the anchors.

As he stood back from the heat of the open gate, the ifs gave way to whats: *What* will the reporters say on Monday morning? *What* damage will result to Corning's prestige? *What* actions should be taken to save face? *What* will be the effect on the decision of the Observatory Council to continue the program?

The decisions were all his. McCauley ordered the ladlers to resume ladling. He would fill the mold and salvage the disk by whatever they had to do, even if it meant laboriously grinding pockets in the back face. He quickly decided on a triage for his own time. He would need all his ingenuity to salvage the disk. The press release, and the challenge of keeping reporters and photographers away, was a problem left to the ingenuity of Quigley and Amory Houghton.

As the ladling went on, twenty-two cores broke loose, slowly floating to the surface of the semimolten glass. As each core popped loose, the vacated area would fill with molten glass. Wilson and Ruocco, the ladlers, filled and poured more and more ladles. The day stretched on, longer than McCauley anticipated. The crews were tired. Wearing street shoes like the rest of the crew, Wilson slipped and banged his head against a ladle. He was knocked unconscious, though he recovered in time to finish the ladling.

It was six o'clock in the evening before the last ladle of glass was poured into the mold. Most of the invited guests had already left. The line of more than six thousand spectators who trooped through the gallery had dribbled down to a trickle. Only those who were nearest the mold had any inkling what had actually happened.

Quigley and Houghton had already prepared press releases and interview comments, which the newspapers and broadcast reporters dutifully reported. "A trifling accident which temporarily suspended the work," the release read, "had occurred during the pouring." Walter Adams, the ranking representative from Pasadena, was quoted to the effect that the cores that had broken loose would not affect the disk. The release was issued quickly enough to be picked up for the early-morning newspapers.

McCauley knew the truth. That evening, when the workers and the crowd went home, he had to deal with with twenty tons of molten borosilicate glass that had twenty-two cores of refractory brick floating on its surface like toppings on a sundae.

When the factory finally emptied, the melting-room crew was standing by, ready to move the filled mold from under the heated igloo to the waiting annealing oven. With the crowd and reporters gone, McCauley had experienced tank men don protective asbestos aprons and face masks, and try to remove the floating cores from the surface of the disk with long-handled grappling tongs. The glass had cooled enough from the original 1500°C melting temperature to become a viscous, heavy syrup. After a few minutes in the extreme heat of the open gates to the igloo, the men with the tongs would stagger back for breath. It was impossible to remove the cores.

The next attempt was to use long iron rods to break off the tops of the floating cores. Perhaps, McCauley hoped, the remainder of the cores would be embedded in the surface no deeper than the layer of glass that would ultimately be ground off the disk in making the concave surface. The effort to break up the cores was more successful than the effort with tongs, but the surface was far from a pretty sight. This wasn't the perfect disk the telescope needed.

By ten that night the crews had broken up as much of the cores as it seemed possible to reach. The glass and core pieces stood proud of the original mold height, so work crews had to build up the perimeter of the mold with insulating brick, lest the heated cover of the annealing oven press down on the cores and further embed them in the glass. With the new row of bricks in place, McCauley allowed himself a break while the mass of glass slowly cooled down to the temperature at which the long controlled cooling of the annealing would begin. It was his first break since the operation had begun at eight in the morning.

He hadn't eaten since before the pouring began. Hostetter, who had spent the afternoon and evening with visiting VIPs, joined him at the Athens lunch counter, a popular Corning spot. The room was empty except for the waiter. McCauley, famished, ordered a roast beef sandwich.

"Sorry gentlemen, no beef."

"Ham and eggs?" McCauley asked.

"No ham."

"Bacon and eggs?"

"No bacon."

The crowds, more people than Corning, New York, had ever seen, had eaten everything. The two men settled for eggs and coffee.

When McCauley got back to the blowing room, the disk was cool enough for its trip to the annealing oven. He pushed a button and the screw hoist slowly lowered the mold and glass below the floor. Crewmen on the hand-cranked windlass moved the heavy load along the tracks until it was centered beneath the annealing kiln. McCauley thought to himself that in an hour the disk would be inside the oven, hidden from the world. The long day would be over.

He pressed the button on the hoist, the screw hoist slowly turned, and the disk started upwards. It was halfway on its journey when the overload switch opened and the hoist stopped. With the switch reset, McCauley tried lowering it a short distance, then raising it again. The jammed lift wouldn't budge. Sunday night was no time to call the Whiting Corporation in Chicago Heights, Illinois, for service.

To maintain the temperature of the disk, McCauley and the crews filled in the open space between the top of the mold and the bottom of the annealing oven with Sil-o-Cel. The temperature controllers for the annealing oven were set to hold the interior at 500°C to postpone the cooling as long as possible. It was all they could do. McCauley and the exhausted crews went home.

The early editions of the morning papers were already out. The big headlines and photographs of the pouring had crowded the usual stories about Mussolini's alliance with Austria and Hungary, and Japanese troops in Manchuria, off the front page.

He had told them not to wait up, but McCauley's whole family was there when he got home Sunday night. They had read the late edition of the *Evening Leader* and heard a stream of rumors about the disk. They remembered how ebullient he had been in the morning, and through most of the day. A quiet man, he had been thrust into the limelight and had enjoyed it. Now his face was gray with exhaustion and frustration, as he sat down at the dining room table where he had worked so many evenings while his children did homework.

"I know what happened," he said. "We just have to see that the bolts won't do that."

18

Salvaging Hopes

Even before Walter Adams and Francis Pease brought home the news that the two-hundred-inch mirror had been successfully cast, the barrage of publicity from Corning galvanized Pasadena. No one had been willing to admit it openly, but ever since the dark days of the GE experiments, much of the project had been held back—partly out of fear that the telescope *couldn't* be built and partly to avoid the inevitable publicity and questions that would follow each major decision for the project. After the hoopla in Corning, it was inevitable that questions would be directed to Pasadena: How long will it take to finish the telescope? How far will the telescope see? Where will the telescope be sited? What will it look like? Who will get to use it?

Hale and Anderson had half answers for the reporters' questions. When Hale had said in 1928 that the telescope might take five or six years to finish, the astronomers and engineers close to the project thought him excessively pessimistic. Six years later, if pressed, Anderson and Hale would suggest 1939 as the date when the telescope would be finished, but the date was made of whole cloth. The real answer, which reporters wouldn't welcome, was that they didn't know how long it would take to grind a mirror, to build a telescope, or to put together an observatory. They hadn't announced where it would be sited because the Observatory Council was reluctant to sign leases or even announce a site decision until they knew they could build the telescope. Fundamental questions about the design of the telescope hadn't been agreed on, and no plans had been made for the mount and observatory, because the Observatory Council and the various design committees hadn't been convinced that they would be able to build the telescope. And while the original proposal to the IEB had specified that the Carnegie Institution would work closely with Caltech in the design and ultimately in the operation of the facility, until they were reasonably sure that the telescope would be built, no one from Caltech or the Observatory Council wanted another round of negotiations with John Merriam.

It would be months before anyone knew how the disk would emerge from the annealing process, and whether it could be salvaged, but the Corning officials and Walter Adams were convinced that the success of the ladling process meant that a mass of glass as large and as complex as the two-hundred-inch disk *could* be cast. Hale wrote to Arthur Day expressing his gratitude and admiration for Corning's achievement: "If the mirror proves a success, it is not likely to be repeated very soon." Privately McCauley, Houghton, and others at Corning assured Hale that if the disk couldn't be salvaged, Corning would pour another.

With the disk seemingly assured, it was time for decisions.

The committee on site selection had been accumulating data for years. John Anderson, the de facto chairman of the committee, used his own old-boy network to call on the geology department at Caltech for help.

Geology at Caltech had begun with a Carnegie Institution program, headed by Harry O. Wood, who had worked for Hale's National Research Council during the war. Anderson collaborated with Wood to develop a seismometer sensitive enough to detect small local shocks. Charles Richter (of Richter Scale fame) and Göttingen-trained seismologist Beno Guttenberg later came to Caltech as the core of a geology faculty, and the new Department of Geology absorbed the Carnegie Institution Seismological Laboratory. When it was time to assess potential sites, Anderson turned to his friends at the Geology Department, who were using the successor instruments to the seismograph he had helped develop.

A team of geologists, headed by J. P. Buwalda, assessed the sites on the shortlist and reported that Horse Flats, north of Mount Wilson, was too close to the San Andreas Fault. They found no problems with Table Mountain, east of San Bernardino, or with Palomar. The long-term seeing-test data from Table Mountain and Palomar were roughly equal, but Hale had favored Palomar from the beginning, even when there were some indications that under the best of conditions, the seeing at Palomar did not equal the remarkable seeing at Mount Wilson.

In March 1934, Hale, Adams, and Anderson traveled up to Palomar with Dan Tracey, the forest ranger at Mount Wilson, in the Pierce-Arrow touring car the Mount Wilson Observatory had bought before Einstein's first visit. Hale had been up the mountain before, and Anderson had been several times to check the monitoring equipment on the mountaintop. This time the three of them were looking for a location to site the biggest telescope in the world.

The mountain had changed since 1903, when Hussey wrote the report that captured George Hale's imagination. Once a source for timber to build missions, by the turn of the century Palomar, within a few days' stage of Los Angeles or San Diego, had become a popular summer resort, with seasonal hotels and a tent city that blossomed each summer in Doane Valley. The trail that ascended the west shoulder of the mountain, called the "Nigger Grade"* by the locals, and only later renamed Nate Harrison Grade, took a full day for a team, so there were halfway camps where travelers could rest before the ascent. Those who survived the deerflies in June, mosquitoes in July, and no-see-ums in August could enjoy the unspoiled splendor of acres of summer ferns and azaleas; meadows of dense grass, wildflowers, and butterflies; and the forests of big-cone spruce, white fir, and California black oak.

The automobile brought distant resorts in range of a one-day drive from Los Angeles and San Diego, putting an end to the Palomar resort era, although Bailey's Palomar Lodge held on. A few intrepid souls liked to challenge the switchbacks and hairpin turns of the Nate Harrison Grade in automobiles. The grade took its toll in gears and overheated engines. The real challenge was the downhill run. The favorite technique was to tie a large tree to the rear bumper and drag it downhill; the right-size tree would provide the perfect brake. At the bottom the local Indians took the trees for firewood. It wasn't a road built to haul telescopes up a mountain.

By the 1930s cattle ranching had taken over much of the mountain. The Mendenhall Ranch, on the southern side of the mountain, was the largest. Mendenhall had been around long enough to work through breeds, from mixed-breed, white-faced "California cattle" to Herefords and finally to polled Aberdeen Angus steers that thrived on the high-meadow grass. To the northwest, a sheep-farming effort by a Frenchman named Foussat had failed, leaving behind the name French Valley. Locals attributed the failure to locoweed. There were smaller ranches, including some failing efforts at apple orchards. William Beech and his wife kept a cabin high up the slope in the French Valley. Beech had run a weather station and monitored some instruments for Caltech for five years.

Adams, Anderson, Hale, and Tracey found the Nate Harrison Grade reminiscent of the old toll road on Mount Wilson. From the grade up over the top and down the other side to the Oak Grove Ranger Station, they looked at potential sites for an observatory, finally settling at a place near the center of the plateau, at an altitude of 5,500 feet. From the site they could see north to the San Bernardino

---

* The grade was officially named after Nathan Harrison, a black man who called himself "the first white man on the mountain." Harrison had come to Palomar as a slave in 1848, to work at a mining claim in Rincon. No one knew how he survived year-round on the mountain; his only apparent source of income was tips from travelers grateful for the buckets of water he brought for their struggling teams. "Uncle Nate" died in 1920, at the age of 107.

peaks and south as far as the Coronado Islands off the coast south of San Diego. At night, the loom of light of Los Angeles was barely visible on the horizon. San Diego, though closer, was also scarcely visible. It was hard to imagine that the light of either city would ever be a problem. The Caltech geologists came back to map the character of the terrain and the underlying geological structure and to determine the exact latitude of the site. The engineers would need the latitude for the final design of the equatorial telescope mounting.

If there had been any doubts about Palomar as a site before that last trip, there were none after. The Observatory Council agreed that the telescope would be sited at Palomar and gave Henry Robinson the task of negotiating to buy up the needed land. Word of the plans got out to the ranchers on the mountain before Robinson made his initial approach. Mendenhall in particular, who had heard about the budget of the project, was set to hold out for an enormous sum, despite the depression plunge in land prices.

In 1931 forty square miles of land next to the site, including some private land, had been declared the Cleveland National Forest, part of an effort by the National Forest Service to control forest fires. Caltech wanted a portion of the National Forest land for the observatory. Anderson and Robinson explained their plans, and Guerdon Ellis, the forest supervisor, was enthusiastic about Caltech's proposal: "Inasmuch as I can conceive of no higher use to be made of these 40 acres of land than that which you request, the use of the land is practically assured the Observatory forever."

The bargaining with the local landowners went on all summer. San Diego, eager to have the research facility in its county, agreed to build an all-weather road up the slope to the site of the observatory. The final deal was signed on September 21, in the Beeches' weather-beaten cabin on a slope of the French Valley. Five men, representing Caltech, the ranchers, and San Diego County, sat around an old-fashioned wooden table. An early fall storm was raging outside, so they worked by the light of a kerosene lamp, until the lamp ran dry and they had to use candles.

The agreement, signed at 3:00 A.M., provided that when Kenneth and William Beech were handed a check for $12,000 that was being held in escrow in San Diego, 120 acres from different ranches would be transferred to Caltech. The government would transfer an additional 40 acres, making the observatory site a total of 160 acres. San Diego County would begin work on a road up to the site at the earliest possible time. The ranchers had already agreed between themselves that the Mendenhalls would continue to have grazing rights on the mountain. Cave C. Couts, a descendant of an early pioneer on the mountain, and the Beeches, provided the rest of the land.

The San Diego newspapers lapped up the story, from the storms and kerosene lamps to the announcement that "with the closing of the deal Southern California was assured a scientific institution which will comprise one of the wonders of the world." By the end of the year a Civilian Conservation Corps (CCC) camp had been built in the Doane Valley, west of the proposed observatory site, to house the workers for the County Road. San Diego had already named the road the Highway to the Stars, and proposed changing the name of the mountain to San Diego Mountain. Outraged citizens wrote poems to the Oceanside newspapers demanding that the old name stay. John Anderson, on behalf of Caltech, pointed out that since Palomar was the name in use on topographic maps, changing it would create confusion. San Diego gave in and agreed to leave the name alone. Instead they started printing new publicity maps for the county, with the site of the observatory prominently marked.

In 1934 there was still no design for the telescope, only a collection of specifications, ideas, and requirements.

From the earliest talk about a big telescope, Pease had favored a fork mount, like the mounting on the sixty-inch telescope. The advantage of the fork mount was simplicity: Because the tube of the telescope was held only at one end, the fork permitted the telescope to point to all of the sky overhead. Pease had designed most of the one-hundred-inch telescope. His early drawings and model were the departure point for the project, and the models that were shown at the National Academy of Sciences and brought out for curious reporters were based on his drawings and early models. As late as February 1931, when a reporter asked about the design of the telescope, Hale answered that "the fork type [mount] . . . seems the most promising."

But there were problems with a fork mount. If the telescope tube were supported only at one end, while the other end carried the weight of auxiliary mirrors, mechanisms to interchange the mirrors, and a cage big enough for an astronomer—achieving the required rigidity in the tube became an impossible engineering task. At the tolerances required for the telescope, even a massive tube of steel girders behaves like a hollow noodle cooked *al dente*. Hold it at one end, and the other end droops. Making the forks long enough to support the tube in the middle would substitute drooping forks for a drooping tube. The fork mount also presented problems for the use of some of the alternate foci astronomers had requested, and the concentration of the entire weight of the telescope on a single set of bearings troubled the engineers.

There were two alternatives, which pivoted the tube in the middle. Early on, Hale had urged consideration of the so-called "German" mount, made popular by Zeiss in the telescopes they had designed and built. The German mount—also called the "Victoria" mount after a 72-inch telescope in British Columbia—put the tube of the telescope on an extended shaft, with a counterweight on the other end of the shaft. It was popular for small telescopes, but a counterweight would effectively double the load on the bearings for the tube of the telescope, making the mount impractical for a telescope as large as the two-hundred-inch. The other possibility was a form of yoke mount like the one on the one-hundred-inch telescope. The yoke could be supported at both ends, dividing the bearing load for the telescope. The disadvantage of the yoke was that it blocked the tube of the telescope from dipping low enough to view the North Polar region of the sky. Hubble and Humason were already chafing at the inability of the one-hundred-inch telescope to photograph or take spectra of galaxies in the north polar region. No one wanted to build another telescope with that limitation.

Russell Porter, who had never before worked on a large telescope, had played with some maverick ideas. In the telescopes he had designed for the Stellafane amateur astronomy center, and in drawings he had done for the "Amateur Telescope Maker" column of *Scientific American,* Porter had come up with telescope designs unlike anything anyone had ever seen before. Some telescopes kept the observer in a fixed location indoors, with the moving parts of the telescope outdoors. One design, his Garden Telescope, looked like a decorative sundial, with parts of the mounting done up in Art Nouveau scrolls.

As early as 1918 Porter proposed a split ring for the north end of a yoke. The split ring would allow the tube of the telescope to be lowered enough to view the polar region, while still dividing the weight of the telescope between bearings at the two ends of the yoke. Porter patented the idea in 1918, before he began work on the two-hundred-inch telescope. When he got to Pasadena, he began sketching ideas for a mounting for the two-hundred-inch telescope, using variations of his split ring.

Porter, who was older than the others, and whose hearing difficulties left him the odd man out at meetings, had a hard time convincing people that his ideas deserved attention. He was an outsider, without the formal academic credentials and big-telescope experience of the Mount Wilson and Caltech staff on the project. He had also been brought in by Hale at what others in the project considered a too-high salary. Porter's initial appointment was for a year, and his official design assignments were small projects like the telescope used to measure the seeing at sites, and architectural details of the Astrophysics Laboratory and Optics Lab on the Caltech campus. Frustrated, Porter threatened to go back to Vermont. Hale encouraged him to stay on. When others ignored Porter, Hale would ask him privately to sketch up various designs. Porter's memos were pencil sketches and handprinted notes rather than the typed memorandums with eight carbon copies that characterized most of the official communications of the project.

By 1934 the chief design question was whether the telescope should have a fork mount or some variation of the split-ring design. Porter's sketches, amalgams of his own ideas and the suggestions from brainstorming sessions, gradually evolved into a giant ○ or horseshoe, with the outside of the horseshoe serving as a bearing surface to support the telescope as it turned in right ascension to match the motion of the earth. Francis Pease, who probably came up with the idea independently, had used a form of split ring in one early sketch he did of the three-hundred-inch telescope that had been talked about in 1921. As the design group settled on the horseshoe, Pease was generally credited with the design. Porter, angry that he was not given proper credit, became even more isolated from the others working on the telescope.

With the main design agreed on, it was time to move from sketches to engineering drawings. From the earliest days Hale and Anderson had turned to the Caltech engineers for advice and designs. Theodor von Karmann in aeronautical engineering and Romeo Martel in mechanical engineering, stars of their departments, and the mathematical physicists Harry Bateman and Paul Epstein had all had been consulted on the final design for the ribbed back of the glass disks. They were asked to recommend both personnel and ideas for the telescope mounting.

Romeo Martel recommended a young engineer named Mark Serrurier, who had come to work for Caltech in November 1932. Engineering jobs were tough to find in 1932. The aircraft companies were at a standstill, and public authorities were cutting back to limited maintenance programs. The design program for the telescope soon had its pick of young graduates, eager for the chance to work on a major project. Serrurier had been eager to work in Southern California. The movie industry would have been his first choice, but he was willing to take any work he could get. He did some consulting on the new Golden Gate Bridge in San Francisco. The new telescope seemed an exciting project. Serrurier was too young to realize how impossible the assignment was.

Even with the fast focal ratio of $f/3.3$, the main telescope tube for the two-hundred-inch telescope would be fifty-seven feet long and weigh close to 250,000 pounds. Early in the design process Hale wrote: "The first point is to study the tube, which should be *extremely* rigid and capable of carrying . . . the heaviest attachments without injurious flexure." Dr. Ross, the designer of the corrective lens for the prime focus of the telescope, calculated that effective use of the lens required that it be fixed within 0.040 inch (approximately one millimeter) of the optical path of the telescope, and that the position of the lens could not move more than 0.020 inch during an exposure. What this meant in engineering terms was that for the duration of a photographic exposure, typically one hour, the alignment of the tiny lens and a mirror fifty-seven feet away could not vary by more than a half-millimeter— all while the enormous telescope tube that held both in alignment was moving to track an object. To complicate the design, the light path for the Coudé focus required a slit on one side of the telescope tube, so the light could be bounced to an auxiliary mirror.

In his engineering courses or his work on the Golden Gate Bridge, Serrurier had been given design problems for skyscraper and bridge frames that were allowed to sway by feet. Quick calculations showed that the truss sections and cross-bracing in his engineering texts wouldn't work for the telescope tube. Nor would the massive riveted-girder construction that had been used for the sixty- and one-hundred-inch telescopes. If the telescope tube were massive enough to hold its alignment by brute force, the ends would droop under their own weight, and the telescope would be so massive that it would be impossible to design bearings and a drive mechanism that could move it smoothly.

Day after day Serrurier sat in his office in the basement of the astrophysics building, playing with rulers and pencils to model the stresses on a tube structure. No idea worked, until one day Martel came by and suggested that the tube didn't really have to be so rigid that the mirror and the lens at the focus would never move in relation to one another; what mattered was that they remain *aligned* within a half-millimeter. Imagine, Martel suggested, a tube that sags at both ends, but in such a way that the ends remain precisely parallel to one another. The lens at one end would still be perfectly aligned with the mirror at the other. In engineering terms Martel had switched the problem from flexible stress to sheer stress. Serrurier listened, sketched the idea on paper, then went back to work with his pencils and rulers. The solution wasn't immediately obvious, but at least he was off dead center.

Sinclair Smith had joined Anderson and Pease as a part-time Caltech employee. Smith was a brash young researcher in physics and astronomy, until he spent a year at the famed Cavendish Laboratories in Cambridge, where James Chadwick and his associates were probing the secrets of the atom during the 1920s. He returned "much matured," and Anderson recommended him for the lab staff at Mount Wilson. Smith had worked with electronic detectors and controls, so his special project became the drive mechanisms for the telescope.

Large telescopes like the one-hundred-inch and the sixty-inch were equipped with clock drives that turned the telescope synchronously with the sidereal rotation of the heavens, so that a star would appear to stand still in the telescope during a long exposure. Astronomers had long known that the apparent motion of the heavens is not simple. The rotation of the earth is not perfectly regular, and the apparent motion is also affected by atmospheric effects, gravitational influences from the moon, and other factors which could be modeled in equations. Smitty took on the challenge of duplicating not just the simple rotation of the earth but the other minuscule motions, so the motion of the telescope would come as close as possible to mimicking the apparent motion of the heavens, minimizing the demands on the observers for hand-guiding the exposures.

He had the assignment of finding out all he could about photosensitive devices, time standards, servo controls, and other devices that might be useful to control the telescope. Like others who had pledged their time to the project, he soon discovered that the telescope could be all-consuming. Although Caltech was paying only half his salary, the project was eating up most of his time, cutting into the time he had free for research. And even as Anderson brought the engineers into the project, some basic and seemingly insoluble problems remained.

First the rough engineering calculations showed that the bearing for the split-ring horsehoe would have to support a weight of close to 1 million pounds—five hundred tons. The big telescopes on Mount Wilson had used drums of mercury and floats for their polar axis bearings, which produced a smooth motion along with a then-unrecognized danger to the staff from the quantity of mercury. The weight and size of the two-hundred-inch telescope precluded that solution. The consideration of other choices—ball bearings, roller bearings, or a plain bearing—were frustrating enough that discussions of the bearings, Anderson recalled, "usually resulted in a headache." The bearing for the horseshoe, which would carry most of the 500,000-pound weight of the moving portion of the telescope, would be the largest journal bearing ever built. The load would deform any kind of ball bearing, and while Pease put roller bearings into his drawings, even his own calculations showed that roller bearings would require some 22,000 foot-pounds of torque to overcome the friction and move the telescope. Moving a telescope against that much friction would require huge electric motors; the resultant vibrations would be impossible to isolate from the telescope. The weight of the horseshoe would also ultimately distort the rollers, which would then introduce wobbles into the motion of the telescope.

Even if a bearing could be built, no one was sure that a structure as large as the horseshoe could be built with the rigidity the telescope required. Pease and Porter weren't engineers, but their preliminary calculations showed that no matter what material or structure was used to build the horseshoe, the open end would deform slightly as it turned from one extreme to the other. A sag of one-sixteenth inch in the forty-six-foot diameter horseshoe would be enough to leave the telescope axis unacceptably misaligned.

Still, the horseshoe design solved so many other problems that the unresolved issues were shunted aside. Caltech was a cocky institution.

The engineers and scientists who worked on the project, some with little commercial experience or background in the strict hierarchy of an academic department, weren't easily daunted by challenges that stodgier souls would call "uneconomical," "impractical," or "impossible." And after the apparent success of the mirror casting in Corning, even Hale, Adams, and Pease—who had lived through the birthing pains of other big telescopes—were confident that the two-hundred-inch telescope *could* be built. Their optimism was contagious for the young scientists and engineers who worked on the project. Getting a job during the depression was an achievement. When the job was at Caltech, working on the biggest scientific project ever undertaken, with a committed budget, it was hard not to feel that there was nothing you couldn't do, including solving problems that the books and experts said were insoluble.

John Merriam, at the Carnegie Institution of Washington, had been chafing at the publicity given the project ever since the grant for the two-hundred-inch telescope was announced. Every time a newspaper or radio report appeared based on the GE press releases, Merriam wrote George Hale to protest that proper credit had not been given to the Mount Wilson Observatory for their contributions to the project. Hale or John Anderson would write back, explaining that they had nothing to do with the GE publicity and that they had tried repeatedly to persuade GE *not* to publicize their work on the project. Merriam brushed the fine points aside. He knew only that adequate credit hadn't been assigned to *his* institution, the Mount Wilson Observatory.

Mount Wilson—still the site of the largest operating telescope in the world—had stayed in the news. Edwin Hubble, with his tall stature, tweedy attire, omnipresent pipe, and affected British accent, was an ideal subject for the newsmagazines. He liked the attention, liked to be photographed with visiting celebrities, from Albert Einstein to movie stars, and the magazines liked to run photographs and stories on what they called Hubble's "law" of the expanding universe. The favorable publicity wasn't enough for John Merriam.

By early 1934, after Lowell Thomas's broadcast made the telescope a household word again, Merriam was so angry at what he saw as slights of his institution that he appointed a committee "to give study to questions touching cooperation with California Institute in furtherance of the 200-inch telescope project." Merriam named Walter Adams as chairman; the members were Hubble and Seares from the Mount Wilson staff, Fred Wright from the Carnegie board, and "Dr. Hale if he desires to associate himself with the committee." Merriam had a way with words that could provoke even a peacemaker like George Hale.

Merriam pulled other strings as well, sending Arthur Day, the director of the Carnegie Institution Geophysical Labs in Washington and a vice president of the Corning Glass Works, to talk to Max Mason at the Rockefeller Foundation. Day explained that he was concerned that the original spirit of the grant for the telescope was not being observed. The casting of the mirror seemed to increase the confidence of the Caltech people, and Day worried that the telescope might not be open to anyone else. Mason, reluctant to interfere, agreed that overlapping staffs of the Mount Wilson Observatory and Caltech were a good idea—a harmless concession, since Anderson, Sinclair, and Pease were already overlapping—and that Day should keep him informed if it appears that the "spirit" of the grant was not being observed. Mason marked his memo to the file on the meeting PERSONAL lest it fall into other hands and start a stir.

Merriam's committee was a sham. He knew what he wanted: "As the greater weight of authority relative to matters that touch questions of astronomical study and operation lies with the Carnegie Institution, I assume that this contribution by the Institution may at least equal in significance the use of funds available to California Institute and the contribution made by California Institute in the general scientific and engineering sense." Merriam's concern was the same as it had been in 1928, when he almost aborted the project: He wanted credit, especially *public* credit, for the Carnegie Institution.

George Hale was too tired to fight. His health was faltering. With the mirror apparently successfully cast at Corning, the design process in full gear, the site picked, and negotiations underway for the actual observatory site, he was marshalling his energy, selecting which meetings he could attend, eager to see the project through to completion. He was frequently fatigued, more often than not confined to his dark room, reluctant to waste time on matters that did not contribute to the progress of the telescope. He had begun jotting autobiographical notes, recalling his childhood and early years.

There was also a minor scandal at the Mount Wilson Observatory that Hale wasn't eager to publicize. A bookkeeper, James Herbert, had embezzled $13,000 by forging Adams's signature. When Adams reported the theft to Hale, Hale reported that Herbert was an "untrustworthy thief" who had previously stolen securities entrusted to him, and that he had only silenced the earlier episode out of concern for Herbert's wife and children. Herbert wasn't prosecuted because Merriam feared that "radical elements" in California would pillory the Carnegie Institution for paying a bookkeeper so little as to drive him to theft.

When Hale heard that Merriam had been lobbying Max Mason, he took another train ride to New York to meet with Mason. Hale explained once again the circumstances that had led to the grant, Merriam's effort to derail it, the actions of Elihu Root, and the overruling vote of the executive committee that had finally forced the matter. Merriam, he explained, was not *opposed* to the telescope. He just wanted to "proceed along different lines." Far from urging stronger cooperation between Mount Wilson and Caltech, he had actually objected that Adams, the director of Mount Wilson, was "too close" to the Caltech people.

When he got back to Pasadena, in April 1934, Hale answered Merriam's charges in a sharply worded statement, pointing to the minutes of decision making by the Observatory Council; the de facto membership on the council of Adams, the director of the Mount Wilson Observatory, and his record of attending every meeting; the fine working relationship between the Mount Wilson staffers on the project and the personnel at Caltech; and the use of lenses and other equipment developed as part of the two-hundred-inch project on existing telescopes at Mount Wilson, which had greatly contributed to the research of Hubble and others. "In the future," Hale wrote, "unless effective cooperation between C.I.W. [Carnegie Institution of Washington] and C.I.T. [California Institute of Technology] is prevented by the action of C.I.W., the advantages to C.I.W. will greatly increase." Elihu Root congratulated Hale privately on the statement and the progress of the project, commiserating that "an atmosphere of disgust, suspicion, and personal dislike and resentment" will stop the best work anywhere.

With the depression giving no signs of abating, with 13 million unemployed, with the pitched hatred of the weekly radio broadcasts of Father Charles C. Coughlin and Gerald L. K. Smith drawing even bigger audiences than *Amos 'n' Andy*, and with members of the American Student Union on college campuses openly demonstrating for Communist causes, the Observatory Council's concerns about publicity for the project were exactly the opposite of Merriam's craving for public recognition and credit: "In periods of social unrest subversive tendencies may at any moment make their appearance; and in a crisis there might not be time to convince an ignorant public that large sums of money flowing from the great foundations have always been used to useful public purpose."

John Merriam had been too consumed with institutional pride to see what was happening outside his Washington office.

19

Revelation

At the end of May 1934 the two-hundred-inch disk was cool enough to view. George McCauley wasn't willing to predict what he would find when he lifted the cover of the oven.

After the crowds finally left Corning the night of the pouring, it had taken McCauley and the Corning millwrights and electricians, with advice from the manufacturer, three days to change the lubricants on the screw hoists and retune the linkages so they were able to lift the disk into the annealer. By then the temperature of the disk had fallen to 300°C. McCauley raised the temperature to 530°C, at which annealing normally began, held it for thirty days, then set a rapid cooling schedule, dropping the temperature of the glass by eight degrees Celsius per day.

With the reporters gone and the attention of the fickle public shifted back to Hitler's newest threats against Austria and the trial of the accused Lindbergh kidnapper Bruno Hauptmann, McCauley enjoyed the breathing room. He knew the cooling schedule he set, approximately ten times faster than the annealing schedule they had calculated for the disk, wouldn't produce a satisfactory disk, but with orders for other large disks backed up, he was anxious to free up the annealer. The rapid cooling was an experiment.

When he opened the annealing oven, the surface of the two-hundred-inch mirror disk was an ugly mask of scars and chunks of refractory brick, the remnants of the floating cores. McCauley wasn't concerned about the surface. What mattered was the quality of the glass, how well it had annealed. With a polarimeter and a light source, McCauley tested the disk, shining the light source through the glass and measuring the residual strains in different sections of the disk. The strains were approximately ten times what he would have expected for a properly annealed disk. McCauley eagerly reported the good news to Pasadena. If the brief, rapid annealing had reduced the strains to that point, it meant that the full annealing schedule would produce a disk free of strains.

Enjoying the freedom from outside attention, McCauley began a salvage effort. His tool was a long chisel that had been fabricated in the Corning blacksmith shop. Another man worked alongside him with a portable sandblasting machine. They wore breathing masks and goggles, but after hours of digging and blasting, their ears and hair were filled with fine refractory dust. It took most of a day to dig and blast the remnants of the brick cores out of the surface of the disk. They then vacuumed the surface clean, and the disk was put back into the casting oven and slowly heated for five days, until it reached a temperature of 1100°C, hot enough for the surface to remelt and gradually smooth itself. When the last traces of the broken cores were gone, the disk was again consigned to the annealing oven for another annealing cycle.

It had taken everyone a long time to unwind from the mad whirl of the casting day. A few reporters still called, and Hostetter, flushed with the attention of that day, had begun intercepting calls, even queries from Pasadena, to answer them himself. Although he had little to do with the technical work on the disk, Hostetter was quick to use the first person plural, or even the singular, in claiming credit. He told the producers of the "March of Time" that he had supervised the pouring, and made sure their broadcast scripts were filled with references to "Dr. Hostetter." McCauley silently fumed. He wasn't a man to complain, so friends and even family assumed his sullenness and immersion in work was his way of accepting the responsibility for the accident with the cores.

For months McCauley worked late at the dining table, reviewing the design of the mold anchors and studying how they could have been built. He went over tables of heat-resisting alloys. When he found a promising alloy, he had Ralph Newman test a sample by heating it in a laboratory furnace. There had been no decision to cast another disk, but McCauley was determined to have the designs and procedures right. He hadn't been cautious enough before. That wouldn't happen again.

A young draftsman named Walter Smith got the assignment of drawing up new rod designs and mold sections. Smith had come to Corning with an EE degree from Ohio State to take a job as an electrician's helper. His mother and grandmother liked Pyrex pie plates; they thought Corning sounded like a great place to work. Smith was glad to have any job. He had worked on the mirror project earlier, laying out the routing of the monorails that carried the heavy ladles from the glass tank to the mold, then had gone back to the usual design jobs, working on machinery that would be used to mold baby bottles, pie plates, or laboratory flasks. With the crowds gone, the mirror project was another job. Smith had been working on some design problems for a glass dance floor when he was asked to work on the mold anchors. The only difference was that the drawings for the mirror project were done to the unusual accuracy of $\frac{1}{128}$ inch to impress the need for accuracy on the mold makers.

The high-temperature alloy McCauley had selected for new anchor bolts, Calite B-28 from the Calorizing Company, was too hard to shape by milling or forging. Since the anchor rods could not be threaded in a machine shop, Walter Smith drew up plans for rods that would be cast with hooks on the end to attach to the frame that supported the mold. McCauley wanted springs on the anchor shafts to maintain tension as the anchor bolts expanded and shrank in the heat of the mold. The final design needed two springs on each shaft. Walter Smith looked at the specifications and concluded that auto valve springs would do the job, saving the time and money of a special order. He found the smaller ones off the shelf at Rhodes Buick, the local dealer. No nearby dealer had larger springs in stock, so Smith rummaged for what he needed at Maxx Russ Auto Supply, an infamous junkyard where new, overstock, and used parts—some of dubious origin—shared the boxes and bins. The final design was exactly what McCauley wanted—cautious, overbuilt, using reliable stock parts wherever possible.

In late September 1934, two months after the disk had been returned to the annealing kiln for a second cooling, it was again cool enough for inspection. The surface scars were gone now, but there were fine checks in some of the lower edges of the ribs. McCauley attributed those to a too-rapid reheating when the surface was remelted; the heat had not been maintained long enough fully to melt and heal the lower extremities of the disk. The clear face of the pale yellow Pyrex showed the missing core holes. Grinding those pockets into the disk would not be an easy job, and the checks in the adjacent ribs were a reminder that a slip of the grinder could fatally crack the disk.

In the relative quiet of the spring and summer, McCauley and his colleagues weighed the question of whether to pour another disk or to salvage this one. With the new engineering work they had done, McCauley estimated that the chances of successfully casting a new mirror were about the same as the chances of repairing the damaged disk. Neither was certain. In a process as complex as the pouring of a mirror disk, something could always go wrong; the larger the casting, which meant the longer the process, the more likely an accident. The same caution applied to grinding out the pockets in the back of the first disk; a workman with a grinder might discover the *fatal* check in a glass rib in the last pocket he ground.

Whichever disk they ultimately used, McCauley had to face the problem of removing the disk from the mold, cleaning away the mold cores from the back, and crating it for shipment to California. The lack of experience at maneuvering a twenty-ton disk of glass suggested that having an expendable dummy to practice the various operations would substantially increase the chances of delivering a usable disk in one piece.

In the end, the decision was too important for anyone except Amory Houghton to make. Houghton concluded that the cost of making a new disk wasn't that much greater than the cost of an attempted salvage, and that it was much in the interest of Corning's reputation as a company to supply a disk to the exact specifications of the Observatory Council rather than a salvaged disk. The Observatory Council agreed. Hale wanted this telescope to be as perfect as they could make it. No formal announcement was made in Corning or Pasadena, but McCauley quietly began plans to cast a second mirror. In addition to the new mold anchors, there would be one other change in the procedure: They would cast this mirror without an audience.

On the riverbank behind A Factory, the Corning millwrights erected a new structure of steel columns and beams, smaller but similar in construction to the steel building GE had built for its work on fused quartz. Four ten-ton hand-operated chain hoists were suspended from the beams. A careful crew could use the lifts to maneuver forty tons of glass, steel, and brick. At the end of September workmen used the hoists to lift the first two-hundred-inch disk from its mold onto a support of heavy timbers, left over from the dismantling of a section of railroad trestle. There, the two-hundred-inch disk, the beautiful geometry of its back ruined by the filled-in pockets where the cores had broken free, was locked away from prying eyes.

The 3A melting tank of 715-CF Pyrex had been used through the summer to cast other large mirrors, including an eighty-two-inch mirror for the McDonald Observatory of the University of Texas, auxiliary mirrors for a large telescope Heber Curtis was planning at the University of Michigan, and two sixty-inch mirrors for Harlow Shapley at Harvard. Curtis and Shapley, whose debate had inspired the two-hundred-inch telescope project, both visited Corning, inquiring about the progress of the mirrors, and prying into secrets about the two-hundred-inch mirror.

By October the Texas and Michigan disks had either been consigned to the smaller annealing oven or sealed away in Sil-o-Cel awaiting a free annealing oven. The big casting oven was cleaned out, the steel flooring of the big mold was rebuilt to accommodate McCauley's new mold anchors, and the masons went to work building another jigsaw puzzle of bricks to form the pockets and ribs. There had been no publicity about the decision to cast a new mirror, but rumors had been rife. The new official policy was that the company would talk openly about what *had* been done but make no arrangements for exhibitions or news statements about future plans. Leon Quigley, in the publicity office, did what he could to deflect queries, but with Harlow Shapley carefully leaking what he knew to the press, it soon became impossible to deny the rumors. Houghton and Quigley also thought there should be some documentation of a successful effort, so that the only record of the Corning participation in the project wouldn't be photos and stories of the failed pouring.

In late October, Quigley got permission to bring a Pathé News team to the factory with their cameras and power plants. Pathé agreed in advance that they would release no film or printed statements until after a new disk had been successfully cast. The gaffers and crewmen strung heavy cables throughout the blowing room at the west end of A Factory and positioned their cameras to focus on the tank openings. McCauley, trying to be accommodating, watched and explained the various steps in the pouring process. When the equipment was finally ready, the director gave the shout, "Action!"

"What action?" McCauley asked, explaining that the fires hadn't been lit in the tank.

One of the cameramen, eager for some "motion" for his motion pictures, asked to have the mold under the casting oven raised and lowered a few times.

McCauley shrugged and pressed the start button for the screw hoist. The cameraman shouted, "Faster!" McCauley explained that the screw hoist was operating at its full speed of two inches per minute. The cameramen looked at one another and at McCauley. From the newspaper reports of the earlier pour, they expected something more exciting than a heavy platform moving so slowly the motion would be imperceptible on film. McCauley suggested that the horizontal motion of the mold platform from the casting oven to the annealing oven might be a little more interesting, but the film crew was still disappointed. The director discovered the special building outside where the first two-hundred-inch disk rested on its support of trestle timbers. He suggested that the four hoists could be used to move the disk with enough speed to show up on the film.

McCauley refused. He wasn't willing to risk unnecessary movement of the disk for publicity. The compromise was that the Pathé cameramen would run their cameras while the workmen *pretended* to move the disk. Then with the hoists disconnected from the disk, the cameras would take footage of the chains moving up and down. After a day of filming, the crew left with little to show, and McCauley could settle down to the real business of pouring another disk.

At the end of October the 3A tank was refired, and the men and equipment were reassembled. The only change to personnel in the pouring crew was the addition of two experienced ladlers from the Blue Ridge Glass Corporation of Kingsport, Tennessee. McCauley had wanted to use an extra crew of ladlers for the first two-hundred-inch disk. Wilson and Ruocco, who had ladled the glass for the earlier disks and were proud of their work, insisted that they could handle the longer workday for the much larger disk, and McCauley had yielded to their arguments. But fatigue had stretched the pouring time for the ill-fated disk and contributed to a minor accident. This time McCauley wanted another crew to share the ladling. The stakes were high enough to risk hurt pride among the workers.

The second disk was poured on Sunday, December 2, 1934. There was no special church service that morning. A tiny group of guests was present, including Heber Curtis, A. G. Ingalls of *Scientific American*, E. P. Burrell and Henderson from Warner & Swasey, Dr. & Mrs. Francis Pease, and twenty-five representatives of the press, including Howard Blakeslee, science editor at the Associated Press, and William L. Lawrence of the *New York Times*. Max Mason, who had missed the first pour, came up to watch, along with Warren Weaver from the Rockefeller Foundation. George Hale had been invited as well but was in another of his periodic bouts with the demons. He was well enough, however, to fire off memos urging that the audience be limited lest its presence again cause problems.

The operation went like clockwork. Hickman and Harris, the new ladlers, filled the ladles with molten glass; Wilson and Ruocco emptied them into the mold. The rest of the crew, now practiced at the operation, opened furnace and oven doors, manipulated the switches on the overhead monorail track, sprayed the ladle rims, filled spray tanks, emptied ladle skins into the special wheelbarrows, and returned the skins to the tank. A total of fifty-six men—carpenters, tinsmiths, laborers, construction workers, millwrights, pattern makers, pipemen, masons, electricians, and foremen—worked on the operation. Their hourly wage ranged from $0.40 for laborers to $1.25 for the experienced mason who rebuilt the openings to the glass tank during the pour. Joe Ruocco, the skilled ladler who poured the glass into the mold, was paid $0.48 per hour for eight hours of work, without a lunch break, that day. His share of the $6 million project was $3.84.

By two o'clock in the afternoon the work was done. The cores had stayed in place. The two visiting ladlers spent the afternoon seeing the sites in downtown Corning before they went back to Tennessee.

That evening the disk was transferred to the annealing oven and sealed in. Although the pouring had not been publicized and wasn't open to the public, everyone in Corning knew what was happening. As a signal that all was well, the Glass Works whistle went off at 11:35 P.M.

This time the annealing schedule was not an accelerated test but the real thing. After an initial stabilizing period at 500°C, McCauley began the actual cooldown on January 21, 1935. The optimum annealing schedule required that the disk cool by exactly 0.8°C each day for ten months. A crew of attendants was assigned to work eight-hour shifts, twenty-four hours a day, seven days a week, manning the controllers for the nichrome heaters in the kiln. Every three hours one of the ten controllers had to be reduced by 1°C, an operation that took approximately ten seconds. The attendants had nothing else to do.

McCauley lectured the young men who were hired that if they missed even one adjustment of the controllers—each attendant faced a maximum of three ten-second operations in an eight-hour shift—they would ruin the entire disk. Knowing they might guess that a single missed adjustment would not imperil a twenty-ton mass of glass for the entire annealing period, McCauley drove or walked down to the 3A tank every morning to check the dial settings and thermocoupler readings on the control panel. On the weekends he would take his young children with him, stopping off before or after church on Sundays. He never missed a single day. He often would leave notes on the dials, reminders to the attendants of the importance of the precious object in their care.

The melting tank and ladling crew stayed busy pouring auxiliary mirrors for the two-hundred-inch telescope, including some tricky oval mirrors for the Coudé focus and a special disk for the Bell Telephone Labs. By midsummer the telescope mirrors had all been poured, and the fires in the 3A tank were extinguished. McCauley was pleased that they had received an order from a "blue chip" company like Bell Telephone in addition to the telescope orders, and that the procedure had become so routine that they could confidently pour mirror after mirror. But no matter how routine the operation became, he couldn't stop worrying that something would still go wrong, that some accident, some unpredicted disaster, would befall the twenty-ton mass of glass sealed away in the dome-shaped annealing oven. Checking the controllers each morning was his way of fighting off the anxiety. He was a winemaster, making daily visits to his cave, reassuring himself that his slowly maturing, priceless treasure would emerge intact and perfect enough for the telescope.

Left alone the astronomers and engineers might have gone on refining their designs for the telescope forever. By late summer of 1935, with work on the site underway, George Hale concluded that it was time to start building a telescope. From the beginning, he had assumed that Warner & Swasey, with its vast experience, would build the telescope. The company had kept in close touch with each step of the design, and had worked on the models that were exhibited at the National Academy of Sciences. But as the initial sketches began turning into engineering drawings, it became clear that Warner & Swasey did not have the facilities to build the huge telescope mounting. Indeed, few companies did.

Hale had turned to a shipbuilding yard for the mounting of the one-hundred-inch telescope. He tried again for the two-hundred-inch, asking Homer Ferguson, the president of Newport News Shipbuilding and Dry Dock Company and a member of the Board of Trustees of the Carnegie Institution, if he would be interested in the job. Newport News had pioneered welded construction for ships. The project Hale had in mind would be the largest precision-machined welded structure ever built, but Ferguson declined, citing the backload of the shipyards and observing that the welded steel structures Hale wanted were so large that if they were built in the West, the only place with facilities big enough to anneal them was at Boulder Dam. Ferguson couldn't recommend a company that could undertake the work.

Hale was too ill to manage the construction of the telescope himself. He also knew that there was no one on the staff with the experience and time to get it built. John Anderson would have his hands full supervising the figuring and polishing in the optics lab, when the mirror disks arrived. Francis Pease and Russell Porter had no experience supervising large-scale construction. And Walter Adams had a full-time job running Mount Wilson. The engineers on the project, like Mark Serrurier, were too young and inexperienced to undertake the overall management. Hale turned to Max Mason, at the Rockefeller Foundation, for advice.

Without hesitation Mason said he knew the best man in the world for the job, although he doubted that they could recruit him. Mason had met Clyde S. McDowell at the University of Wisconsin, where McDowell, already a Navy officer, had gotten his doctor of science degree. During World War I, McDowell had been in command of the Naval Experiment Station in New London and had recruited Mason to do research on acoustic devices for antisubmarine warfare. After the war McDowell had gone on to supervise the navy yard in Samoa and then become yard manager at Pearl Harbor, working his way up to the rank of captain. During the postwar disarmament, he had been a strong advocate for both applied and theoretical research, arguing that the war had been "won by *research.*"

Hale trusted Max Mason's advice. In October he and Mason met with Captain McDowell, who was then stationed at the Navy Office of Inspector of Machinery at the New York Shipbuilding Corporation in Camden, New Jersey. McDowell was quick to offer his suggestions for the telescope project. From what Hale and Mason told him, he thought the design was not far enough along to put out to contract. Navy practice, he explained, was not to enter into contract for any portion of construction until the general features were agreed on. Hale, impressed with McDowell's quick grasp of the problems of building the machine, and his no-nonsense, take-charge manner, asked McDowell if he would be willing to serve as coordinator and planner of the construction if he could be temporarily released from his navy duties. McDowell said he would be interested if the details could be worked out.

In less than a week McDowell wrote that he was willing to retire from the navy, passing up the chances for promotion to admiral, to take charge of the building of the telescope. He asked for an "honorarium" of $12,000 per year, substantially higher than any salary paid at

Caltech. Hale balked, offering $8,000 and traveling expenses. McDowell held firm, and Mason finally had to persuade Hale to consider comparable commercial salaries and the fact that it wasn't a salaried position but only a short-term assignment for McDowell.

Within a month McDowell moved into temporary quarters at the Atheneum, the Caltech faculty club. He was introduced around as "Sandy," but there was no mistaking his background. He immediately began sending out flawlessly typed memorandums, with multiple carbon copies, to everyone working on the project. Oral authorizations, he noted in an early memo, might prove expeditious, but "in all cases such action must be confirmed as soon as possible in writing." The memorandums were followed by organizational charts. On McDowell's charts only Hale and Anderson were above him. Everyone else reported to McDowell.

Sandy McDowell was ambitious. He had had a promising career in the navy, with a good chance for promotion to flag rank, but the Bureau of Ships, in a period when isolationism and minuscule defense budgets reduced American shipbuilding to a standstill, wasn't enough. In New London during the war he had directed a staff of the brightest physicists and mathematicians he could locate, working with virtually unlimited wartime budgets. He had done a good job in New London, forcing reluctant scientists to meet deadlines, making decisions that others feared, and ultimately getting the job done by producing workable submarine detection devices before the end of the war eliminated the need.

Building the two-hundred-inch telescope, the biggest scientific project in the world, was a challenge McDowell couldn't resist. He hit Pasadena like a whirlwind, firing off a barrage of memorandums, asking each person on the project for timetables, status reports, and flowcharts of project progress. He wanted clear lines of reporting, channels of communication, and specified procedures for every eventuality.

Francis Pease, who was in Corning for the pouring of the second disk when McDowell arrived in Pasadena, was astonished at the changes in the working atmosphere in Pasadena when he returned. Before McDowell came, the various project committees used their meetings, usually once a week, to catch up on formal reports and to prepare minutes for Hale, who missed most of the meetings. The real work took place in informal bull sessions, in an office, a lab, or under a tree with box lunches in the partially landscaped area outside the buildings on California Street—wherever two, three, or more scientists and engineers could get together to bat around ideas before they went back to their calculations and drawing boards. Now, suddenly, McDowell wanted formal minutes at every meeting. He would sometimes ask each person present to report what they had worked on during the week. Pease, who had been through the birth pains of a large telescope and had worked on this one for a decade and a half, didn't like the forced efficiency of the new regime. He never got over his discomfort with McDowell.

By January 1935, barely a month after he arrived in Pasadena, McDowell was back on the East Coast. In a whirlwind trip, he saw Max Mason at the Rockefeller Foundation, Vannevar Bush at MIT, Hannibal Ford of the Ford Instrument Company, GE officials in Schenectady, Warner & Swasey in Cleveland, Hostetter at Corning Glass, and Bethlehem Shipbuilding in Quincy, Massachusetts.

McDowell searched out the contacts he had developed in his navy years. Vannevar Bush had done research on automatic guiding mechanisms for guns, and Hannibal Ford had built servo controls for gun turrets. The problems of mounting and guiding the telescope, for McDowell, were only a variation of the problems of controlling naval guns. John Anderson had to point out the difference: In guiding an astronomical telescope, the guide star might be as faint as the fifteenth magnitude, visible only in a large telescope, and stars, which had to be held on the slit of a spectrograph for hours, really weren't comparable to the targets of the naval guns.

McDowell wasn't surprised by Anderson's response. He was used to what he thought of as the hair-splitting of scientists. He saw his responsibility as cutting through the fine distinctions, urging, and even forcing mutual comprehension between the scientists and real-world engineers.

One of McDowell's earliest moves was to hire a retired army man from Pasadena, Col. M. L. Brett, to serve as resident superintendent at Palomar. Brett, a West Point graduate and former ordnance officer, had been a personal aide to the secretary of war in 1918. He was an asthma sufferer and eagerly accepted the position because he thought the mountaintop would be good for his health. His brother was a general who had served in Southeast Asia, and Brett did his best to carry on the family tradition, setting up his headquarters at the William Beech cabin east of the observatory site with his personal "striker," a young man who cooked, kept house, and polished the knee-high boots the colonel wore each day.

The conditions on the mountain were primitive. There was no telephone, an old gasoline-driven generator for power, and all supplies had to come up by the rugged Nate Harrison Grade. Communications were via mail and visitors from Pasadena. Even so, Brett didn't have to advertise for labor. In 1935 an unconfirmed rumor of work was enough to prompt men eager for jobs, or with goods to sell, to hitch a ride up to the old Nate Harrison Grade to the peak to seek out the colonel. A young man named Ben Traxler, whose radio repair shop had failed in the depression, got a ride up the mountain with a hardware merchant eager for a new account. When Traxler mentioned his radio experience, Brett suggested the possibility of opening an amateur radio station on the mountain for regular communication with Pasadena. Traxler knew that an amateur license couldn't be used for regular traffic, but he needed a job enough to agree with the colonel's ideas. Two weeks later he was told to report to the mountain with long underwear, warm bedding, and any tools he would need.

A few of the newly recruited workmen slept in the back of the colonel's cabin. The others slept in a bunkhouse nearby, called the "Boar's Nest," with an attached kitchen and mess hall. Traxler tried a tent his first night, thinking he would enjoy the privacy, until he discovered the cold of nights on the mountain. By morning he had moved into the bunkhouse.

Each morning Colonel Brett lined the workmen up in a semblance of an inspection, so he could march back and forth in his freshly shined boots and sharply pressed trousers, barking orders for the day's work. The orders from Pasadena arrived in a steady stream. McDowell expected the work on the mountain to run like a Navy SeaBee operation, and he had time for even the smallest procedural details: All mail from Palomar to Pasadena was to go in locked mailbags, to be unlocked at Palomar by Colonel Brett and at Pasadena by Miss Gianetti; spring faucets were to be used in all showers and the washhouse to conserve water; meal expenses were to be held under $1.25 per worker per day. Some of the crew cleared brush from the eventual building site with a small Caterpillar tractor, others started on digging and pipe laying for a well with a holding tank in the lower valley. The plans called for a pump station to lift the water three to four hundred feet to a 1-million-gallon tank and water tower, initially for the concrete work on the observatory and eventually to supply the observatory.

While Colonel Brett's crew started on site work at the top, a road crew from San Diego started on the new Highway of the Stars up the mountain. The flurry of New Deal alphabet agencies had provided for road projects like this one, and word soon got out that Basich Brothers, the contractors, would establish a WPA camp to house the roadworkers. To an old cattle farmer like George Mendenhall, the whole New Deal smacked of Reds and Communism; the WPA was nothing less than government-funded bands of revolutionaries. "Owing to the general lack of control over these men, I am forced to refuse permission for any of these men to cross the French Valley at any time," he wrote in protest.

The problems with Mendenhall were smoothed over and by August 1935 everyone was ready to celebrate the beginning of work. George Hale received an invitation to the ceremony but was too ill to attend. Everyone else—Walter Adams, John Anderson, Francis Pease, Sinclair Smith, and Mark Serrurier—from Caltech and Mount Wilson enthusiastically trooped down to Palomar. Mendenhall, Beech, and some of the other property owners on the mountain came up for the festivities, joined by civic officials from San Diego and the reporters they were able to drag along. Poor Hale, bedridden in his darkened room, had to rely on reports from others. "I attended with considerable pleasure and amusement the party yesterday on the top of Palomar Mountain," Adams wrote. "It was very much of a love feast."

20

Swept Away

The old A Factory at the Corning Glass Works sat close by the bank of the Chemung River, hemmed in by the tracks of the New York Central Railroad. It was a beautiful site. The river was wide enough to afford a perspective across to the hills in the distance and swift enough to keep itself clean. From the riverbank, a visitor could see the hillsides above the factory where so many of the Corning employees lived, and the Monkey Run that ran down off the slope into the Chemung. The river was part of life in Corning. Its changes signaled the seasons, from the low water of late summer to the wild flow of spring runoff. Old-timers could read the river the way an experienced sailor can read the sea.

From time to time the river misbehaved. When the spring runoff reached a peak, the Chemung sometimes overflowed its banks. If that weren't enough, after a sustained period of rain the Monkey Run, which carried runoff from the hills into the river, would surge over its channel, and the floodwaters would find their way into A Factory, extinguishing furnace fires in the cave level and leaving behind a layer of mud that would take weeks to clean up. The worst flood anyone could remember had been in 1918. There were still marks on the walls of the factory caves to show the height the waters had reached.

McCauley took some ribbing for his excessive caution, but when the controllers and transformers for the annealing oven for the two-hundred-inch disk were installed, he had made sure they were on a raised platform several inches above the high-water mark. The entire casting and annealing setup was designed so that the only equipment below the high-water mark were the bases of the four lifting screws, which were shielded with watertight steel cylinders, open at the top, and filled with oil to lubricate the screws and protect them against the inevitable dust from the factory cave. Even without the protective shields, the hoist screws would only be vulnerable to flooding in the lowered position, when the disk was being moved into or out of the annealing oven.

For most of June 1935, the carpenters and millwrights had worked on a crate for the mirror. The shipment plans had been debated in Corning and Pasadena. Hale's initial preference was shipment via the Panama Canal, but track clearances on tunnels and underpasses were too low to move the disk to a dock in New York, Baltimore, or Philadelphia, which left Albany as the only seaport they could reach from Corning. Even if they could get a direct cargo ship from Albany, via the Panama Canal to Los Angeles or San Diego, shipping by sea would involve the extra handling of the disk at Albany and a West Coast port, in addition to the initial loading onto a railcar at Corning and the final unloading in Pasadena.

The alternative was to ship the disk by rail from Corning to Pasadena. The 120-inch disk had been shipped successfully across the country by rail in a crate, shielded and supported by heavy timbers. A 10-foot-diameter mirror disk was a large cargo, but it required no special routing or handling. The two-hundred-inch disk was another matter. Even if it were suspended vertically in a well car, so that the bottom of the disk would be only inches above the tracks, the crated two-hundred-inch disk would require tunnels and overpasses with a height of eighteen feet. The normal east-west routes through Kansas City didn't have the clearance. The New York Central Railway authorities thought it *might* be possible to find a route through Chicago or St. Louis that would accommodate the disk, though it would take months of planning by their schedulers and the schedulers of other rail lines.

The crate builders, told that their cargo was close to irreplaceable, that it had to fit under tight railroad bridges and tunnels, and that it had to be movable by crane, spent months engineering a crate. Here, too, the Caltech engineers got into the act, sketching a metal drum, constructed of half-inch and quarter-inch boilerplate, reinforced with heavy channel and angle sections. The American Bridge Company, in Elmira, was engaged to fabricate the larger sections from hot steel. Corning millwrights and carpenters fabricated the balance of the crate from sheet and angle steel, with felt, rubber, and cork cushioning designed to hold the disk rigidly. As each section of the complex crate was finished, it was test-fitted to the original two-hundred-inch disk in the temporary steel building on the riverbank.

Besides the crating work, the summer seemed quiet in Corning. There were still back orders for disks, but with the two annealing ovens occupied with the two-hundred-inch disk and an 86-inch mirror blank for Heber Curtis at the University of Michigan, McCauley directed all efforts at smaller mirrors that didn't need the big ovens. In addition to some specialized mirrors for solar telescopes, the publicity around the pouring of the big disk had generated an unexpected flurry of orders for one-tenth-size replicas of the two-hundred-inch mirror. Caltech wanted one for a mockup of the telescope to test designs. The other orders were from individuals and institutions that apparently reasoned that what was good enough for a two-hundred-inch mirror was good enough for a

20-inch one. In fact the ribbed back, necessary for the bigger mirror, served no purpose on a smaller one, except to present a headache to the Corning engineers: The tiny mold cores were too small to enclose metal mold anchors. McCauley concluded that they would need a "sculptor" to carve the mold out of solid blocks of insulating brick.

The first week of July saw steady rain in Corning. The ground turned sodden, and the normally tame Monkey Run filled to its banks. Some puddling showed up here and there in the caves under A Factory, but it was no more than the usual sporadic flooding from spring runoff. McCauley was glad he had taken the precaution of building the annealing oven, where the precious two-hundred-inch disk was halfway through its annealing schedule, well above the highest high-water mark. Still, his family noticed his nervousness as the rains continued. It was only at the end of the week, when the rain tapered off, that he relaxed again.

Then, on Saturday, the rains returned to south-central New York State. The downpour was steady and heavy, from early afternoon through the night. On Sunday morning McCauley's son Jim woke up in his cabin in the Finger Lakes to cries of "Help!" from a neighbor whose boathouse had floated away. The lake had risen six feet overnight. In Corning the Monkey Run became a raging torrent. With the ground of the hills above the glassworks saturated, the runoff coursed down to the river below, overflowing the banks of the Monkey Run. The Chemung overflowed the dikes along the bank in front of the factory and covered the roadbeds of all three bridges over the river. It was the worst flooding in forty years.

By Monday morning, when the early shift showed up for work, the factory yard on Walnut Street was a shallow lake. To get into the factory McCauley had to skip from one "island" to another. He ran down to the caves, where the controllers and thermostats for the annealing ovens were installed on their raised platform.

Water was rising on the floor of the caves, coming in from the overflow of the Monkey Run to the south. On the north end of the factory, closer to the river, the water was rising even faster. A quick inspection revealed that the river had risen over the drainage ports in the concrete dikes; the holes that normally drained the caves into the river were instead bringing the river into the caves. By midmorning the entire floor was covered with water, and the level was rising rapidly.

McCauley got the factory foreman to issue an all-hands call for men to build a restraining wall of sandbags, brick, and concrete to isolate the section of floor below the controllers and transformers for the annealer. With dozens of men working, the wall went up quickly. But as the double line of sandbags cut off the flow across the floor, the hydraulic pressure of the rising water in the rest of the factory forced jets of water up through cracks in the concrete floor under the annealer and controller.

An emergency call to the Corning Fire Department brought men and pumpers. When someone on the phone mentioned that the effort was needed to save the two-hundred-inch disk, the fire department sent every pumper it had, filling the passageway outside the west side of the factory with red fire engines and a maze of black hoses. Alas, there was no place to pump the water except back into the river. Even with every pumper the fire department could supply, they couldn't get ahead of the steadily rising water. Down on the floor, electricians, masons, laborers, and scientists worked side by side, standing in water up to their knees as they wrestled with sandbags, bricks, and mortar. McCauley remembered the Pathé news cameramen a half year before, who had been so anxious for action shots.

All Corning's men and equipment were no match for the floodwater. As the water level marched steadily toward the high-water mark left from the Great Flood of 1918, McCauley knew the battle was lost. The only hope was to cut off the electrical power to the transformers before they shorted out and somehow move them to dry ground and reconnect them in time to maintain the annealing schedule on the glass disk. The disk was down to 370°C. If the power were cut off, the insulation of the annealing oven would limit the rate of cooling for a short while. How long they could go without power without introducing strains into the glass—twelve hours, twenty-four, forty-eight—was anyone's guess.

The laborers who had been building walls of sandbags were put on air hammers on the floor above the transformers. Other workmen cleared away the production equipment for molding nursing bottles in the factory area above the transformers while the hammers pounded at the heavy reinforced concrete floor. The call went out for more jackhammers, but rising water in the compressor room cut down the air supply, limiting the number of hammers they could keep in operation. The water in the cave began creeping up the sides of the transformer cases. If the water reached the lids, it would flood the interior of the transformers, displacing the oil and rendering them useless. Replacement transformers would take weeks, maybe months, to order. During that time the disk would cool uncontrollably.

McCauley had the laborers concentrate all the jackhammers on a single hole. A crane was rigged overhead, ready to lift the transformers to safety. In the cave below, electricians on the scaffolding disconnected the reactor units that controlled the flow of current to the oven. They got sixteen reactors out before the water was too high to work. Just over their heads, as many men as room permitted hammered at the concrete floor. As soon as a man tired on a jackhammer, another took his place. No one had to be reminded of the consequences of failure. Everyone there had seen the world rush to Corning to witness the casting of the great glass disk. This was the piece of glass that had made Corning famous.

In the midst of the chaos, McCauley did some rapid calculations and concluded that one transformer could supply enough current to complete the annealing schedule; there would be no margin for error and no backups, but it would just work.

The water was lapping the lower edge of the transformer lids when the hole in the floor was finally large enough to lift the first transformer to safety. The bottom edge of the lid was already wet. A few more minutes and the water would cover the top, saturating the transformer. A cheer went up as the crane hoisted the first transformer clear. Through the hole in the floor, men could see oil seeping out of the other two transformers as the water lapped over the lids and flooded the windings inside.

When the first transformer—the only one they had rescued in time—was lowered to the floor, McCauley saw that the oil gauge on the side had been broken off. A sample of oil drawn from the hole was contaminated with water. The other two transformers were already under water.

It was morning outside. McCauley and many of the workers had worked all night.

All day Tuesday reporters called Corning, eager for stories about the fate of the big disk. McCauley, too busy to take calls, told the publicity office that there had been no damage to the disk, that in fact the water had never been within five feet of the disk. That was all he said, and all they printed.

On the factory floor, laborers suspended the waterlogged reactors and controls outside a heated glass tank to dry. Even the one salvaged transformer couldn't be used as it was. McCauley had a pump hooked up to pump the oil out of the bottom of the transformer case through a centrifugal separator that would separate off the water; the oil was then pumped back into the top of the transformer.

McCauley finally went home Tuesday evening. He had been working without a break for thirty-six hours. Too anxious about the disk to sleep, he worked with his little drafting board at the familiar round table. He guessed that it would take at least another twenty-four hours before the reactors, controls, and transformer were dry enough to use. It might be as much as another thirty-six. In seventy-two hours without power, the temperature of the disk would drop seventy degrees, to 300°C. Would the glass survive that interruption to the annealing schedule?

All day Wednesday, McCauley went back and forth from his calculations to the cleanup and salvage work. No one in his family had ever seen him so jittery. It took forty-eight hours to get the two flooded transformers out of their position. By then the water in the caves was receding, draining the transformers. Plant electricians estimated that they could be dried with current applied to the primaries, but it would be a long, slow operation.

The question of what to do with the annealing schedule was

McCauley's to answer. There were no precedents for restoring an annealing schedule. At best it was a black art, as much guesswork as science. The only thing clear was the consequences. Too much heat or too little, heating too fast or too slow, could leave strains in the glass, destroying the usefulness of the disk. McCauley didn't say much about his real worry: Could the disk survive the sudden drop in temperature intact? Or would they open the annealing oven, in November, to find the glass cracked?

Thursday, when the transformers and reactors were dry, and power was restored, McCauley raised the temperature of the disk to 370°C, where it had been before the power was interrupted. He had decided to hold it there for five days, then resume the original schedule. His new schedule was a guess. All McCauley knew for certain was that they had lost eight days from the annealing schedule, and that after six months of controlled cooling, dropping by a carefully measured 0.8°C per day, the temperature of the disk had dropped seventy degrees in three days.

For the next five months, every day when he checked the thermocouples on the oven, he thought about the flood and wondered about the disk hidden inside the annealing kiln.

In October the eighty-six-inch disk for Heber Curtis at the University of Michigan was cool enough to come out of the smaller annealing oven. The oven had gone through the same power outage and sudden cooling as the bigger oven with the two-hundred-inch disk. The disk emerged intact—it had not broken during the power outage—but it was so filled with glass faults that McCauley classified it R.R.R. (reject requiring replacement). Curtis wasn't disturbed by the news. He suggested that they make the replacement a ninety-eight-inch disk.

Curtis, who would get a bigger disk, might be sanguine. McCauley wasn't. The eighty-six-inch mirror was made from the same batch of glass, in the same 3A tank, that had been used to cast the two-hundred-inch disk. The glass had looked fine when they poured the first two-hundred-inch disk, but in August, when the tank was cooled off after the last of the big mirrors had been poured, the cullet that was broken out of the tank contained streaks of devitrification. The streaks seemed to mark the outlines of the ladle skins—suggesting that the glass that had been returned to the tank during the ladling, to keep the glass level up so the subsequent ladling efforts could reach the molten glass, had been contaminated with some impurity. What if the impurities also affected the two-hundred-inch disk?

By mid-October, the natural cooling rate of the insulated annealing oven finally matched the rate that had been maintained by the electric controllers. The disk was cooling itself without help. The day-and-night guardians of the controls were moved to other jobs, and the power was turned off. On October 17, McCauley lowered the screw hoist carrying the disk one and a half inches, increasing the rate of cooling. The day of unveiling was rapidly approaching.

After the disasters of the early publicity, McCauley knew he had to· explore the unknown alone. On October 25, a Friday, he waited until all office and laboratory personnel in A Factory had gone home for their dinners, then went down into the factory cave alone. With the heaters in the annealing oven turned off, the cave area was dark, so he carried a flashlight. At the disk he stood motionless, listening to make sure he was alone in the cave, before he pushed the hoist's down button. When the disk had lowered just enough for his slim 150 pounds to fit, he stopped the hoist and crawled up onto the still uncomfortably warm glass, keeping his thumb off the flashlight button until he was entirely inside the oven.

When he switched on the beam, he saw a strip of clear, solid glass. The disk hadn't broken.

Then, as he moved the beam of the flashlight, the light came back in scattered reflections, as if from broken glass. "No!" he thought. "Not this one, too!"

McCauley crawled toward the reflections. The surface of the disk looked as if some Paul Bunyan had embedded a huge crowbar several inches into the glass while it was still soft, then waited ten months to pry it loose. His mind raced, wondering what could cause scars like that. Could it have been the power failure during the floods? Was the disk ruined?

He scanned the disk with his flashlight, looking for more wounds, then crawled over the surface toward each broken reflection, supporting himself on his knees and left hand, while he traced over the glass with the flashlight in his right. He found a second, a third, a fourth— nine in total. He sat down at the ninth scar, exhausted. Years of planning, two pours, almost a year of annealing, and it came to this. He felt as if his life had come to an end with nothing accomplished.

As he sat on the disk, his eyes roved upward. The cover of the kiln had been fabricated to hold a layer of insulating brick over the surface of the disk, matching the insulating qualities of the mold underneath. Above the scar in the surface of the glass, he could make out a welded junction of beams in the cover. Could this have been Paul Bunyan's crowbar?

The physicist began calculating. When the molten glass disk was raised into the oven, the heat of the glass would make the side of the cover closest to it expand. The layer of insulating brick would shield the upper layer of the cover from the heat, and the uneven expansion would make the cover bend downward toward the glass, perhaps far enough for the steel beams to touch the surface. The explanation *almost* fit. But there were only three welded areas on the lid. How would they account for nine scars on the disk?

The physics was easy. Common sense took longer. McCauley finally realized that in his anxious search for scars he had crawled around the disk three times. He had "found" the scars on each lap, magnifying the problem by three. Small consolation; even three scars, in what was supposed to be a perfect disk, were three too many.

As he maneuvered around to climb out of the kiln, he saw the head and shoulders of Ralph Newman peer under the oven cover. Newman, who had worked on the mirrors from the beginning and had every right to be in the kiln area, was surprised at the vehemence of McCauley's order to maintain "absolute silence" about what he had seen.

With the kiln closed, that night McCauley thought through the options. It was clear that the Observatory Council should not be expected to pay for another casting. The telescope project was already behind budget and schedule, and as much as the Corning Glass Works had reveled in the attention of the world when they cast the disk, McCauley wasn't eager to preside over the announcement of a Corning Glass Works failure. A second option was to reheat the disk to level its surface, then reanneal it. But the experience with the first disk, which had been reheated to smooth the rough surface left by the floating cores, hadn't been encouraging. The reheating had introduced fractures to the ribbed supports. And reannealing meant another year in the oven, with the risks that entailed.

There was one other option. The disk had deliberately been cast deeper than required. What if the surface were ground down far enough to remove the surface scars? McCauley calculated that the resulting disk would be thinner than the original plans had specified. Would it be too thin? The only evidence on the rigidity of mirrors with ribbed backs was based on much-smaller mirrors. The test results from Pasadena had all been good. The mirrors held their shape better than the designs had anticipated. The two-hundred-inch mirror would have active mirror supports in the pockets in the back, helping it hold its shape against gravity. If the mirror supports worked, the thinner disk might even be better, because it would react faster to temperature changes. As he calculated and considered the alternatives, McCauley was convinced that grinding the mirror down was the only choice.

McCauley wasn't the only one thinking about the mirror. After the orgy of publicity at the pouring of the first disk, the telescope, and astronomy, had become a regular topic for newspapers, radio broadcasters, and the weekly and monthly magazines that supplied news and entertainment to Americans. Reporters, prompted by their tickler files, called constantly for progress reports. "In December, you said the annealing would be finished in October. How soon can we see the disc?" Every potential disaster, like the flood, brought another round of calls from reporters who hoped for a fresh angle on the mirror.

In the depths of the depression, the telescope had become a staple of the news, a symbol of scientific and technological progress. When Edwin Hubble's search for larger red shifts led to the idea of an expanding universe, journalists seized on both the connections to the famed Albert Einstein and the loose ends in Hubble's evidence. Few reports on Hubble failed to assure their readers that the great two-hundred-inch telescope, still under construction, would solve the dilemmas. The humorists, too, had made the telescope project grist for their columns in magazines like *The American*. After the "minor problems" with the first casting were announced, Robert Benchley sarcastically explained to his readers that what had gone wrong was that "40,000 pounds of glass is too much glass." The telescope disk became a theme for Benchley, who picked up on each announcement of another mirror being cast at Corning: "If you ask me, they have got started making gigantic glass lenses up at Corning and can't stop. And California is being made the sucker."

All summer and fall McCauley, usually an easygoing man, had seemed tense to his colleagues and family. He took his wife and children down to see the disk but didn't seem enthused. He had never been a man to discuss what troubled him, and his wife and children assumed he was anxious about the disk, or bruised by the barbs of Benchley and others who took the disk to task.

In fact McCauley enjoyed Benchley's barbs. He would buy each issue of *The American* and read the tidbits out loud to his family and friends. Though he did not relish the thought of Benchley, or of the other journalists who circled Corning like vultures, seizing on the news of the flaws in the surface of the disk, what really worried him was closer to home.

McCauley had never been greedy for fame. He was a quiet, churchgoing man with a strong sense of justice. Ever since the world had descended on Corning to witness the pouring of the first disk, J. C. Hostetter, the director of development and research at Corning, had been taking personal credit for the successful disk. Hostetter was not a scientist. His own role was more that of a product manager; his style and background, as a New York–club sort of fellow, suited him well to the job. Hostetter invited reporters to interview him, directed the Corning publicity office to route queries to him, and held court during the two castings as if they were *his* operations. In public statements, and in stories passed along by his assistant George Maltby, Hostetter did all he could to leave the impression that *he*, more than anyone else, was responsible for the achievement of casting the disk.

By late summer the bitterness between them was strong enough that McCauley began avoiding Hostetter. The friendship between their families evaporated. In a small city like Corning, where people share the same church and social activities, the situation became uncomfortable.

In October, when McCauley reported the scars on the disk at a secret meeting with Amory Houghton, Houghton asked McCauley to show Hostetter the disk before a final decision was made. Hostetter had been sending periodic reports on the disk to George Hale, and Houghton was planning a round of discussions, like the initial talks at the University Club that had begun the Corning work on the telescope. The important element, all three men agreed, was that the Observatory Council and the Rockefeller Foundation should be informed quietly, before newspaper reporters sensationalized the story.

The University Club meetings were not to be. Early Friday morning, November 1, an earthquake rumbled through the Finger Lakes area of New York State. It was close enough to be felt in Corning, and by midmorning reporters were telephoning and telegraphing the press office at the Corning Glass Works for news on whether the disk had been harmed. Leon Quigley assured them that the disk was fine and had not been affected by the tremor. When persistent reporters began asking, "How do you *know* the disc is all right?" it was clear that the world, beginning with the Observatory Council and the Rockefeller Foundation, had to be told the state of the disk.

Although Hostetter had been eager to have all earlier reports about the disk come from him, this time he deferred. On Monday morning Amory Houghton, the president of Corning, told George McCauley he was going on a business trip. He was to buy a ticket only as far as Chicago. No one in Corning, not even his family, was to know the final destination.

21

The Journey

Marcus Brown grew up on his father's chicken farm in the California foothills. Like most farmers he was a good mechanic, and he eventually got a job driving and repairing a delivery truck for the Mount Wilson Observatory, where he made a good salary for the 1920s—one hundred dollars a month. Much of his job was picking up instruments and optical devices at the Optical Labs on Santa Barbara Street and driving them up the tricky mountain roads to the observatory, so he spent a lot of time in the lab, watching the opticians and workmen as they patiently labored over the glass-grinding machines. Brown— friends called him Brownie—had little formal education, but he was patient and careful with tools. As he watched the men work on the glass machines, he decided that polishing glass was exactly the right work for him.

John Anderson interviewed him. A job in the optical shop would mean a big cut in pay, Anderson explained, and Brown would have to start as an apprentice, taking orders and doing the lowest work in the shop. Undeterred, Brown took the job. His first tasks were routine and boring. He would clean up around the machines, or stand over a disk for a full shift, slowly pouring a slurry of carborundum and water onto the edge of a glass disk while the grinding machine turned under the direction of a more senior worker.

Whatever the task, Brown was enthusiastic. Many men, lured by the appeal of steady indoor work, took jobs at the Optical Lab only to discover that they couldn't stand the routine, the confinement in a windowless room, the painfully slow process of grinding glass. Brownie seemed to thrive on it. He had unlimited reserves of the essential temperament for an optician—*patience*. Hour after hour he seemed not to notice the droning noise of the machines, the occasional screech of abrasives against glass, the progress that often couldn't be measured after even days of work. Where most men were too eager to move from one step to the next, rushing the switch to finer grits of carborundum or beginning the final polishing stages before the last residual scratches had been ground off the surface, Brownie was the consummate craftsman, never rushing the glass.

Anderson was a good judge of men. Before long he assigned Brown responsibilities of his own, including the grinding and polishing of a twelve-inch mirror. Brown did a good job. He couldn't get enough of the optics lab. At night he studied optics on his own. On his days off he built a grinding machine from water pipe and the insides of an old thresher from his father's farm. He was one of those men who had found his calling. Brown made no secret of his ambition: He wanted to grind the mirror for the two-hundred-inch telescope.

In the optics lab as at the observatory, there was a distinct gap between the men who worked on the machines, and the trained opticians—often astronomers—who designed and tested the mirrors and lenses. But Anderson took notice of the curly-haired man with glasses. After Brown had worked in the optics lab on Santa Barbara Street for three years, Anderson asked him if he would move over to the new optical laboratory on California Street.

The new lab was a huge building, designed in Russell Porter's characteristic Art Deco style, with stone latticework on the ends to conceal the lack of windows. McCauley described the vast interior space as "large enough to enclose five six-room bungalows, with space enough above their gabled tops for a roof garden containing 5 badminton courts with side lines for spectators." The machine shop next door assembled an impressive array of grinding and polishing machines for the optics lab. The smaller machines were familiar, modeled after machines at the Mount Wilson shop, but there were also big machines, of welded steel, large enough to hold the two-hundred-inch mirror and tilt it to a vertical position for testing. Brown started on the smaller machines, grinding test mirrors that had arrived from Corning. The mirrors would never be used in a telescope. They were figured solely to test the Pyrex disks and the grinding machines, to determine grades of carborundum and tentative grinding schedules, facing block materials, and procedures for holding the ribbed-back disks in the new grinding machines.

While Marcus Brown tested mirrors in a temporary laboratory in the basement of the astrophysics machine shop, the machinists upstairs worked on the big grinding machine. The machine was a challenge. It would cradle the largest and most valuable piece of glass in the world. SKF Industries proudly proclaimed in advertisements that the two 51-inch-diameter bearings they had provided for the 17½-foot-diameter grinding table were "the most accurate giant anti-friction bearings ever produced."

The machine shop was also turning out models for the engineers who were close to a final design for the telescope and the observatory. Russell Porter had finished a design for the observatory building itself, including the preliminary engineering for the dome. Up on Palomar workmen were leveling the site and excavating the foundations. While the flurry of work went on, Brown and a few men ground and polished test disks that would ultimately be discarded. From time to time he would look up at the the majestic door, two stories high, at the far corner of the lab. Brown was waiting for the day when that door opened.

George McCauley arrived in Pasadena on the Santa Fe Chief. He had traveled in secret, picking up the ticket to Pasadena in Chicago so no one in Corning would know his ultimate destination. He had a private Pullman compartment, and tried to concentrate on sketches and data for his presentation to the Observatory Council. The Santa Fe seemed to conspire against him. Just outside the city limits south of Chicago, as his train passed a northbound freight, a wrecking crane on the freight train broke free, hitting the leading diesel of the Chief. McCauley arrived in Pasadena two hours late and exhausted with anxiety.

His meeting was at the Astrophysical Laboratory on Thursday afternoon. He was ready with his sketches of the disk, measurements of the location and magnitude of the scars he had found, calculations, and tables of data to convince the council that the scars would not affect the finished mirror if the surface thickness were ground to 3⅝ instead of the previously planned 4 inches. George Hale wasn't there— he was too ill to attend—but McCauley anticipated a tough meeting with Anderson, Porter, Pease, and the others.

Anderson studied McCauley's data and passed it around before giving his opinion that there would be no problem with a final thickness of 3⅝ inches for the surface of the disk over the ribs. Anderson had already tested smaller ribbed disks that Brown had figured. They were more rigid than the preliminary engineering had predicted. With the project moving into its eighth year, everyone was eager to start the most important part of the project—grinding and polishing the mirror. Arthur Day, who had been notified of the condition of the disk, had met in New York with Max Mason and Sandy McDowell. They too had given their go-ahead.

Later that afternoon McCauley was taken to see Hale. Hale, confined to his darkened room, brightened at the news that the disk had survived the annealing intact. He said he was eager to see the grinding begin.

On the way back to Corning, in Chicago, McCauley received a telegram from Corning warning him that a "well-known science editor" was going to try to intercept him for a scoop on the mirror. He managed to evade the reporter, but it was obvious that news on the disk couldn't be delayed much longer. The reporters were now joined in their queries by a book author, David Woodbury, a friend of Russell

Porter's from Maine, who had begun interviewing anyone who would talk to him about the telescope project.

On December 8, 1935, with George Hale's approval, the press was invited to see the disk lowered from the annealing oven. Hale urged that they be shown the bottom of the disk, instead of the face, both because it would be more interesting and because he wanted to quell possible rumors about the scarred face. But the newspapers wanted the whole show. Photographers were allowed to erect lights around the oven and along the tracks that led from below the oven out to the temporary building on the riverbank. The room was bitterly cold, heated by portable "salamanders." When McCauley gave the order, the four screw hoists took twenty-four minutes to majestically lower the disk into view. When the surface of the disk was almost in view, a quick measurement showed that one edge might strike a radiator along one wall, so the operation halted while workmen with torches removed the radiator and its supporting hooks. The delay increased the suspense for the crowd of reporters and photographers, none quite sure what he would see.

When it was finally lowered into place, the surface of the disk was dusty, but observers could still see the the ridges and pits, some several inches deep. When the disk was within an arm's distance of a special balcony that had been built for the reporters, McCauley and Hostetter fielded questions and posed for photographs on the disk. "No effort had been made to get a perfectly smooth surface," they explained, since the face of the disk would be completely reshaped in the grinding and polishing stages. The pits, they reported, were handmade, deliberate efforts with the sandblaster to explore the consistency of the glass. The headlines in the newspapers the next day, Monday, December 9, set off a wave of demands for a public unveiling.

There was plenty of news in the United States at the end of December 1935, most of it depressing. The weather had been terrible: Floods and storms had followed the summer drought. The Midwest had begun to turn into a dust bowl. Father Coughlin and his "Golden Hour of the Little Flower" were at their height; his anti-Semitic radio broadcasts generated so much response that he received more mail than anyone else in the United States, including the president. In Pennsylvania employers were deducting thirty-three cents per week from the pay of children to indemnify themselves for the $100 fines imposed for working the children ninety hours per week. The average steelworker's clothes caught fire once a week. The average family income, excluding those on relief, was $1,348 per year, which typically supported a mother, father, and one or two children.

In the fifth year of the worst depression anyone could remember, the successful casting of the great mirror, the triumph of American technology, was the kind of good news many Americans were eager to celebrate. Corning yielded to the demands. The first public unveiling was scheduled for Sunday, January 26. By then Corning millwrights, using the first two-hundred-inch disk as a model, had finished the steel crate for shipping the disk. The first disk was rolled out of the steel building on the riverbank, to be stored on a platform of heavy timbers, and the second disk took its place. The disk was nestled into the steel crate, swung upright with the four corner chain lifts, and the masons began the tedious work of chipping the mold cores out of the ribbed back. McCauley wanted to display the disk with the scarred surface hidden, and its best face—the ribbed back—showing.

The cores were only half out by January 26. A blizzard made roads in the Corning area nearly impassable. Still, two thousand people showed up to see the great symbol of American progress.

Two days later the rest of the cores had been removed. The disk was again open for public viewing on February 2. This time more than five thousand visitors came to view the disk. An observer described the audience as "filled with awe and puzzlement"—awe no doubt at the sheer size of the immense disk and puzzlement at just what this great waffle iron of glass would do. No matter how many press releases Leon Quigley put out, all but a few newspaper reporters called it an "eye" or "lens." That the huge mass of glass was not to see *through* was a concept counterintuitive to all but the amateur astronomers and others familiar with telescopes. Most reporters chose to ignore the careful wording of the press releases.

On February 6 the press photographers were invited to take photographs of the disk before it was crated. One photographer, Robert Richie, stayed on long after the others left. Richie tried every possible angle to get the photographs he wanted, including lying on the floor and standing on ladders. Late in the day he finally had the angle he wanted, but announced that the photo needed human interest. Would McCauley be willing to pose?

McCauley agreed, and Richie asked him to climb up to the central hole in the disk, some nine feet off the floor, and pretend he was measuring the stress in the disk with a small polariscope. After much tinkering with the lights and camera angles, Richie finally snapped his photograph. The resulting photo, with McCauley looking very tiny in the middle of the great disk, was widely distributed, winning national recognition for Richie, even wider fame for the telescope disk, and a temporarily sore back for McCauley.

The cascades of publicity for the as-yet-unveiled disk, and the much reproduced photographs, brought a barrage of queries from all over the country. What had been an object of curiosity for those in central and southern New York was suddenly a focus of national attention. Schoolchildren, mayors of small towns, hucksters looking for a promotion, and reporters who had rarely covered anything more distant than their local county fair wrote to find out when and where they could see the great disk. Told there would be no more public viewing, they asked when the disk would be shipped to California. Where would the route take it? What cities would it visit?

The mid-1930s had seen a wave of building projects, many with the kinds of superlatives that had been the triggers for national pride a decade before. Radio and the wire services had also tied the sprawling nation together. The nightly news and the familiar entertainment programs were shared from coast to coast. Still, hearing about the new Golden Gate Bridge on a Lowell Thomas news broadcast, or seeing UPI photographs of Hoover Dam in the newspaper, didn't evoke the same pride as the possibility of actually *seeing*, up close, a great and unique achievement of American technology. In the eight years that the telescope project had been underway, none of the superlatives had been diminished. It was still the biggest and most expensive scientific device ever planned. The mirror disk—the largest and heaviest piece of glass ever cast—was the first tangible evidence of the telescope. Americans, in small towns and big cities, in rural hamlets that had never seen any telescope except in a Sears Catalogue, wanted to see the triumph of American achievement.

The railroads, aware of the popular interest, besieged the Corning Glass Works and the Observatory Council with requests to carry the disk. The Argonaut Steamship Line offered to ship the disk by sea, guaranteeing delivery in twenty days. Teamsters offered their services. Glen M. Wiley, the "designer and builder of Wiley Whirleys," offered to truck the disk across the country, assuring Hale and Anderson that he could provide the maximum publicity for the disk. Hale scribbled on the letter, "This man must think we have the point of view of a three ring circus!" Wiley, not easily discouraged, wrote highway departments in every state on his proposed route for permission to carry the oversize cargo.

The final route was shaped as much by preferences at Corning and Pasadena as by questions of clearance for the huge cargo. George Hale, impressed with the Santa Fe's handling of the 120-inch disk and the proximity of its station in Pasadena, favored it over the Southern Pacific and Union Pacific. Nate Cole, the traffic manager at the Corning Glass Works, preferred the "Big Four" route out of Corning, with the New York Central taking the shipment from Corning to Cleveland and handing it off there to the Cleveland, Cincinnati, Chicago & St. Louis Railway. The Erie Railroad and the Delaware, Lackawanna & Western were turned down because the Erie route involved the switching yards at Chicago, which George Hale—a Chicago native—thought might involve shunting too rough for the disk. The Delaware, Lackawanna had low-clearance problems on portions of its route.

Although too sick to take part in the day-to-day decisions on the telescope, Hale took a special interest in the shipment. When F. E. Williamson, the president of the New York Central, wrote that they planned to ship the mirror as the sole cargo on a special train, Hale, not eager to pay a special fare, wrote back that a car in an ordinary train would be fine. Williamson later explained that there would be no additional cost for the special train. McDowell suggested sending a dummy cargo, as large as the disk, ahead of the shipment, to make sure of clearances. Hale, the kind of man who complained when too many pencils were missing from his secretary's desk, tried to take advantage of the competition of railroads to negotiate down the cost of the shipment. The western roads offered a rate of $1.80 per cwt. The eastern roads wanted $3.43. Hale wanted the whole shipment to go at the lower rate and tried to get an additional discount on the grounds that the disk would be insured by Lloyds of London, so the railroads would need no special insurance.

Insurance was a tricky question. The railroads insisted on a replacement value for the bill of lading. No one had ever put a price tag on the mirror. Corning had sent some interim bills for time and material, which had totaled less than $300,000. If asked, McCauley and the bean counters could have come up with the incremental cost for the second mirror. But that was back when a portion of A Factory had been set aside and a large melting tank was kept filled with 715-CF Pyrex for the manufacture of telescope mirror disks, and a special crew, including experienced ladlers, had been assembled for the pours. The tank had since been converted to regular production of baby bottles, and the ladling crew dispersed. Whatever the cost in man-hours and equipment, McCauley, who had sweated through the bobbing cores, floods, and earthquakes, would have been the first to say that there was no way to get insurance for the "luck" that Corning had needed to cast the disk.

As details of the train journey leaked out, the railroads began issuing press releases about "the most precious single shipment ever handled by the American Railroads," and the extraordinary precautions that were being taken to crate, secure, and guard the disk on its journey. Sandy McDowell and John Anderson arranged a one-hundred-thousand-dollar insurance policy for the disk with Lloyds of London for a premium of nine hundred dollars. The railroads raised the possibility of dignitaries and scientists riding with the disk. McDowell, picking up Hale's penuriousness, wrote that he had no objection to the use of a special train or to railroad people and others accompanying the disk if "no additional charges would be involved."

The schedulers for the New York Central worked with the schedulers of the other lines for months to map out a route that would accommodate the disk. Heavier cargos had been shipped across the country by rail, including parts of the enormous turbines at the Hoover Dam. The challenge for the disk was its extreme fragility and size. On its side the disk would be too wide for most railbeds, so it had to be carried vertically. But even on a modified New York Central Rail-

way well car, with twenty-six-inch-diameter wheels instead of the standard thirty-three-inch wheels, and with the crate suspended in the well
so that the bottom edge was only five and a half inches over the rails,
the top of the crate would tower almost eighteen feet over the
roadbed—higher than many tunnels and underpasses on the cross-
country route.

It was March 1936 before the railroads and the millwrights building the crate were ready. The Chemung River, as if sensing the
impending drama, acted up again. On March 9, when the disk was
loaded onto a low-slung trailer provided by an Elmira contractor, the
river was already at the floor level of the factory caves. The journey
from the building on the riverbank behind A Factory, to the New York
Central spur that led into the Corning yards, a distance of one-quarter
mile, took thirty hours. The floodwaters were rising so quickly that
many of the men who supervised the loading and lashing of the cargo
were knee deep in water as the trailer started its journey. When the
cargo wouldn't clear the corner of one building, masons were called in
to dismantle the bricks. Two sharp turns required teams of men with
crowbars to hand-steer the trailer.

The next day, despite bleak weather—at least it wasn't still raining,
and the river had started to subside—a crowd of townspeople, newsmen, and movie cameramen from Fox, Paramount, Pathé, and Hearst
showed up to witness the arrival of a huge NYCRR wrecking crane
and the engine pulling the special well car. The car had been fitted
with beams, rubber padding, and bolts to hold the crate. Hoisting the
crated mirror, swinging it to a vertical position, and lowering the crate
into the well of the railcar took most of a day. There were no mishaps,
and the reporters—some of whom had come from as far as New York
City—were left to fill their stories with clichés about the mirror fitting
"like a glove" into the well and comments like, "It appears as if nothing
short of an earthquake could jar the valuable contents of the package."

The car disappeared from public viewing for a week in the Corning railyard, while the millwrights fitted tie rods and braces to support
the crate. The weight of the mirror had sagged the crate closer to the
rails than the anticipated 5.25-inch clearance, so gum rubber cushions
and wooden planks were substituted for the sponge rubber pads that
had originally supported the crate. Spring bolts increased the loading
on the welded rails under the disk to a compression of approximately
60,000 pounds, which McCauley had calculated would be enough to
dampen any bounce in the load. The millwrights finished by fastening
steel plates, thick enough to deflect a bullet, to the sloping braces that
held the mirror upright. The car was moved out of the factory yard to
the NYCRR switching yard, just east of Walnut Street, and a crew
from the Corning paint shop painted the crate bright white and let-
tered a huge PYREX and a smaller "200" TELESCOPE DISC" and
"CORNING, N.Y." in bold black block letters.

The assembled train was probably the shortest ever to set off across the country. Behind the steam engine, which flew the twin white flags of a Special, was only the tender, a single boxcar with the lifting sling and accessories for unloading the disk, the well car with the disk, and a caboose for the train crew. Every rail line on the route had to agree to George Hale's orders for the journey: The train was never to exceed twenty-five miles per hour and was to move only during daylight. At night it was to be parked under bright illumination and heavy guards.

On March 25 the assembled train went to the NYCRR scales at the north yards with McCauley, a crew of millwrights, and a photographer riding on the well car. The weighing was a formality. The NYCRR had insisted that the car be ballasted against tipping with a load of heavy rails, so the 65,000 pounds of the crated disk was only half the weight of the loaded car. That night, the train was back on the Walnut Street siding, ready for a noon departure the following day. The Baron Steuben and other Corning hotels were again full with reporters and visitors.

March 26 dawned a chilly, bright day. A crowd gathered early, creating a workday traffic jam. There was no ceremony scheduled, so the newsmen searched for their stories. One sharp-eyed reporter found graffiti on the white sloping panels that protected the disk. Someone had scribbled, "I'd like to hear from you," and given an address. Under it was a penciled, "Come up'n see me some time—Kitty."

Representatives of each of the rail lines that would be carrying the precious cargo—the New York Central; the Chicago, Burlington & Quincy; and the Santa Fe—had spent the final week of preparations in Corning. They would ride the caboose with the trainmen, special railroad detective Fred Phillips, and four railroad security agents who were entrusted with guarding the disk.

The waiting crowd hushed when NYCRR trainmen Phil Barrows raised his hand. Before he could signal the engineer, someone called out, "Where's Murphy?" A search began for the jovial representative of the CB&Q, who had been the butt of jokes all week. Murphy was finally spotted on the run from the Athens Hotel, his breakfast in his hand. The crowd was still laughing when Murphy joined the others in the caboose, Barrows gave the signal, and the train slowly pulled away toward Watkins Glen.

McCauley felt a lump well up in his throat as the train eased over the grade crossings. It was exactly two years and one day since he had supervised the pouring of the first telescope disk, with the world watching. McCauley was still a young man, at the peak of his career. As a scientist, he was ready for whatever problem came along. He knew there would be other projects, other challenges. He also knew this glass disk would always be his masterpiece.

★　　★　　★

For fourteen days the four-car train wound its way across the United States, its journey chronicled daily in the newspapers and on the radio. Wherever it went the crowds were larger than anyone had anticipated, larger than any train had ever drawn. As interest built, even the increasingly garish stories about the trial of Bruno Hauptmann were crowded off front pages by reports of the stately journey of the great disk.

The first big city scheduled was Rochester, New York. A crowd turned out at the main station, only to discover that the closest the train would come to the city was across the Genesee River, north of the Ballantyne Bridge. Hundreds of people hurried to the bridge. Deputy sheriffs were hardly able to control the traffic.

Wherever the train went, crowds were waiting. Outside Buffalo, the train switched to the Delaware, Lackawanna & Western tracks, because the NYCRR route lacked adequate clearance. The DL&W route had two bridges with marginal clearance. As the train came to a near halt before the Bailey Street bridge, crowds pressed forward, and newspaper photographers snapped pictures of the clearance engineers and trainmasters giving the signals and the train inching forward. The crowd cheered when the top of the crated disk cleared the bridge by five inches.

The train averaged eighteen miles per hour during the day. At night, it was parked under guard. At Cleveland, Sergeant D. H. Simonson of the NYCRR, part of the guard crew, told the newspapers, "If anything happens to this tonight, I might as well start running."

Some in the crowds turned away disappointed, when all they could see was the shields and the signs painted by the Corning signpainters. For others it was enough. The train arrived in Indianapolis on a Sunday afternoon. Highways leading to the station were jammed to a standstill with cars, and the mob on the streets was so dense that the yardmaster had to call in city police to clear the tracks. Police estimated the crowd at ten thousand. The next day the train went through Terre Haute at noon. Schools and colleges were dismissed for the day so the students could see the disk.

At Charleston, Illinois, the disk was parked overnight under a heavy guard of railroad police. Murphy, the CB&Q man who had boarded late in Corning, wrote, "I estimate everyone in this town of 17,000 will see it."

Eighteen-below-zero temperatures and a heavy snowstorm didn't stop the science class of Cameron High School, in Cameron Junction, Missouri, from walking to the railroad tracks for a glimpse of the newest wonder of science. On the open plains, where the train rolled through tiny towns at twenty-five miles per hour, men and women lined the tracks, classes from one-room schools, millworkers, farmers, mailmen, housewives, and the unemployed. Grandfathers perched small children on their shoulders, so that some day they would be able to tell *their* children that they had seen the great "eye."

Everyone insisted on calling the disk an "eye" or a "lens." Even the Burlington Route, celebrating its role in transporting "one of the most precious single objects" ever moved, from St. Louis to Kansas City, titled its self-promotion brochure, "Transporting the World's Largest Telescopic Lens."

In Kansas City, where the cargo was to be transferred to the Santa Fe Railroad, the scheduled route led under the Broadway overpass. The official height would have cleared the disk easily, but the orders were that every questionable overpass was to be checked. The train slowed to a halt, then crept forward so the clearance engineer and trainmaster could measure the actual clearance.

The officials of the four lines had agreed that the final gauge of clearance would be the thickness of a pocket Bible. The crowd hushed as the trainmaster held the tiny Bible over the crate. It wouldn't clear—frozen ground under the tracks at the overpass had heaved—so the train backed off and waited under guard while local routing agents devised a detour. That low overpass cost four hours of zigzag switching through and around Kansas City yards. No one kept track of how many trains had to be held up or rerouted to provide clear tracks for the special freight that for two weeks enjoyed priority over every other train in America.

As the train approached the Rockies, and again in the Cajon Pass between Barstow and San Bernardino in California, a local section crew rode a scooter car ahead of the train, searching the track for stones or debris from rockslides. The portion of the disk crate that hung below the level of the well car, less than six inches above the track, was as vulnerable as the top of the crate.

Outside Albuquerque, the train took a one-hundred-mile detour to avoid a questionable highway overpass. At Albuquerque a coal spout hung over the main line at sixteen feet above the track. The entire yard was shut down while the Special was routed through on a side track. Later, at the entrance to the Johnson Canyon Tunnel, the clearance over the westbound track measured only sixteen feet; the eastbound track showed eighteen feet. So, explained the freightmaster, "We went through on the wrong side," though only after orders had been issued to halt or reroute all eastbound trains for the day.

Near Gallup, New Mexico, the train crossed the Continental Divide at an altitude of 7,300 feet. The disk was almost two thousand feet higher than it would be on Palomar. As it descended from the Cajon Pass into San Bernardino, a day away from its destination, a reporter from the *Los Angeles Times* wrote: "Looking like an armored train ready for battle, the mirror special came into the Santa Fe yards just as the sun was setting. First a little old-fashioned locomotive puffing with importance ... then the mirror, and last the all important caboose that always contained more railroad officials than train crew."

On Friday morning, April 10, 1936, the train arrived at the little

Lamanda Park station in East Pasadena. It had taken sixteen days to travel three thousand miles. Hundreds of thousands of people had seen the train, making the journey to stations or to trackside in snow, rain, and cold. It looked more like *The Little Train That Could* than the most valuable cargo ever shipped.

Francis Pease was at the station in Pasadena to officially receive the cargo, along with Ferdinand Ellerman of the Mount Wilson Observatory, G. W. Sherburne of the astrophysics machine shop, and Marcus Brown. A crowd of more than five thousand gathered at the station to view the arrival, and police had to be called in to keep the area around the railcar clear as the crowd pushed close for a chance to touch the wondrous object that the nation had been watching in its journey across the country. By the time reporters showed to take photographs of the mirror before it was unloaded, Edwin Hubble had arrived and posed near the car.

McCauley had sent a detailed description of the crating and loading procedure and instructions for the unloading to George Hale. The crowd watched as workmen used acetylene torches to cut away the six steel rods that held the disk upright and the fifty heavy bolts that held the metal shields in place. When one plate buckled as it was cut away, the crowd let out a collective sigh. Pease inspected the crate before it was lifted off the railcar. He found the names of dozens of railroad men scrawled on it, but no damage.

A few hours after the mirror, the "Barstow Hook," the largest railroad crane the Santa Fe operated, arrived at the station. The crane was rated to lift 150 tons; it was tested on a 116,000-pound locomotive before it was swung into place over the disk. It took five hours to lift the 20-ton disk, in its 15-ton crate, off the car and load it, horizontally, onto a 40-foot-long rubber-padded trailer rig provided by the Belyea Truck Co., which advertised that it could ship any cargo anywhere. The truck, escorted by Pasadena police officers on motorcycles, and followed by workmen in cars and boys on bicycles, drove through the heart of Pasadena to California Street and the astrophysics buildings at Caltech. For one turn greased steel plates had to be placed under the wheels of the truck. Movie cameras on trucks recorded the entire journey.

George Hale, too ill to leave his bed for the station, kept track of the journey of the disk from his bedroom, taking telephone reports as the disk arrived in San Bernardino, Pasadena, and finally at the optical shop. He sent a wire to McCauley to announce the safe arrival. From the quiet of his room, Hale contemplated the progress of telescope making. He remembered when the forty-inch-diameter objective lens of the Yerkes refractor, the first large telescope he had shepherded into existence, and a marvel of its era, had arrived at the Yerkes Observatory in 1897. The Yerkes lens, he pointed out in a note to a friend, would fit inside the hole in the center of the great two-hundred-inch disk.

At the optics lab, the tall doors—which had been built to accommodate the disk—swung open and an overhead crane lifted the crated disk off the trailer. When the crane rolled back into the laboratory, the doors were shut behind it.

Little did anyone imagine that it would be eleven years before those doors opened again.

22

On the Roll

The safe arrival of the disk put the telescope project in high gear.

Crews had worked all winter on the mountain, clearing the site and laying pipes for the water supply. Hale's original plan was to build the entire observatory, except for a small pumping plant and some wooden cottages for observers and caretakers, in a single building. As the plans proceeded, it soon became obvious that they would need far more facilities, from a separate machine shop and utility building to housing for the large work crews.

Not only was Colonel Brett a martinet, but he ran the workers' mess hall at a profit and served a horrible meal once a week so the men would appreciate the rest of the meals they got. Whether the situation was routine or an emergency, the colonel liked precise procedures. When a carpenter suffered a heart attack while making concrete forms, the colonel lined everyone up for a lecture on procedures and responsibilities before sending the thirty-five-horsepower Caterpillar tractor down the road to clear a way for the doctor and ambulance to get through two feet of new snow that had begun to turn to ice. Ignoring the colonel's orders, two men raced down on the Caterpillar, smashed the county gate that had been left locked, and pulled the ambulance up the road to the colonel's cottage. They were too late. The worker died. In the morning, they had to clear rockslides off the road to get the ambulance back down the mountain.

Despite the isolation, celebrities were brought up the mountain to see the early work. Herbert Hoover, only a few years out of the presidency, visited Caltech in Pasadena and was brought to Palomar, where he stayed over in Colonel Brett's house. Sandy McDowell's wife came along to supervise the dinner preparations for the former president and to hide the colonel's whiskey bottle of "salad dressing."

In the spring McDowell hired a Caltech graduate engineer named Byron Hill to supervise the construction work on the mountain. Hill had been working for the Metropolitan Water District in Banning, California, as a concrete specialist. He was young, confident, and energetic. They agreed to pay him the respectable salary of three hundred dollars per month, plus an unfurnished cottage with heating equipment and hot water.

Colonel Brett, wearing his uniform and tall boots, greeted the new man outside his cottage. Byron Hill had brought his collie, Mike, with him. Mike sniffed the colonel's boots, found them satisfactory, and lifted his leg. The colonel kept a straight face, but he and Byron Hill never got along.

By then Pasadena was generating reams of plans. Engineering students and draftsmen were given sketches and told to produce working plans. The Caltech students who drew up the plans for the utilities on the mountain were precise and careful draftsmen; they drew the power and water lines straight and true. When crews went out to dig, they would find solid rock where the lines were supposed to go. Byron Hill knew engineering professors at Caltech and realized that they were sometimes astonished when engineers in the field, or even their own children, could make the leap from drawings to working models and machines.

Crews of Caltech students came up for the summer to dig the footings of the dome and the telescope. Beneath the acidic topsoil they discovered decomposed granite, which had to be blasted and hauled away to get down to bedrock solid enough for the foundations. The only heavy equipment on the mountain was the ancient Caterpillar tractor and a fifty-year-old air compressor. Much of the work was done by hand with wheelbarrows, two-wheeled concrete buggies, and hand winches. In the middle of the circle of dome excavations were separate excavations for the piers to anchor the telescope. To reach bedrock for the piers, they had to dig twenty-two feet. A Caltech student with limited blasting experience, a cavalier attitude toward insurance company regulations, and a gung-ho attitude set the charges to blast the boulders out of the decomposed granite.

Each step of the early construction was a negotiation between the site realities and the plans that came up from the drafting tables at Caltech. The huge one-million-gallon water tank and water tower had been planned to provide water for the concrete work. But filling the caissons with solid concrete would have required a daily fleet of trucks making the rough climb up the steep, winding Nate Harrison Grade, a dirt road that was impassable during much of the winter and spring. Instead Byron Hill framed the caisson excavations with wooden forms, filled the bottom and eight to ten inches of the outside edges with concrete, dumped decomposed granite inside, then poured concrete over the top. They could draw what they wanted at Caltech. Byron Hill knew concrete.

Hill took over the mountaintop like a whirlwind. In his rakish leather jacket and aviator sunglasses, with the cockiness of a recent

Caltech graduate, he leapt into the work, pushing the students and workers, revising plans, improvising when equipment broke down or wasn't available. The concrete was mixed in a one-cubic-yard mixer, and poured by hand from two-wheeled concrete buggies. The students who did much of the work made a game out of it, racing across the site in timed runs. At the peak of the concrete pouring that summer, every person on the mountain, even the cooks, was called to help with the buggies. Colonel Brett, though nominally in charge, chafed at the pace and Hill's nonmilitary style. When they began fighting openly, Brett was called to Pasadena for discussions. The story told on the mountain afterward was that the colonel was given fifteen minutes to resign.

The Caltech students came and went. The men working on the mountain, a small band that had wintered over in the old barracks of the cattle ranch, turned into a cohesive group. They called themselves "zombies" and on Saturdays trooped down to Captain Bolin's store halfway down the mountain where they would buy Don Leon wine. Byron Hill, in charge after Colonel Brett's resignation, took a dim view of drinking and carousing but realized that in the isolated conditions allowing some partying was the only way to hold on to good workers.

To supplement the weekly mailbags from Pasadena, McDowell authorized a radiotelephone. Ben Traxler got a Class I license to operate the unit. A red button on the transmitter at Palomar would bring up the carrier and ring a bell in Pasadena, letting Hill make his reports and requests. McDowell took the calls himself in Pasadena. He liked to acknowledge each statement to show he was on top of every matter. His barked "Yup" would trigger the voice-operated transmitter in Pasadena, cutting off Hill in midsentence. The interruptions, together with the ignition noise on California Street in Pasadena, ensured that Pasadena heard little of what was reported from Palomar. Hill, content to carry on with as little interference as possible, welcomed the arrangement. He could later claim that he had reported his plans before he went ahead.

By the end of the first year of work on the mountain, cottages had been built for Hill and a few others with families. Mary Marshall became the schoolteacher. Her husband, Harley Marshall, kept account books and arranged public relations for the tourists who were already finding their way up the mountain to see the world's largest telescope.

Beyond the bare site for the two-hundred-inch telescope, workmen were well along on the construction of a smaller dome. Inside, workers from the Santa Barbara Street optics laboratory of the Mount Wilson Observatory and machinists from the astrophysics machine shop at Caltech were putting together a telescope unlike any the world had ever seen.

Bigger telescopes, like the two-hundred-inch, let astronomers see farther into space. For Hubble and Humason, a bigger telescope would let them measure the red shifts of ever-more-distant spiral galaxies, building more evidence for the cosmologists. The big telescopes bit off a tiny portion of the heavens with each exposure or spectrograph, probing deep into one small area of the sky. But even as they designed and built ever-bigger telescopes that could probe ever deeper into the unknown, astronomers dreamed of another sort of instrument, a camera that would allow them to photograph huge segments of the heavens in a single exposure. A camera that could photograph a relatively large swath of the sky, recording celestial objects down to very dim magnitudes, would serve as a "finder" scope for the big instrument, exploring promising areas of the heavens and producing plates that would let the astronomer identify targets for study with the big telescopes. The survey camera could also search for new and rare celestial objects, covering areas of the heavens in a few dozen plates that might take a lifetime to explore with the narrow field of view of the larger telescope. When Hubble began his search for novas in the Andromeda galaxy with the one-hundred-inch telescope—a search that ultimately led to the identification of Cepheid variables and the end of the "island universe" debate—he had to confine his research to three promising areas of the galaxy, because the field of vision of the one-hundred-inch telescope could not take in the entire Andromeda Nebula.

The problem of building a wild-field camera had stumped opticians and astronomers for years. The mirror of a fast, wide-field camera must be steeply curved. But with traditional paraboloid mirrors, the more steeply curved the mirror, the smaller the area of sharp focus in the center of the field of view. The area around the edges is distorted with coma, so the stars look like teardrops instead of sharp points.

In 1929 Walter Baade was sent on an expedition, with an optician named Bernhard Schmidt, to photograph an eclipse of the sun from the Philippine Islands. Schmidt had lost one arm playing with a pipe bomb of his own design at the age of eleven, and had later eked out a living and money for brandy by polishing telescope mirrors for amateur astronomers in an abandoned bowling alley. He was finally hired by the Hamburg Observatory as their optician. Single, an Estonian by birth, a pacifist by conviction, and eccentric enough in his habits and politics to attract police attention, Schmidt worked alone in his basement workshop, dressing in a cutaway coat and striped pants to grind and polish mirrors and lenses with his one good hand.

On the ship on their way back from the Philippines, Schmidt casually announced to Baade that he had an idea for a telescope that would photograph enormous areas of the sky with a single exposure, with pinpoint images from edge to edge. The idea, he said, was to use a deep, spherical mirror. Spherical mirrors are relatively easy to shape—the first step in the shaping of a telescope mirror is to bring the surface to a spherical figure—but Baade knew that spherical mirrors smear the light of distant objects, distorting the images. The answer, explained Schmidt, was to put a thin correcting glass in front of the mirror. The ripples in the surface of the correcting lens would be so subtle that they would be invisible to the naked eye, but they would be sufficient to correct the aberrations of the spherical mirror. Baade agreed that the idea was fantastic. If Schmidt could do it, the telescope would be phenomenally useful. But how would anyone grind such subtle shapes into a glass plate?

Back in Hamburg, Schmidt procrastinated. When Baade and the director of the observatory pressured him, Schmidt said he wasn't ready. He needed time. He finally asked Baade for some physics handbooks, put on his cutaway jacket, and retreated to his basement vault. He worked for forty-eight hours, taking only cigars and brandy as nourishment, to create a thin fourteen-inch-diameter disk of glass. His method was elegantly simple. He sealed the disk on top of a pot and drew down a vacuum inside the pot, distorting the surface of the thin disk while he polished it flat. When he released the vacuum he had the shape he needed.

Schmidt and Baade tested the new camera by photographing the Neuer Friedhof, miles away, from a window of the observatory. The photograph recorded the names on the tombstones and the fine edges of leaves of the trees in the cemetery.

In 1931, when Baade was offered a job at the Mount Wilson Observatory, he brought along a print of the photograph he and Schmidt had taken of the cemetery. John Anderson was impressed. Just when "it is said that there is nothing new under the sun," Anderson reported to Hale, Baade had brought an idea from Germany "which comes pretty near being new. . . . The trick is to use a spherical mirror. In a plane roughly parallel to the mirror surface, but passing through the *center of curvature*, is placed a thin sheet of glass—really a very weak negative lens so figured that it distorts the incoming parallel light just enough to correct the spherical aberration of the mirror."

It hadn't been requested in the original grant from the IEB, but Anderson and Hale agreed that a Schmidt camera, with its wide view, would be a perfect compliment to the two-hundred-inch telescope. Russell Porter began working on a design. There was already a pressing need for the new camera.

In 1934 Fritz Zwicky and Walter Baade began to collaborate. They had discovered that some stars could explode with extreme violence, and coined the name *supernova* to describe the phenomenon. Supernovas were so rare—the last one in the Milky Way was at the time of Tycho Brahe—that Baade and Zwicky were forced to study the remnants of the few supernovas that had been seen by naked-eye observers, like the Crab Nebula. Zwicky wanted a wide-field Schmidt camera to search for supernovas. When he asked George Hale to allocate money for the camera, Hale said it sounded like prospecting for gold but approved the funds. While Porter was working on the design, Zwicky wandered into his office, made some suggestions, and before long was telling everyone that Bernhard Schmidt had shown *him* the photographs of inscriptions on the gravestones on a visit in 1935, and that the idea of building a Schmidt camera was actually *his*.

Opticians at the Mount Wilson optical labs figured the twenty-four-inch mirror and eighteen-inch correcting plate for the new Schmidt camera. The Caltech astrophysics machine shop built the tube and mounting. Russell Porter's design looked like an eighteenth-century cannon mounted in a tuning fork. The correcting plate was at the muzzle of the cannon, protected by shutters that opened and closed to control the exposure of the film. The film holder inside the camera was reached through a door in the side. The machine shop built film holders that would cut the photographic emulsions in a circle and bend the emulsions so their curvature exactly matched the spherical focal curve of the camera.

As the eighteen-foot-diameter dome went up at Palomar, experienced Mount Wilson workmen came up to install the new telescope. Jerry Dowd, the master electrician from Mount Wilson who had wired the electric controls for the solar telescope in George Hale's laboratory, did the wiring. Dowd had retired from the observatory to a ranch, until he heard about the new work going on at Palomar. He couldn't resist the opportunity to wire another telescope, even though he was only paid the same sixty-seven and a half cents per hour that the regular construction workers received. Ben Traxler got the job of helping the mechanics and technicians from Pasadena do the installation and learned the tricks of slip-couplings for the rotating dome and the complexities of wiring for telescopes.

By the fall of 1936, the Schmidt camera was in operation. One of the cottages on the mountain had been rushed to completion for Zwicky's use, and he began making the trip from Pasadena to Palomar to photograph galaxies, searching for supernovae. Byron Hill met Zwicky on one of his first trips up the mountain, in a heavy snow. Hill had sent the Caterpillar tractor with a chain to pull Zwicky's car. Zwicky wanted rope instead of the chain, and before long they were in a full-scale argument. To Zwicky, Hill was one more "spherical bastard"—no matter which way you looked at him, he was still a bastard. The astrophysicist recruited Ben Traxler as his night assistant on the telescope, and soon Zwicky would signal his arrival on the mountain each night with a shout of "*Wo ist* Traxler?" That slight of protocol, circumventing Hill's authority as superintendent on the mountain, didn't sit well with Hill. The politics of the mountain began early.

Zwicky, in his first films of galaxies in Virgo, found supernovas. He kept up the search, pushing and tugging at the sometimes balky Schmidt camera until its battleship-gray tube was dented with scars. Zwicky still holds the record for the most supernovas recorded by a single observer.

In Pasadena the arrival of the disk, coupled with Sandy McDowell's eagerness to begin the actual construction, put the final stages of the design of the telescope in high gear.

Once the yoke mounting, with a giant horseshoe bearing to support the weight at the north end of the telescope, had been selected, the pieces of the design began to fall into place. Mark Serrurier, who had been given the challenge of designing the tube of the telescope, had fiddled with his pencils until he stumbled on an idea so elegant and simple that engineers who looked at his drawings shrugged and asked, "Why didn't I think of that?"

Serrurier knew that the weight of the mirror in its cell at one end of the tube and the prime focus cage, with equipment and auxiliary mirrors, at the other end of the tube, made it fundamentally impossible to design a telescope tube that wouldn't flex. He kept playing with Martel's suggestion that the tube didn't have to be absolutely rigid, that what mattered was for the primary mirror at one end of the tube, and the correcting lens or auxiliary mirror at the other, to be perfectly *aligned* with each other. If both ends of the fifty-seven-foot-long tube drooped, even as much as one-sixteenth of an inch (an enormous distance at optical tolerances), it would not affect the alignment of the mirror and the prime focus as long as both ends drooped the same amount and remained parallel to each other.

Suddenly the problem was much simpler. Serrurier sketched supports that ran diagonally from the corners of the tube to its central pivot point. Compared to the massive braced girders that had been used for previous telescopes, the supports seemed airy. But the diagonal supports meant that any motion of the ends of the tube would create compression in the supports. Try to compress a rod, as opposed to bending it, and the effect of Serrurier's structure is immediately clear. The ends of his tube would droop by a millimeter or so, but the diagonal struts would keep the two ends parallel and perfectly aligned. The elegant, symmetrical structure immediately earned the name Serrurier truss. In various forms it was soon widely copied for telescopes and other structures.

One major question remained for the telescope: the bearings. The entire weight of the moving portions of the telescope, half a million pounds, would rest on the north and south bearings of the yoke. Francis Pease's drawings still showed huge roller bearings for the north bearing under the horseshoe, but the horsepower needed to move the telescope on roller bearings, and the potential distortion of the rollers from the heavy load, were discouraging. Ball bearings were worse. Even the declination bearings for the telescope tube presented problems. Serrurier's new design, which lightened the tube, still concentrated the full weight of the tube on the declination bearings.

Sandy McDowell, who had been brought into the project to supervise construction, was eager to move from design to actual building. His experience in the navy had been with large welded components, and he wanted to have as much of the telescope as possible built from seamless, welded construction. Especially if the welded sections were annealed—a process of heating and controlled cooling, similar to the annealing of a glass disk—which would remove residual strains in the steel, the telescope would emerge a strong, monocoque structure, with no danger of fastenings loosening or corroding.

Some consultants argued that a welded structure would be *too* stiff, that riveting the members together would permit a flexibility that would slow down or dampen vibrations in the structure. Welding, by increasing the rigidity, would shorten the period of vibrations, increasing the number of cycles per unit of time. The welding advocates argued that vibrations would only be a problem if they didn't get the drives and mount right. They had worked long enough on this project that they *would* get it right. Finally even Pease, the traditionalist among the design staff, came around to the welded construction.

Welded fabrication had another appeal for the project. It was fashionable. In place of the bolts and rivet heads of the Mount Wilson telescopes, a welded telescope would be sleek, smooth, and streamlined, like the new aerodynamic designs that were just being introduced in automobiles and locomotives. An earlier era had marked its technology with the strength of massive raw-steel construction, producing the Brooklyn and Golden Gate Bridges and the massive riveted and bolted frames of skyscrapers. Welding not only offered the technical advantages of fusing the metal fabrications into large, strong structures but created a look as modern as that of the sleek cars that graced the pages of *Life* magazine.

There were few companies with experience of welding large structures, and even fewer engineers who had the experience of designing large structures to take advantage of welded joints. The telescope added the additional requirement that many surfaces of the structures had to be machined to extremely fine tolerances. McDowell concluded that only companies with experience building large hydroelectric turbines and heavy gun turrets combined the two skills. He approached the men he knew: Wylie Wakeman, general manager of Bethlehem Shipbuilding; Homer Ferguson, president of Newport News Shipbuilding; J. F. Metton, president of New York Shipbuilding; and W. W. Smith of Federal Shipbuilding. When they turned him down, McDowell approached the commander of the Mare Island shipyard in San

Francisco, asking whether they could do the final machining on the mounts if someone else did the fabrication. None of the companies he approached had any experience with telescopes.

When the shipbuilders turned him down, he went to companies with experience on hydroelectric turbines and other heavy structures. George Hale had already begun preliminary negotiations with Westinghouse and Babcock & Wilcox, who had both done large welded structures for the Boulder Dam. General Electric had done some heavy welded structures for electrical systems in Russia. Baldwin-Southwark Corp. built locomotives. The American Bridge Company, Warner & Swasey, Budd Steel, and Inland Steel were all interested. Westinghouse sent a vice president, the manager of their huge South Philadelphia plant, and a group of engineers to meet with McDowell. Except for a few government projects, like the Boulder Dam, their business had been slow. The end of 1935 and 1936 seemed an opportune window to keep some idle factory capacity busy.

While McDowell was negotiating to decide who would build the mounting, the preliminary plans were sent to a blue-ribbon list of independent consulting engineers: S. C. Hollister at Cornell, George E. Beggs at Princeton, W. F. Durand at Stanford University, S. F. Timoshenko of the University of Michigan, John Lessells and George B. Karelitz of Lessell & Karelitz in New York City. The consultants unanimously supported the horseshoe design.

By the end of 1935, McDowell contracted for the first large fabricated component, the cell that would hold the mirror, which he ordered from Babcock & Wilcox, who had taken on Baldwin-Southwark as a partner on the proposal because a portion of the machining was so large it would require an oversize boring mill at the Baldwin Locomotive Works. On his trips east, McDowell also held more talks with Westinghouse. Their huge South Philadelphia plant built ship turbines. They had lathes, milling machines, and boring mills large enough to machine the biggest components of the telescope, spare capacity in the plant, and an engineering staff with experience on welded structures. Jess Ormondroyd, the manager of the experimental division at Westinghouse, agreed to send a Westinghouse engineer out to California to work on the details of the mounting.

Rein Kroon was the youngest engineer on the staff at Westinghouse. A Dutchman, he had graduated from the Federal Technical Institute in Zurich, couldn't find a job in Europe, and was hired as a design school student with Westinghouse by Timoshenko and Lessells, both consultants on the Palomar project. Kroon had been too young to enter the Federal Technical Institute when he first graduated from high school, so he had spent a year working as a volunteer in a Swiss factory. Under the Swiss apprentice system, as soon as a worker announced that he had spent long enough in a job, he was moved to a new position. For a quick learner, a year of apprenticeship meant a wide exposure to machine work.

Kroon was newly married, with a one-month-old son, when he was told that he would be temporarily assigned to a project in Pasadena, California. He hadn't heard about the telescope project—the newspapers in Europe were too busy with Hitler, Mussolini, and the abdication of the king of England to chronicle an unbuilt American telescope—and he had to look up Pasadena in an atlas to see where he was going. He carried his son on the train in a basket.

When Kroon got to Pasadena, McDowell had arranged a welcoming party and found Kroon and his family a bungalow on Los Robles Avenue. The next day Rein Kroon, a tall, lanky man with a soft voice and a remarkably soothing accent for a Dutchman, showed up at the astrophysics laboratory, where he was given an office with a drawing board and a stack of sketches and drawings.

Kroon had an odd advantage over everyone else on the project. Almost everyone who had worked on the telescope had some experience as an astronomer and ideas of how a telescope *should* be built. Kroon had never worked on a telescope before and hadn't even seen a large astronomical telescope. He saw every problem of the project as a challenge in pure engineering. It didn't matter if the project was a telescope or a jet engine: Engineering was problems to be solved.

The first problem they handed Kroon was the horseshoe bearing. When he came to Caltech, McDowell had asked around about an idea that had been discussed in the navy—the possibility of using a thin film of oil, under pressure, as a bearing for the weight of the telescope. Pease, pushing for roller bearings, dismissed the idea as impractical. McDowell, who had his doubts about Pease, solicited opinions on the idea from outsiders. J. Emerson, from the navy shipyard at Mare Island, recommended against oil bearings: There were too many variables to control, such as temperature and the viscosity of the oil; it would be too hard to keep out vibrations and tremors; it would require considerable power; and any movement or slippage would wear grooves in the bearing surfaces that would effectively destroy them. As an illustration that the bearings couldn't work, he suggested the experiment of putting a playing card on a spool; no matter how hard a person blows, he cannot lift the card. Like Pease, he preferred the roller bearings.

In fact, the only real interest in the oil-bearing idea came from Guenther Froebel, an engineer at Westinghouse, who thought that if the problems could be solved, oil bearings would be simpler and smoother than roller bearings. No one else wanted the problem of trying to design an oil pressure bearing, so it was given to the new man.

Kroon began by going over the calculations that had already been done. Francis Hodgkinson, the chief engineer at Westinghouse, had done some preliminary work on oil pressure bearings for the heavy rotors in power turbines. The idea of the bearings was that instead of having a metal surface press against metal, the two metal surfaces would be separated by a thin film of oil, pumped into the space between the surfaces. Hodgkinson had calculated the power required to maintain a film in a bearing for the two-hundred-inch telescope, and had come up with six hundred horsepower. Motors that large would need a huge generating plant and would cause enough vibration to shake the entire mountain. Pease looked at Hodgkinson's figures and gloated.

Intuitively Kroon guessed that Hodgkinson's figure was too high. Kroon had a strong background in physics and applied mechanics. The engineering formulas were fresh in his head. He studied Russell Porter's sketches of a possible bearing and realized that Hodgkinson had calculated the power required as if the entire oil flow were through an orifice. In fact the oil would be spread in a film between the two bearing surfaces. When Kroon recalculated the oil pressure as a film flow, the needed horsepower dropped by a factor of one thousand. With his new calculations, the pumps for oil bearings would need less than one horsepower.

Kroon showed his figures to other engineers and scientists in the basement of the astrophysics building. He was greeted with skepticism. He offered to demonstrate that his oil bearings would work, and the Engineering Committee approved the expenditure of fifty-eight dollars for an oil pump to test the idea.

At Westinghouse a test model would require detailed working drawings, approvals of the plans, requisitions for materials and model-makers' time, layers of authorizations, and an interval to schedule the project in the modelmaking shops. Caltech, even with Sandy McDowell's efforts to introduce proper procedures, didn't run that way. The astrophysics machine shop was in the building next door. There were seventy machinists, under the direction of Sherburne, who had been recruited to run the shop from his own machine shop in Pasadena. Sherburne, a first-rate machinist himself, seemed to know the best machine operators and machinists. His shop turned out parts for the Schmidt telescope, new instruments for testing on the Mount Wilson telescopes, and models for the engineers and telescope designers with a minimum of paperwork and bureaucratic hassling.

When Kroon went over to the machine shop late in the afternoon with his sketches, he was introduced to a machinist and told to explain what he wanted.

"When do you need it?" the machinist asked.

"As soon as possible," Kroon answered. Looking at the huge shop, filled with machines and work in various stages of completion, he assumed that it would be weeks, maybe even months, before he had his model.

Early the next the morning, the machinist came over with the model Kroon had asked for. After the formality of work at the Federal Institute in Switzerland and the bureaucracy at Westinghouse, Kroon was amazed. The machinist had worked most of the night to finish the model.

Later that morning Kroon invited the engineers and scientists in the astrophysics building to watch a demonstration. The test model was a three-foot-square steel slab, six inches thick, loaded with lead weights to a total weight of 12,000 pounds, on an inclined plane. Three oil ports in the inclined plane were hooked to a pump and reservoir of oil. The slab was suspended on the equivalent of a hinge on one side so it could be moved over the pads, but the friction of the weight was so great that if the slab was moved from one side of the plane to the other, it remained in place.

Then Kroon started the pump. Oil pumped through the three orifices and spread into a thin film under the weight. When the pump reached full pressure, Kroon nudged the heavy weight to one side. It swung over, then back and forth. By measuring the amplitude of the swings of the pendulum, Kroon could measure the effectiveness of the oil film as a bearing surface. At the optimum setting, the weight oscillated freely from one side to the other, almost frictionless.

Even Pease, who had been the most skeptical about oil bearings, came around after Kroon's demonstration. The clearance between the weight and the inclined plane was only a few thousandths of an inch, but it behaved as Kroon's calculations predicted: A film of oil a few molecules thick could support hundreds of thousands of pounds of telescope. The coefficient of friction was so low that a fractional horsepower motor would easily move the telescope. Kroon's experiment was expanded to a large weighted bearing, carrying the estimated five-hundred-ton load of the telescope. Moving it at the normal driving speed required fifty foot-pounds; Kroon calculated that a ball bearing would require thirty thousand foot-pounds of driving force to move the same load.

Kroon's final design for the horseshoe bearing used four cavities in each pad for the oil orifices. Limiting the flow through the orifices from 600 to 300 psi (pounds per square inch) provided additional stability. In less than a week, Kroon finished his calculations; his sketches and figures, some on the back of an envelope, went to a draftsman. The bearing problem, which had haunted the telescope since Pease's earliest designs fifteen years before, was solved.

Next Kroon tackled the design of the horseshoe. If the telescope was to move smoothly, the bearing surface had to be stiff enough to remain round as the horseshoe tipped from resting on one edge, around through the bottom of the ○ and over to the other side.

The earliest ideas for the horseshoe were for an open trusswork design that would be riveted together in the field, the construction that had been used for the one-hundred-inch telescope. To remain rigid with riveted construction, the horseshoe would have required such heavy plates, and such closely spaced stiffening diaphragms, that the whole horseshoe design was almost abandoned. The strength of a welded construction from solid plates made the horseshoe possible again. The examples of the penstocks and cylinder gates of the huge Boulder Dam showed that massive structures could be welded. Still, the design was tricky. The loads on the horseshoe would put a combination of compression and shear loads on large plates of steel. The completed horseshoe would be some forty-six feet across, far too large to ship as a single unit, so the design also had to permit fabrication in sections small enough to move by rail, ship, and ultimately truck to the mountaintop. The challenge was to design internal stiffening members that would retain the needed stiffness under load and still be arranged in a way that would allow the internal structure of the horseshoe to be welded.

Kroon knew that it was possible, with enough calculation, to arrive at an analytic solution to the design. But he also remembered from his days as a machinist apprentice how often an analytic design by an engineer had made no provision for the difficulties of manufacture. One afternoon at the cottage on Los Robles, while his three-month-old son played on the floor, Kroon cut out sections of cardboard and glued them together in different configurations. He tested each model with tiny weights, measuring the deflection of the cardboard panels with a ruler. By the end of the afternoon, he had an internal design that met all his requirements. When he calculated the stresses the next week, they worked.

Kroon found the style of work in California, where an engineer could take rough drawings to the machine shop without review, authorization, or certification, both refreshing and productive. Westinghouse had sent him to California for six months. In the first two months he had solved two of the thorniest design problems. Even Pease was delighted to see more of the design details directed toward Kroon.

Serrurier's tube design, with his diagonal trusses, had solved the problem of building a relatively lightweight yet rigid tube for the telescope. Kroon was handed the problem of how to pivot the tube on bearings. Ball bearings could carry the load of the tube—only a fraction of the load on the horseshoe bearings. But any play in the mounting or strains passed to the bearings would distort the tube, and the careful alignment of the optical system. Theodor von Karmann, from the Aeronautics Department, had been studying the problem of the declination trunions, and had his men calculating the stresses.

Kroon had never seen a big telescope. He had no idea how tubes were usually pivoted in a telescope. But as soon as he saw the problem, he had an idea. The spokes of a bicycle wheel are stiff in tension and compression, though flexible in bending. A light bicycle wheel remains round even with a heavy rider going over a big bump. A ring of spokes, running from the telescope tube to the bearings, he reasoned, would locate the tube firmly. This problem was too tough to model in cardboard, so he calculated the loads and drew up sketches for the draftsmen. Once again his solution looked too simple. Von Karmann, the famed Caltech professor, looked at the thin spokes and asked, "Did you calculate the buckling loads?"

Kroon, awed that a world-famous professor was reviewing his sketch, hesitantly pointed out that the tension on the opposite spoke would balance the compression, maintaining the system in alignment. Von Karmann agreed, and yet another solution from the gentle Dutchman found its way into the telescope.

Kroon's gentle manner served him well. Though an outsider among a crew of Caltech astronomers and engineers, he was accepted as one of the team, invited on the camping trips in the desert led by Russell Porter, where the nighttime naked-eye star observations were interrupted by coyote howls. Kroon was in Pasadena when the mirror arrived, and like the others he felt the sudden glare of the publicity spotlights on the project he hadn't even heard of a year before.

His last project was the south bearing of the telescope, which would share the load of the entire fork with the great horseshoe bearing. He spent the Easter weekend "monkeying around at home" in the little Los Robles cottage, trying to think of a design that would create no forces to disturb structure. When he got stuck in his monkeying, he would practice music or play with his infant son. Finally, what emerged was a ball on oil bearing pads. The ball design would work under the enormous thrust loads of the south bearing, and could be powered by the same pump system that served the pads for the great horseshoe.

With the six months up and the design problems solved, Rein Kroon packed up and moved back to Philadelphia. Like the hero of a Western, he had come to town, cleaned up the troubles, saddled up, and ridden on, to other towns and other problems. He never saw the telescope.

23

The Endless Task

John Anderson had trouble sleeping the night the disk arrived at Caltech.

The glass disk had been unloaded from the Belyea Brothers truck onto a heavy timber easel strong enough to support three times the weight of the disk. On Easter Sunday, Marcus Brown and his workmen removed the front and back of the steel packing crate, revealing the marred face of the disk and the beautiful honeycomb structure of the back. Like the Corning officials when they had exhibited the disk to the public, "Brownie" put the disk on the easel with the back showing, although in his case it was only because the initial work was scheduled for the back of the disk. Brownie fretted for a long time to make sure the disk was secure. The circular sections of the crate were still in place around the edge of the disk, and the overhead crane that ran the length of the optics shop was left attached to the slings that fastened to two points on the crate. Anderson came over to check the easel and supports for the disk when it was finally secured in the optics lab.

In the middle of the night Anderson began to wonder what would happen if an earthquake hit Pasadena. From his work on seismographs, he knew how frequent earthquakes were in Southern California. What if an earthquake jostled the easel while the disk was still hooked up to the overhead crane? The thought was enough to get him to bolt from bed and drive to the optics lab. There was no earthquake that night. The disk was where they had put it. It was only the first of many nights that John Anderson lay awake worrying about the great mirror.

In 1935 and 1936 there were plenty of applicants for work in the optics lab. The pay wasn't great: starters got forty cents an hour; the most experienced men ninety-five cents an hour. But when word got out about a steady job, working indoors, men lined up. Some had no work; others wanted better work. Mel Johnson, a young mechanic, was willing to give up a one-hundred-dollar-a-month salary for the promise of steady work. Brownie warned each applicant with the same speech: "Glass won't ever do what you expect. It's human. It has as many moods as a movie star, and no two of 'em are alike. You've got to know them all. And remember this: If you don't know what you're doing, don't do anything till you find out."

The men listened, nodded, and went to work on a trial project. Most left after a few days or a week. It took special talents to work on optics, to stand up to the routine. Previous career was no indication of who had the temperament. The men Brownie hired included a failed insurance salesman, a man who had worked on a garbage truck, and a pump jockey from a filling station.

Work began every day at 8:00 A.M. Everyone who worked in the optics lab changed into a white shirt and trousers, cotton hospital uniforms, and canvas sneakers. Sometimes, depending on the work, the men needed a clean uniform each day. At lunch they changed out of the uniforms, ate from brown bags, and played handball outside and then changed back into the uniforms for four more hours of the routine. There were no breaks. Brownie rolled Bull Durham cigarettes and sometimes chewed Bull Durham while he worked, and others picked up the habit. A few men were interested in optics, and Anderson agreed to give evening classes to explain the mysteries of the mirrors and lenses. Otherwise the only times the lab was quiet enough for conversation was during lunch or while the men showered and dressed. The favorite topic was guns and hunting. Anderson was a hunter and had once shot a polar bear. But the gap between Caltech professors and Mount Wilson astronomers—the men who wore suits and ties—and the workers who actually ground the glass, was too great for small talk. Russell Porter, at heart still an amateur telescope maker, sometimes worked on his own projects in the optics lab. He was too deaf for much conversation.

The obsession with cleanliness in the optics shop was more than many men could stand. The floors were swept and washed daily. A worker rolled a magnet over the floor daily, sometimes several times a day, to pick up even tiny specks of metal. If a speck was found it was put into an envelope, and the search began for the culprit machine. Was it a chip off a gear? Abrasion of some metal part that no one heard because of the noise of the grinding machines? A foreign speck off the shoes or uniform of a careless worker? Whatever the cause, it had to be found. A speck of metal under a polishing tool on the surface of a disk could make a scratch that might destroy months of work.

Mostly the deadly routine got to men. The machines were huge, driven by electric motors big enough to power a large lathe or milling machine. Despite the size of the machines, glass can only be worked slowly. Removing millimeters of glass can take months of slow grinding. In the later stages of work, polishing a disk to optical tolerances of fractions of a millimeter can take years. Men used to seeing progress in their work, accustomed to a finished product at the end of the day, would leave in frustration. Week after week each day was exactly like the one before—tending a machine, performing a routine task, like feeding carborundum into a funnel-hopper on a grinding tool, sloshing the carborundum-water slurry over the face of the disk, or even sitting perched, for hours at a stretch, on a scaffold platform under the big disk-grinding machine, greasing a drive gear by hand to make certain the drive mechanisms did not gall. The unchanging routine, coupled with the fear that a single lapse could destroy a priceless disk, was more than many could stand. Marcus Brown said, "Time is worth less than glass around here."

Before the disk arrived Anderson and Brownie tested the big grinding machine with smaller pieces of glass. Brownie had also tried the lift with loads heavier than the mirror blank. But no amount of testing could assure everyone that the procedures and machines could safely move and work the heavy disk. From the easel where it had been stored on arrival, the disk first had to be lowered, face down, onto the surface of the big grinding machine. As the grinding and polishing proceeded, moving the disk on and off the machine, and from the horizontal grinding position to a vertical testing position, would become a regular weekly procedure. Brownie took the controls of the crane himself for the first moves. Every step was tentative, like a father holding his newborn child for the first time.

Before Brownie and his crew could begin the actual work of shaping the disk into a parabolic mirror, the two faces of the disk had to be true: absolutely flat and parallel to one another. He started with the backside. The ribs and pockets in the back would gall on a grinding tool, so the pockets and spaces between the ribs were filled with wooden tables, built so their surfaces were two inches below the level of the glass. The gaps were then filled with plaster of paris. The initial grinding was done with a seven-ton cast-iron disk, half the size of the mirror disk. The surface of the grinding tool was covered with glass blocks, held in place with wax. The actual abrasive was a slurry of carborundum and water, poured between the grinding tool and the surface of the disk as the tool rotated in a slow epicyclic motion that covered the entire disk.

Grinding glass, even a relatively simple task like truing the faces of the disk, is a job for the patient. What takes but a minute to describe took months on the grinding machine. Hundreds of pounds of carborundum and thousands of gallons of water went into the slurry. The glass blocks on the face of the grinding tool wore out rapidly and had to be replaced. The surface of the disk had to be washed completely each afternoon, lest water left on the disk etch the glass. When the back was ground flat, the plaster of paris and the tables were cleaned out and the disk was turned over.

The face of the disk was a sorry sight. The scars that George McCauley had discovered in the annealing oven looked like jagged wounds. One, near the center, had been excavated with a laboratory sandblaster to a depth of almost five inches, at a point where the disk would reach its maximum "dish."

Anderson had already decided that the top two inches of glass, including much of the "scar tissue" of the annealing oven wounds, would be ground off the disk before the actual shaping of the mirror began. The face of the disk would end up four inches thick instead of the originally planned five and one-half. The loss of rigidity wouldn't be a problem if the edge- and back-support system worked, and the thinner disk would have substantially less temperature inertia.

The grinding was tedious. The coarse carborundum grated as the iron grinding tool turned. The iron, in turn, reverberated the noise into a screech. Conversation was impossible. From the first days of work, Brownie calculated that they would need five tons of carborundum to remove the two and a half tons of glass on the surface of the disk. Five tons was a lot of screeching. Anderson came in daily to examine the disk. He was spared the screeching of the machine, but he had bigger worries.

Once the train carrying the disk left Corning, life settled into a kind of normalcy at the Corning Glass Works. There were still some smaller disks to cast for other observatories, which kept George McCauley busy. He was also much in demand for lectures on the casting of the disk.

For months tension had been brewing between McCauley and Hostetter. Hostetter had taken the limelight when the press returned to Corning for the unveiling and the shipment of the disk. He was good with reporters. He wasn't a scientist and had done no research; his job was an administrative position that brokered between the client and the research and production people. But when he spoke about the disk, the most famous product Corning had ever produced, he gradually shifted his voice from *we*, meaning himself and McCauley, to the first person singular. What reporters heard was that Hostetter, with the *help* of McCauley and others, had produced the great disk. Hostetter's assistant, George Maltby, officially a reports collector (which corresponded roughly to a timekeeper for different projects), spent much of his time promoting his boss.

McCauley, normally an easygoing man, resented Hostetter's open grab of credit. The disk had been McCauley's obsession. He had stayed awake nights worrying, had gone every day for a year to personally check the annealing. He had worked side by side with the workmen when the flood threatened the disk. And he had taken on the task of explaining the wounds in the surface to the Observatory Council.

By Christmas 1935 the two men weren't speaking. When a Christmas card arrived at the McCauley household from Hostetter, George McCauley sent back a note saying that under the circumstances he could not reciprocate. They went on that way for months. The breaking point came early in the summer of 1936, when an invitation arrived asking McCauley to speak in England on the casting of the disk. Hostetter had intercepted the invitation and accepted, *for himself*. McCauley went directly to Amory Houghton. Under the circumstances, he explained, he couldn't continue to work at Corning. Houghton knew George McCauley and his family from the Episcopal church they all went to each Sunday. He assured McCauley that his work was much appreciated. Everyone, Houghton said, was exhausted with this project. He suggested that McCauley take his family on a long vacation as soon as school was out. In the fall, when you get back, Houghton said, "We'll talk about it."

McCauley was packing the car for that vacation when he got an emergency call from California on Friday, July 3.

The caller was John Anderson. He asked McCauley to please be in California on Monday morning. Anderson didn't say why he wanted McCauley there, but it wasn't hard to guess. It also didn't surprise McCauley that Anderson said so little on the phone. From the first meetings in New York City, they had agreed that communication about the project, to the extent possible, would be oral and in person, to avoid headlines and publicity.

Only Houghton knew McCauley's destination when his chief scientist caught the late-afternoon Erie train for Chicago. There McCauley boarded a twelve-passenger United Airlines Boeing for the flight to Burbank, California.

The boyish enthusiasm of a physicist on his first plane ride competed with apprehension about what he might find in Pasadena. From the airport at Cicero, on the west side of Chicago, the plane bumped its way up to a flying altitude of 12,000 feet. McCauley watched the view from the window until they crossed the Mississippi, then settled back to read a new bestseller, *Gone With the Wind*, thinking the title particularly appropriate for his first plane ride.

The plane stopped in Omaha, Cheyenne, Laramie, Salt Lake City, and Las Vegas. The repeated ascents and descents, the sensation of flying through the Rockies instead of over them, and a rough passage through a thunderstorm outside Las Vegas left McCauley with an earache that swallowing and the ministrations of a stewardess with a swab of ointment couldn't cure. The earache persisted for six months, an unpleasant souvenir of his first plane trip, but the flight to Burbank had taken only twelve hours, twenty-seven hours faster than even the extrafare trains. He was able to spend Sunday afternoon with his brother in Los Angeles and show up rested for the Monday-morning meeting with Anderson.

They met at the optical shop and put on white coveralls and the required white cotton shoes in the anteroom. While the workmen stood back, Anderson and Marcus Brown showed McCauley a series of fractures that had appeared as the grinding tool cut down into the glass.

McCauley examined the fine fractures with a magnifying glass. He knew the disk had been thoroughly annealed, and while he couldn't explain what caused the fractures, he assured Anderson that the checks would disappear by the time they ground down to the final shape of the mirror. No one, including McCauley, had any experience with Pyrex castings that large, but McCauley's confidence was enough to assuage Anderson's worries about the disk.

As Anderson and Brown proudly showed off the optical shop equipment to McCauley, Brownie explained that the biggest problem they had encountered in the work so far, aside from the unexpected fractures, was that the glass blocks they were using on the face of the grinding tool were breaking. The face of the tool was a deep lattice, made by welding sheet stock in a crisscross pattern. Glass blocks four inches square and two inches thick were fastened into the openings in the lattice with molten wax. The spaces between the blocks allowed the carborundum-and-water slurry to flow freely. The disk worked well, but when they heated the wax to fasten in a new block, many of the blocks broke from the thermal shock.

McCauley offered a solution: Pyrex blocks. When Anderson protested that new blocks would take too long to make and cost too much, McCauley assured him that the Corning Glass Works would turn them out in no time and that the saving from having no more breakage would make up for any additional cost. McCauley flew back to Chicago and got back to Corning on Wednesday at noon. The entire trip and consultation had taken only five days.

The A Factory made good on his commitment to produce Pyrex blocks, which were on their way to Pasadena within a week. That done, he took up Houghton's suggestion of a long vacation. He packed his family into the Hudson and set off cross-country. They visited family in Missouri, stopped at Mount Rushmore, listened to George's story of his flight through the Rockies, and put up with his hobby of photography, which required sudden halts of the car, followed by a lengthy ritual of setting up his tripod and camera, then burying himself under a black cloth until he got everything just right to snap his plates. In California they visited the University of California and Caltech. McCauley sneaked in meetings with Robert Millikan and Ernest Lawrence. To McCauley's children, who had grown up with the long nights and constant anxiety over the disks, it seemed a real vacation. For the first time in years, they had gone weeks without hearing a word about the disk.

It was only when they got back to Corning in the fall, and saw Am Houghton greet George McCauley at church like a long-lost friend, his exuberance and enthusiasm out of all proportion to the short time McCauley had been away, that George McCauley's wife and children realized that the visits to Berkeley and Pasadena, and the private meetings with Lawrence and Millikan, hadn't been idle sightseeing. He had been exploring new jobs. But at work the next day, McCauley learned that Hostetter had left Corning and the Glass Works.

George Hale tried to follow developments on the telescope, but by 1936, his afflictions had become incapacitating. In the fall of 1935 he was suffering attacks of vertigo so severe that he had to cancel trips abroad and to the East Coast. The attacks would "come suddenly, without warning" and lay him up for weeks. Even the darkened room at the solar lab, where he had once sought solace from his demons, wasn't enough to ward off the vertigo. When the spells hit, Miss Gianetti, who had shifted duties from being Hale's private secretary to general secretary for the Palomar project, protected him from all visitors and calls. John Anderson had been accustomed to making oral reports to Hale, keeping him up to date on various aspects of the telescope. Francis Pease, Walter Adams, and others who had known Hale from Mount Wilson, would come to visit during the quiet periods between attacks. They would find Hale as alert as ever, a bundle of energy as he fired off questions about one detail or another of the project. When McCauley went to Pasadena in November 1935, to report on the state of the disk as it emerged from the annealing oven, Hale asked him and Anderson exactly what thickness of glass would remain if the wounds on the face of the disk were ground away. "If further defects do not appear below 3⅜-inch depth," Hale later reported to Max Mason, "and if no magic mirror defects develop," he was confident not only that the mirror would be satisfactory, but that the thinner mirror would be "better."

But as the year wore on, Hale realized that his lucid periods were fewer and farther apart than ever before. He missed the arrival of the disk, missed the party to inaugurate work on the mountaintop, and wasn't strong enough to visit the optical shop to see the disk in place. When problems developed or decisions had to be made, George Hale couldn't be counted on for the wisdom and experience that had guided so much of the telescope building.

Hale knew that his withdrawal from the project left a vacuum. John Anderson would increasingly be concerned with the figuring of the mirror. Sandy McDowell was efficient at arranging and supervising contracts and pushing along the work on the mountaintop, but he had no experience in astronomy or with telescopes, and his aggressive management style was increasingly resented by the scientists on the project. Walter Adams had his hands full with Mount Wilson, and Millikan, though he had never taken the title of president of Caltech, was running the institution, his time taken with fund-raising, faculty raiding, and aggressive promotion of the school. There didn't seem to be anyone who could step into George Hale's shoes.

To complicate the question, in November 1935 Max Mason announced his retirement from the presidency of the Rockefeller Foundation. Mason had taken over at the Rockefeller Foundation just after the grant for the telescope had been announced. He had presided over the reorganization of the foundation and boards, consolidating all science activity in the Rockefeller Foundation. The payments to Caltech for the telescope came first from the IEB, which had made the grant, and then from the GEB, which had taken over responsibility for the grant after the IEB funds were depleted. But even as the checks came out of those accounts, it had been Mason and the men he hired at the Rockefeller Foundation, like Warren Weaver, who supervised the grant. Mason had fended off the interference of Merriam at the Carnegie Institution, handled some of the trickier negotiations with GE, recommended McDowell to supervise the construction work on the telescope, and served as a sounding board for Hale.

In 1932 Hale had even asked Mason to be chairman of the committee on design for the mirror of the telescope. It had been a largely symbolic position, but Hale felt that Mason—as a theoretical and experimental physicist—could pull the work of Pease, Day, McCauley, and others together. It had been an unusual step to ask the president of the foundation supporting the project to participate actively in the project. Usually grant recipients do all they can to keep the funding source at arm's length, lest the leverage of funding influence a project. But Mason's counsel had been so valuable, and his authority as a scientist and administrator of science and head of the Rockefeller Foundation so effective, that Hale treated him as a member of the Observatory Council in all but name.

Mason's retirement gave Hale an idea. In March 1936 Hale suggested that Mason come to Pasadena and accept dual appointments as a research associate at Caltech and vice chairman of the Observatory Council. Confidentially it had been agreed that with George Hale's health rapidly failing, Mason could serve as the chairman of the council in case of Hale's absence or incapacitation. Mason agreed, and by October 1936 he was settled in Pasadena, and already getting McDowell's memos on pay scales, organization charts with side boxes and arrows, copies of memorandums on contract details, and showing copies of memos on myriad other details. Mason still consulted with Hale whenever Hale was able to receive visitors, but before long he wrote to Ray Fosdick at the GEB that the work was going well "without Hale."

Mason had been chairman of the Observatory Council for just two months when another crisis threatened the entire project.

24

Crisis

George McCauley had been looking forward to the calm of a Corning autumn. There were still a few back orders of astronomical disks to fill, but the crews were experienced, the procedures practiced, and the technical details established. Although they were still made under the auspices of the research department, telescope disks had become a production process at Corning. McCauley, who had developed the processes and refined the technology, looked forward to answering the mountains of mail he had received, writing up articles for the journals on the engineering and physics issues involved in making the disks, and responding to the many requests for lectures on the disks, especially the now-famous 200-inch disk. His expectations were short-lived.

At the optics shop on California Street, Brownie and the crew were steadily grinding down the surface of the big disk. Anderson and Brownie had been buoyed by McCauley's calm assurances during his visit in early July and expected that at any point the grinding would finally work through the last of the troublesome checks and fractures. Some weeks the work went well. When Brownie and his crew washed the disk down on Friday afternoon, in preparation for Anderson's careful examination, there were times when many of the troublesome fractures of the week before had been washed away in the slurry of carborundum, glass, and water.

But even in the best weeks, Anderson would find new fractures, deeper in the glass, especially in a dark stria that showed up halfway between two of the pour points. He began sending McCauley regular reports, often with photographs of the fractures they had discovered.

The photographs wrecked McCauley's plans for a quiet fall. He had believed his own assurances to Anderson early in the summer. There hadn't been comparable troubles with the other disks, including the 120-inch disk. What had been so different with the two-hundred-inch disk?

The continued reports and photographs got him worrying. Maybe this disk was different. His first guess was that either the shutoff of heat to the annealing oven during the flood or the contact of the cover plate with the surface of the disk was to blame. Yet his tests with a polariscope after the annealing had showed minimal strains in the disk; as frightening as the shutoff of power during the flood had been, it didn't seem to affect the annealing. And as it became clear that the fractures were distributed deeper in the disk than the surface wounds from the contact with the cover of the annealing oven, McCauley knew he needed a different explanation.

He recruited Ralph Newman, his lab coworker from the days of the first experiments for the mirror project, to work with him. Painstakingly they reconstructed the pouring procedure. The proximity of many of the fractures to the three pour points was suggestive, but the nagging fact that at least three big disks produced by the same pouring process, including the 120-inch mirror, had been successfully ground with no checks or fractures, wrought havoc with every explanation McCauley tried.

By December, Brownie and his crew had used close to five tons of carborundum in the rough surface grinding. More than two tons of glass had been ground off the disk. The surface level at the edges was approaching the final planned thickness of the disk. On Friday, December 11, Brown gave the disk a thorough cleaning. The next morning, Anderson and Max Mason came to the shop to inspect the disk with Brown. There were fewer fractures in the glass than a month before, but there was at least one relatively large check that was not near the flow points. In a Monday letter to "Dear Mac" (McCauley), Mason suggested that it would be useful to simulate the flow procedure of the pour with a model to determine the cause of the fractures.

It was an interesting suggestion. In Mason's own research at the navy underwater research labs in New London, the modeling of flow patterns was an essential test procedure. But water flowing past the hull of a submarine or a surface ship is easier to model than molten glass. Glucose and water, the closest material McCauley knew of to simulate the flow of molten glass, was not temperature-sensitive enough to mimic the behavior of Pyrex.

McCauley tried to appear calm and reassured Mason and Anderson in California that the checks would eventually disappear before they ground the glass down to the final curved surface of the mirror. This time McCauley's assurances weren't convincing. In January, Mason wrote Houghton—the mail was marked PERSONAL and CONFIDENTIAL—to ask the cost and delivery date for another two-hundred-inch mirror. McCauley agreed to go out to California for another look. All discussion of the state of the disk, in Pasadena and in Corning, was kept secret. After the incessant attention of the press during the unveiling of the disk and the journey across the country, both Pasadena and Corning were wary that a reporter would seize on a mention of a

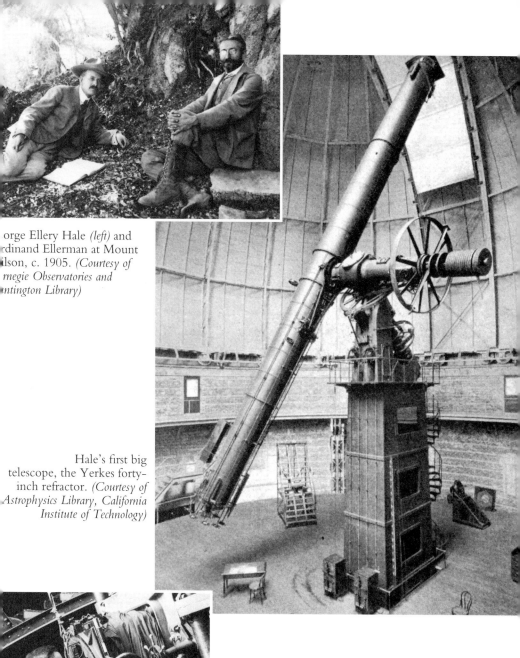

orge Ellery Hale *(left)* and
rdinand Ellerman at Mount
lson, c. 1905. *(Courtesy of
rnegie Observatories and
ntington Library)*

Hale's first big
telescope, the Yerkes forty-
inch refractor. *(Courtesy of
Astrophysics Library, California
Institute of Technology)*

Edwin Hubble observing at the
Newtonian focus of the hundred-
inch Hooker telescope, 1922. The
bentwood chair is traditional for
observers. *(Courtesy of Carnegie
Observatories and Huntington Library)*

Francis Pease's first drawing c
proposed three-hundred-ir
telescope is shown below. T
domes of the hundred- and six
inch telescopes are superimpos
(Courtesy of Astrophysics Libr
California Institute of Technolo

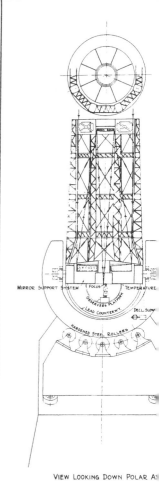

MIRROR SUPPORT SYSTEM · FOCUS · TEMPERATURE

OBSERVERS PLATFORM · LEAD COUNTERWT · DECL SLOW

HARDENED STEEL ROLLERS

VIEW LOOKING DOWN POLAR A

ABOVE: The hundred-inch Hooker telescope at Mount
Wilson, c. 1928. The night assistant's desk, where the observations were recorded on the fateful first light, is in the
foreground. BELOW: Cleaning the hundred-inch mirror
for resilvering. The troublesome bubbles in the blank are
clearly visible. *(Courtesy of Rockefeller Archive Center)*

George Hale at the solar image in the National Association of Science building in Washington, at the time of the great debate. *(Courtesy of Carnegie Observatories and Huntington Library)*

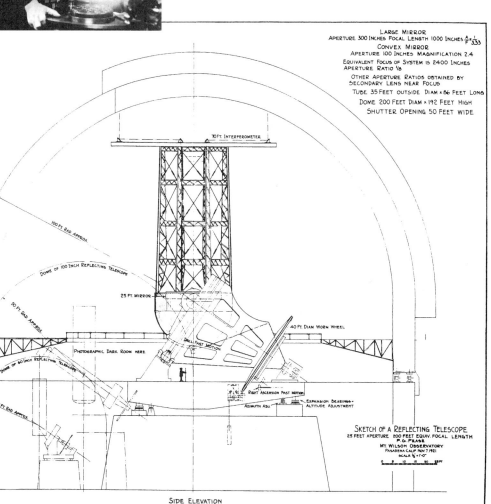

LARGE MIRROR
APERTURE 300 INCHES FOCAL LENGTH 1000 INCHES $\frac{A}{F} = \frac{1}{3.33}$
CONVEX MIRROR
APERTURE 100 INCHES MAGNIFICATION 2.4
EQUIVALENT FOCUS OF SYSTEM IS 2400 INCHES
APERTURE RATIO 1/8
OTHER APERTURE RATIOS OBTAINED BY
SECONDARY LENS NEAR FOCUS
TUBE 35 FEET OUTSIDE DIAM x 86 FEET LONG
DOME 200 FEET DIAM x 192 FEET HIGH
SHUTTER OPENING 50 FEET WIDE

70 FT. INTERFEROMETER

100 FT. RAD. APPROX.

DOME OF 100 INCH REFLECTING TELESCOPE

50 FT. RAD. APPROX.

25 FT. MIRROR

40 FT. DIAM. WORM WHEEL

DOME OF 60 INCH REFLECTING TELESCOPE

DRILL FAST MOTION

PHOTOGRAPHIC DARK ROOM HERE

RIGHT ASCENSION FAST MOTION

AZIMUTH ADJ.

EXPANSION BEARINGS +
ALTITUDE ADJUSTMENT

SKETCH OF A REFLECTING TELESCOPE
25 FEET APERTURE 200 FEET EQUIV. FOCAL LENGTH
F. G. PEASE
MT. WILSON OBSERVATORY
PASADENA·CALIF NOV. 7. 1921
SCALE 1/8" = 1'-0"

SIDE ELEVATION

A. L. Ellis, Elihu Thomson, Walter Adams, and H. L. Watson in front of the furnace at General Electric's West Lynn plant, during the spraying of the first sixty-inch blank. *(Courtesy of Hall of History, Schenectady, N.Y.)*

LEFT: A ladle of glass being poured into the mold in the beehive oven, during the pouring of the first two-hundred-inch disk. *(Courtesy of Corning Museum of Glass)*

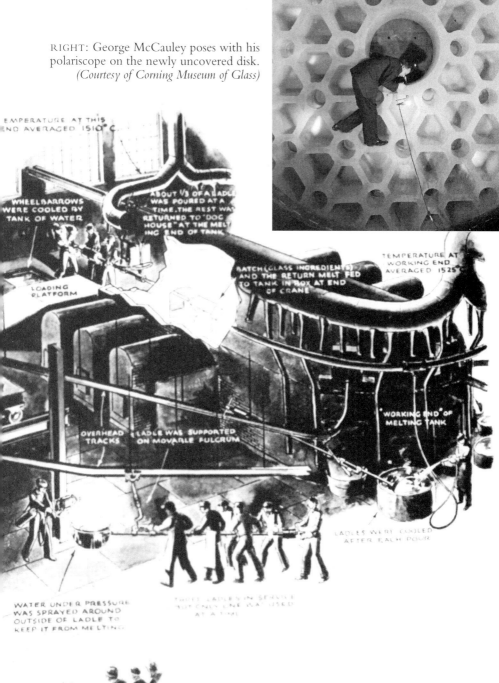

RIGHT: George McCauley poses with his polariscope on the newly uncovered disk. *(Courtesy of Corning Museum of Glass)*

TEMPERATURE AT THIS END AVERAGED 1510°C.

WHEELBARROWS WERE COOLED BY TANK OF WATER

ABOUT ⅓ OF A LADLE WAS POURED AT A TIME, THE REST WAS RETURNED TO "DOG HOUSE" AT THE MELTING END OF TANK

LOADING PLATFORM

BATCH (GLASS INGREDIENTS) AND THE RETURN MELT FED TO TANK IN BOX AT END OF CRANE

TEMPERATURE AT WORKING END AVERAGED 1525°C

OVERHEAD TRACKS

LADLE WAS SUPPORTED ON MOVABLE FULCRUM

WORKING END OF MELTING TANK

LADLES WERE COOLED AFTER EACH POUR

WATER UNDER PRESSURE WAS SPRAYED AROUND OUTSIDE OF LADLE TO KEEP IT FROM MELTING

THREE LADLES IN SERVICE BUT ONLY ONE WAS USED AT A TIME

The pouring of the of the two-hundred-inch disk, as celebrated in the popular press at the time. (*Courtesy of* Popular Science *magazine*)

The disk arrives in Pasadena. *(Courtesy of Astrophysics Library, California Institute of Technology)*

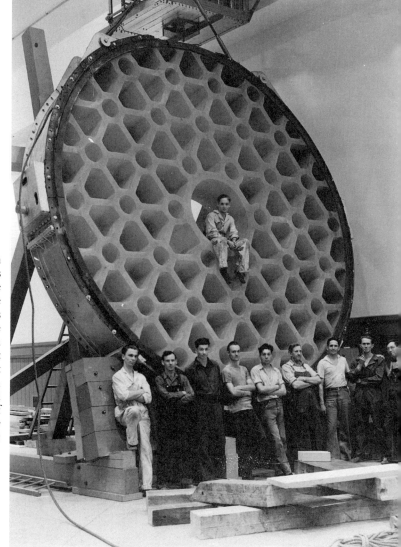

Marcus Brown *(far right)* and his crew, with the disk on its storage easel after it was unloaded from the truck. John Anderson couldn't sleep that night worrying that an earthquake would tip the mirror off the easel. *(Courtesy of Mel Johnson)*

ABOVE: The optics lab from the visitors' gallery, with the test station for the two-hundred-inch mirror and a polishing machine for an auxiliary mirror in the foreground. BELOW: Straightening the rubber skirts around the disk before resuming the polishing. The worker in the background is smoking a pipe; most of the others smoked cigarettes or chewed tobacco. *(Courtesy of Mel Johnson)*

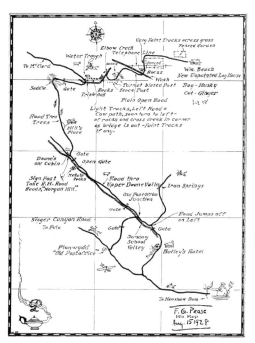

Francis Pease's first map of the route up to Pa[...]mar, made before the grant for the telescope [...]officially announced. *(Courtesy of Sylvia Marsh[...]*

Sinclair Smith, who did not live see the telescope finished. *(Courtes[...] of Olin Wilson)*

BELOW: The mountaintop as construction began. The WPA camp is in the foreground; the wa[...] tower is at the left; the site for the two-hundred-inch telescope is in the middle, with work on t[...] "small" Schmidt camera in the right background. *(Courtesy of Hagley Museum and Library)*

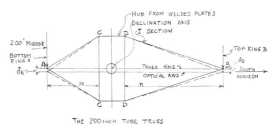

Mark Serrurier's first drawing of the tube truss. In his notes he explained that "dimensions m and n do not change as telescope moves." *(Courtesy of Naomi Serrurier)*

hn Anderson, Captain McDowell, Max
.ason, and Russell Porter inspecting the
rly steelwork. *(Courtesy of Rockefeller
.rchive Center)*

Captain McDowell with
.e one-tenth scale model
.at served as telescope for
generations of Caltech
.ndergrads. The external
piping is for the oil pres-
.ure bearings. *(Courtesy of
Astrophysics Library,
California Institute of
Technology)*

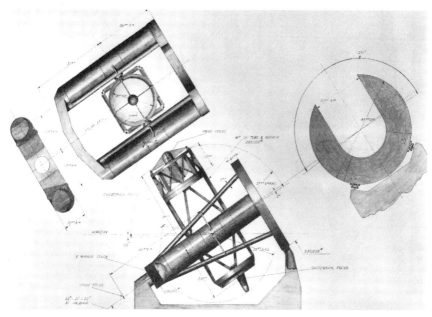

A large-scale drawing of the mounting prepared by the Westinghouse engineers. *(Courtesy of Hagley Museum and Library)*

Westinghouse workers pose inside a section of the telescope tube. *(Courtesy of Westinghouse Corporation)*

Preparing the horseshoe bearing for machining. The holes allow workers inside to weld the internal structures Rein Kroon designed. *(Courtesy of Hagley Museum and Library)*

A bemused Albert Einstein (*front row, center*) on the dais at Westinghouse's celebration of the completion of the telescope tube. *(Courtesy of Hagley Museum and Library)*

ABOVE: Trucking the mounting, here a section of the horseshoe bearing up the south grade of Palomar Mountain in 1938. Two tractors are pulling and one is pushing the heavy load, while a man walks ahead. *(Courtesy of Hagley Museum and Library)*

LEFT: Bringing the sections of the mounting through the hatch into the dome. After unloading the section, Byron Hill said he finally understood what women went through at childbirth. *(Courtesy of Hagley Museum and Library)*

Rein Kroon's bicycle-spoke declination bearings. *(Courtesy of Mel Johnson)*

Ira Bowen, Lyman Briggs of the National Geographic Society, Rudolph Minkowski, Don Hendrix, Robert Harrington, Albert Wilson, and Byron Hill at the forty-eight-inch Schmidt telescope for the beginning of the first Palomar Sky Survey. Harrington and Wilson were the principle observers for the survey. *(Courtesy of Astrophysics Library, California Institute of Technology)*

BELOW: The completed telescope, looking northwest. *(Courtesy of Astrophysics Library, California Institute of Technology)*

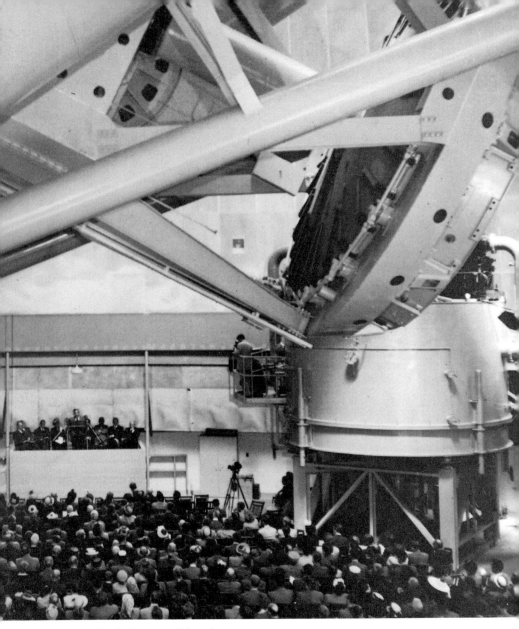

ABOVE: The dedication of the telescope. Raymond Fosdick of the Rockefeller Foundation is speaking. The mirror is at the upper right, covered by its protective diaphragm; the aluminizing tank is behind it. *(Courtesy of Rockefeller Archive Center)*

OPPOSITE PAGE: Russell Port
breakaway drawing of the telesco
(Courtesy of Astrophysics Libr
California Institute of Technol

APPROXIMATE· ·SCALE·

R·W·PORTER·

TWO·HVNDRED·INCH·TELESCOPE·~

Russell Porter's first idea of an observing station at the Cassegrain focus of the two-hundred-inch telescope. Astronomers didn't like the idea of observing from what looked like a child's swing. *(Courtesy of Astrophysics Library, California Institute of Technology)*

Edwin Hubble in the prime focus cage of the two-hundred-inch telescope. The mirror is open below. *(Courtesy of* Life *magazine)*

The two-hundred-inch telescope dome today. *(Courtesy of Jean Muell*

fracture in the disk for banner headlines: GREAT EYEGLASS FLAWED or THE EYE THAT COULDN'T SEE. And Robert Benchley was always waiting to let loose another barb:

> I hate to keep harping on this subject, but what do they *do* with gigantic telescope discs out there in California—eat them? . . . If you ask me, they have got started making gigantic glass lenses up at Corning and can't stop. And California is being made the sucker. I go to California every year, and I have never seen any more glass that you could put a highball into. They must just throw the glass away when it finally, after weeks of publicity, gets out there.

McCauley took another secret trip to Pasadena at the end of January, 1937. He and Anderson pored over the disk, examining the troublesome areas with magnifying glasses. When McCauley wanted to take samples of the disk back to Corning for analysis in their laboratories, Anderson said that it would probably be quicker if they did whatever tests McCauley wanted at the Mount Wilson laboratories on Santa Barbara Street in Pasadena. He phoned the lab, and a few minutes later, he and McCauley showed up with two samples bored from the disk, one of clear glass, the other from a cord that had developed checks.

At Santa Barbara Street, while he waited for the spectroscopists to test the samples on their large-grating spectrograph, McCauley got into a conversation with the opticians who were working on various smaller mirrors. One older optician said that he had run up against checks like the ones McCauley described. On an optics bench, he showed McCauley the face of a mirror that was fully figured and polished, awaiting silvering. After the optician showed him where to look, McCauley could see a fine fracture, approximately an inch long, in the surface of the mirror. The optician assured McCauley that the fracture would present no problems in silvering the mirror and would not impair its performance. McCauley asked for the records on the disk, and found that it was Pyrex and had been made before the switchover to the 715-CF glass that McCauley had used for the bigger disks. It was scant consolation that the two-hundred-inch wasn't the only disk from Corning that had problems.

Before he left McCauley did his best to reassure Anderson that the fractures would disappear. He spent the whole flight back thinking about the results of the spectrographic tests. The two samples of glass were strikingly different. The spectra from the cord of glass with checks contained a higher content of alumina and soda than the sample from the clear glass—a clear indication of contaminated glass. The only possible source of that contamination had to be the refractory lining of the tank. But why did it affect only that disk? By the time he reached Corning, he was at "a new low in spirits." Would the checks continue so deep into the blank that the two-hundred-inch disk couldn't be figured? Would the whole telescope project have to be abandoned for lack of a mirror? The dread of affirmative answers to those questions haunted him.

The only way McCauley could dispel his depression was by work. On the plane, and when he got back, he visualized the processes at work as a newly lined glass tank was filled with glass and fined. The procedure was cautious and precise; he was sure that the contamination wasn't introduced then. That narrowed the possibilities to two points in the procedure: the agitation of the glass mixture against the walls of the tank by the ladles when they were filled, which might have stirred off contamination from the walls of the tank; or drainage down the walls of the tank as the glass level dropped.

Every large disk had been poured with ladles, and the same men had handled the ladles each time. That seemed to rule out the agitation of the glass as the culprit. He went over the list of disks they had produced each year. The big disks cast in 1933, including the 120-inch disk, had all been successfully ground without showing checks. Two 60-inch disks for Harlow Shapley at Harvard had also been successfully ground.

He narrowed his search to the disks cast *after* those successful ones. The first two-hundred-inch disk and an 86-inch disk for Heber Curtis had both been rejected, for reasons unrelated to the tiny fractures. Both blanks were still at Corning. When he got home, he stayed up late at the round table in the dining room, going over the records of the pours, making sure of his analysis.

In the morning he examined the first two-hundred-inch disk, still resting on its bed of timbers in the steel crating room on the riverbank. Knowing what to look for, he quickly found the telltale fractures in the glass. The 86-inch disk that had been cast for Heber Curtis at the University of Michigan, which had emerged from annealing cracked, showed the same fractures.

McCauley had been working on the mirror project for almost ten years. Now, after surviving the floating cores, the first false starts with the wrong glass, the grand public failure with the first mirror casting, a flood, an earthquake, the threats of preachers who had damned the project, and the trials of a journey across the country by rail—it seemed that they were using the wrong refractory material in the tanks, the wrong glass, or the wrong procedure.

It had seemed so simple to copy the ladling procedures of the old window-glass industry. But window glass was ladled at a low temperature, and the tanks were so huge that the glass level scarcely dropped as the ladles of glass were removed. By returning the ladle skins to the tank, he had minimized the drop in level, even for the two-hundred-inch disk, to 10 inches. It was impossible to melt a tank large enough that the amount ladled out for a big telescope disk would not appreciably lower the level of the tank. When the glass drained down the walls there was a risk of contamination from the refractory brick. The older the brick, the more glass that had been recycled through the tank from ladle skins, the greater the risk. What had seemed a terrific production procedure—reviving the old technique of hand ladling—had backfired.

After he received a letter from Max Mason reporting that although they hoped "the glass would behave better as deeper depths were reached" the fractures seemed to be "distributed throughout," Houghton asked McCauley to prepare quotes of a price and a delivery date for a new mirror.

The situation didn't look promising.

When the grant was first awarded, $6 million—the budget figure George Hale had pulled out of the air of his room at the University Club—seemed generous enough for every eventuality. Eight years later, after close to a million dollars had been expended on the GE experiments, and hundreds of thousands more for the series of mirrors from Corning, the machine and optics shops and the astrophysics building in Pasadena, and the site work on Palomar, the grant no longer seemed bottomless. Westinghouse had estimated the cost of the fabrication for the mount and tube at $0.37 per pound, a low, depression-era bid. For one million pounds of telescope mount, it would still come to $370,000, which did not include the cost of transporting the huge pieces to Palomar or assembling them there. The control system for the telescope, and the various eyepieces, spectrographs, electronic sensors, and other auxiliary equipment would also make demands on the budget.

And even as the cost of the telescope mounted, there were new demands for additional instrumentation. The Schmidt camera on Palomar was performing beautifully on Zwicky's searches for supernovas. The astronomers began to discuss the usefulness of a much larger Schmidt camera—as large as opticians could fabricate—to serve as a wide-area survey camera for the two-hundred-inch telescope. Hubble gave the proposed camera his blessing. When he had first searched for novas and Cepheid variable stars with the one-hundred-inch telescope, he had to guess at promising areas and photograph them with the narrow field of view of the big telescope. As he and Milton Humason expanded their search for distant galaxies, they were still using the same technique of *guessing* at the most productive areas to study. The eighteen-inch Schmidt camera had proved that the concept of a wide-field camera worked, but the limiting magnitude and image size of the small Schmidt camera weren't large enough to serve as a survey camera for the two-hundred-inch telescope. Hubble urged that they build as large a Schmidt camera as possible, preferably a fifty-inch $f/2$ instrument.

The astronomers and the Observatory Council all enthusiastically endorsed the proposal, and Max Mason sent feelers to his former colleagues at the Rockefeller Foundation to see if they would approve the expenditure for a large Schmidt camera within the terms of the grant, which allowed for "the purchase of a site, and the construction of an observatory, including a 200-inch reflecting telescope with accessories, and any and all other expenses incurred in making the observatory available for use." Mason was sure he could get the new Schmidt telescope approved, but money spent for a big Schmidt camera came out of the same $6 million budget. Would there be enough left to pour, anneal, transport, and do the preliminary grinding on a new disk for the two-hundred-inch telescope—which might prove no better than the disk they had? How much was a reserve mirror worth in the now-tight budget?

George Hale had been able to discuss questions like that one with Mason. Now Mason had crossed over to the Observatory Council, to being the grantee instead of the grantor. His relationship with his former colleagues at the Rockefeller Foundation was more personal than consultative. Their correspondence, sometimes marked PRIVATE, interspersed comments on mutual friends or fine wines and cognacs and armagnacs into the commentary on projects. When Mason reported on developments on the telescope project, he mentioned that they were still finding fractures at the point when all traces of fractures and checks should have been long gone. But the mention was quickly glissaded over in favor of a discussion of the difficulty of getting a decent 1926 Chambertin.

With no one to ask for advice, and with the pressures of a dozen other aspects of the project pressing on him from all sides, Mason set the figure they would be willing to pay for another disk at fifty thousand dollars, exactly half of McCauley's estimate of Corning's cost, not including overhead or profit. If Corning could quietly produce a new disk for fifty thousand dollars, the council would go ahead with the order. In private correspondence Mason negotiated with McCauley and Houghton, pointing out that the problems with the disk weren't due to the shutoff of power during the flood or the earthquake—the sort of "Acts of God" that are exempted in warranties and guarantees—but were instead striae in the glass from contamination with alumina from the walls of the melting tank.

McCauley's estimate of one hundred thousand dollars for a replacement disk was arbitrary. He had already dismissed two procedures for avoiding contamination in the glass as impractical. A melting tank large enough not to show a significant drop in level from the removal of the glass for a two-hundred-inch disk would cost more than one-third of a million dollars to construct. And pouring the disk in layers, over a period of days or even weeks, would require that the mold withstand the high temperatures for the entire period; it wasn't clear that any refractory brick and any technique for cementing the complex surfaces of a mold could withstand that heating.

Those options were so unpromising that McCauley decided that all future disks would be made by an entirely different process than any they had tried. They would use only glass that had been in contact with unused refractories. A newly lined tank would be filled with the glass mix, heated, and held long enough to fine the glass. The tank would then be cooled, and the glass would be mined from the tank in blocks, with cleanly fractured faces. To make mirror disks, the blocks of pure glass would then be placed on a mold under the beehive oven and heated until the glass sagged into the mold. They had already tried the process with smaller disks. No one knew if it was possible to sag the glass for a two-hundred-inch ribbed disk into a mold.

McCauley could make no assurances about a delivery date. With enough glass on hand for the back orders of telescope disks, and with a growth of orders for commercial products in 1936, the 3A melting tank, which had been used to melt Pyrex for the telescope disks, had been converted back to use for baby bottles. Much of the equipment for the casting and annealing ovens had been dismantled, and many of the experienced personnel dispersed to other projects. Even if they succeeded in producing a mirror by the new process, it would require a year of annealing—with all the perils that entailed—and another journey across the country.

The negotiations went slowly. McCauley's analysis suggested that the glass that had been poured into the mold first should be better than the glass added later. Even in the heated casting igloo, the viscosity of the molten Pyrex permitted only limited homogenization of the glass in the mold, so the contaminants that affected the later pours of the ladles might not have reached down into the layer of glass close to the mold. If he was right, Anderson and Brown should start seeing a rapid decline in the frequency and gravity of the fractures.

All any of them could do was wait and see.

25

Big Machines

The Westinghouse South Philadelphia Works was an enormous factory, acres of space with machine tools as large as any in the world. Their specialty was machining hydroelectric equipment and steam turbines for ships, but American shipbuilding had been slack for years, the union movement of the 1930s had made inroads at the South Philadelphia Works, and the large work force meant a very expensive payroll for Westinghouse. The company already had slack capacity and what public relations firms would later call an "image" problem.

Fortune magazine wrote that Westinghouse wore "seven-league boots" and straddled the "entire market." Its irons pressed gowns in penthouse apartments and red flannels in dude ranch laundries, its generators provided the current to run chippers in Canadian pulp mills and air drills in Arizona copper mines, its motors drove steel slabbing mills and "one-mouse-power" electric razors. But so little of its production was products that the public had seen that most Americans had heard of Westinghouse only in association with broadcasting or home appliances. The contract to build the mounting for what the newspapers now routinely called "the greatest scientific machine in the world" would not only keep machinists and machines busy but would provide marvelous opportunities for publicity that would promote a corporate identity.

Eager as it was for work, Westinghouse was also a company and a shop with a long tradition of doing work its way. The engineers were accustomed to working from large-scale drawings that covered every detail of the proposed work. They would then fabricate, weld, machine, and heat-treat the components, preassemble the turbine on the shop floor to make sure every part fit as intended, and finally disassemble the parts for transport to the shipyard, where they could be reassembled by shipwrights and mechanics. It was a solid, reliable procedure. Since even the largest turbines could easily be accommodated in the factory and foundry buildings, and the workers were accustomed to the procedures involved in assembling the machinery, the extra cost for the conservative approach was modest. Jess Ormondroyd, in charge of the experimental division and with many years' experience on large machinery behind him, was convinced that the procedure was the only safe and reliable way to build large fabrications.

The Caltech engineers had a different idea.

As early as January 1935, when many features of the design of the telescope were still undecided, McDowell heard that Corning had received orders for one-tenth-size models of the telescope mirror—disks with the same ribbed-back design as the big mirror, but only twenty inches in diameter. He proposed that Caltech order one of the smaller mirrors and use it to build an exact one-tenth-size model of the telescope, with every system used on the big one. Not only could the model pin-point design and engineering problems, but it could ultimately serve as a guide scope for the bigger telescope.

The astronomers patiently explained to the navy captain that his idea wouldn't work. The guiding mechanism for the two-hundred-inch tele-scope, like those on the sixty- and one-hundred-inch telescopes, would have to be internal to the telescope, relying on the main optics. Even if an external guide scope were needed, a twenty-inch f/3.3 telescope would not make a good guide scope.

The guide scope idea was abandoned, but McDowell was still intrigued with the possibilities of building a model of the telescope. As the design stages progressed, he had Russell Porter build a celluloid model of the telescope at one-fiftieth scale. By then Corning had begun producing the twenty-inch-diameter replica disks; one arrived at Pasadena for Caltech; and McDowell again took up the model idea, arguing that systems like the oil bearings and the spoked declination-bearing supports could all be tested on the model.

The engineers pointed out that a model wouldn't really test the design, because many of the engineering concepts couldn't be scaled. The loads of a one-tenth-scale model were so much smaller, and the harmonic frequen-cies of the parts so much higher, that deformations, friction of moving parts, or vibrations of the model weren't a useful indication of what would happen with the full-size machine. McDowell was quick to concede their points, but a model, he argued, would test how the various assemblies mated with one another and would provide a working demonstration of the new ideas, like the horseshoe bearing and the oil pads. He ordered the machine shop to build the model. Except that there would be no observer's cage—even Caltech freshmen weren't that small—and that the hydraulic piping ran in external pipes instead of lines within the mount-ing, the small telescope was meant to be an exact working scale model.*

---

* The model was used at Robinson Hall by Caltech undergraduates in astron-omy for many years. It was recently traded to a community college in Corning for a newer fourteen-inch telescope. The community college plans to renovate the one-tenth-scale telescope.

For some problems the model provided valuable data. Although the deflections of the tube in the model would be much smaller than in the actual telescope, the engineers put together a system of mirrors that could accurately measure deflections of 1/100,000 inch, fine enough to prove that Mark Serrurier's tube design would work and that the complex forces in the yoke assembly would cancel one another out.

Westinghouse built its own model of the mounting out of celluloid, using small brass weights to load the model while micrometers and electrical microswitches and microammeters measured the deflection of various components. Stress lines would also show up in the celluloid.

McDowell originally wanted to solicit competitive bids for the mounting, the way he had done with navy projects. The bid from Westinghouse was a minimum of $800,000 and a guaranteed maximum of $1,100,000, far more than the Observatory Council could afford. Hale urged that instead McDowell negotiate a cost-plus arrangement, essentially the same sort of agreement the council had with Corning Glass. The new agreement held the cost down, but the bills from Westinghouse still raised questions.

George Hale, eagle-eyed despite his increasingly frequent attacks, was the first to notice the problem. In January 1936 he read through the correspondence with Westinghouse and noticed what seemed like enormous "miscellaneous" expenditures, including revamping the foundry building ($81,250) and building a new turning rig ($62,500). There were also large expenditures proposed for moving sections of the mounting to the Sun Shipbuilding Company for annealing in their furnaces, and to other Westinghouse plants in Pittsburgh for portions of the machining too large to be undertaken in South Philadelphia. Hale was worried about the expenditures. Between the work in the optics shop and the machine shop, the grading and site work on the mountain, and the proposed contracts at Westinghouse, the expenditures for 1936 were rapidly sliding toward $600,000—one-tenth of the entire grant.

McDowell had pushed for Westinghouse to build the large components of the mount because they had the facilities to fabricate, weld, and machine very large structures. Why, then, did Caltech have to pay to expand the Westinghouse foundry building or build a new turning rig to machine parts of the assembly?

The answer was that Westinghouse planned to preassemble the entire structure in its South Philadelphia plant before it was shipped to California. Just the tube of the telescope was so large that to assemble it vertically would require excavation of the floor of the foundry, the tallest portion of the plant. Assembly of the huge horseshoe and testing of the oil pads would require major modifications to the plant. As the negotiations went on, the Westinghouse metallurgist, Norman Mochel, urged that the annealing all take place at the South Philadelphia plant, where the exact positioning of the various components during the annealing could be controlled. If fabrications were heat-treated in the wrong positions, heat-induced sagging could spoil the alignment so that components designed for precision fits would not match. To anneal the parts in the plant, a temporary furnace would have to be built to accommodate the large components.

Mark Serrurier, sent east to investigate, thought the Westinghouse estimates were based on "meager information and very little detailed study." Sherburne, the machinist in charge of the astrophysics machine shop at Caltech, agreed. Between the faith of the Caltech engineers in their design, and McDowell's trust in his one-tenth-scale model, Pasadena was confident that the mounting did not need to be test-assembled at Westinghouse. The savings in money and time would be considerable. The Westinghouse engineers were equally stubborn and willing to budge on the billing if their procedures were followed. As Guenther Froebel argued, "a maximum investment of fifty to seventy-five thousand dollars now may assure the complete success of an undertaking which has to live through the years." McDowell replied that he was sure that Westinghouse would have preferred to erect and test the huge generators they built for the Boulder Dam at the plant before they were shipped, but just as they could only be assembled and tested on-site, so the telescope could really only be assembled at Palomar. Froebel answered that the Boulder Dam generators *had* been preassembled for testing in the shop.

The negotiations went on and on, and around and around. The Palomar side—McDowell by mail, Serrurier and Sherburne in person, Pease and the others agreeing from Pasadena—were eager to save money and time and confident that their own engineers and staff could put the telescope together. The Westinghouse officials were equally confident that their experience with large welded structures, not the upstart confidence of the Caltech engineers and scientists, should be trusted. Their name would be on the mounting, their publicity officials had already prepared a campaign to attract attention to the Westinghouse contribution to the project, and the last thing they wanted was a glitch and news articles about problems with the Westinghouse-built mounting.

The customer is always right. The Westinghouse officials ultimately came around to a compromise: the tube of the telescope—Serrurier's trusses and Kroon's spoked declination bearings—would be assembled in South Philadelphia, to make certain the boring for the declination bearings was exactly true. The rest of the mounting, including the huge horseshoe bearing, would be fabricated in pieces, predrilled for alignment pins and bolts, and shipped to Palomar for assembly. Although Westinghouse publicity touted the telescope mounting as an "all-welded" structure, a description that sounded modern and advanced, welding the assemblies together on-site was not an option for a precision device like the telescope because the heat from welding could introduce distortions or strains. The sections would be bolted together on the site. Frank Fredericks, a Caltech engineer, would be at Westinghouse during the fabrication, and then at Palomar during the assembly, to make sure it all went together as planned.

Sandy McDowell urged the Westinghouse officials to hold down their publicity about the project. "I know that what Dr. Hale worries about," McDowell wrote, "is misleading or pre-mature newspaper discussion of the various parts of his project."

Like almost every other company that had associated itself with the now-famous telescope, Westinghouse was eager for a part of the glory that seemed to travel with the triumph of technology. The Westinghouse officials commissioned a series of papers on the process of building the mounting, and arranged talks for selected representatives of Westinghouse at sites like Harvard University. Edward Pendray, the director of publicity at the main Westinghouse offices in Pittsburgh, issued a series of press releases about the project, touting the magnitude of the fabrications required, which would only be "handled by a company possessing the necessary equipment and facilities," and the accuracy required for the assemblies, which, though exacting, was "nothing beyond the usual accuracy of Westinghouse practice."

The most demanding tolerance in the fabrication of the tube and mount was the alignment of the declination axis of the telescope tube. The tube of the telescope, in its final configuration, would be twenty-two feet in diameter, fifty feet long, and would weigh 150,000 pounds. The most critical dimension, measuring for boring the declination axis, demanded a measurement accurate to .077 inches over a length of twenty-six feet—a far cry from the accuracy demanded in the optics shop or even the .001-inch accuracy that is more-or-less routine in machine shops.

Yet, what made the mounting fabrication exacting was the sheer scale of the pieces and the uniqueness of the project. By Westinghouse standards, the fabrications were lightweight. The largest unit they had ever built previously in the plant, a turbogenerator installation, comprised a huge condenser that weighed twenty-two pounds per cubic foot, and a generator that weighed fifty pounds per cubic foot. The tube of the telescope, built up of hollow box or tubular sections, weighed only eight pounds per cubic foot.

In many ways the telescope was a relatively easy fabrication project. It was large—the largest structure ever machined—but there was no need for superlightweight design or materials like those used in the manufacture of aircraft. Ordinary mild carbon steel was fine for the telescope, although they did take the precaution of ordering all the steel for each component of the telescope from a single heat batch, to guarantee uniformity of composition. The chief difficulty was finding or building machines large enough to handle the huge sections and devising means to accurately measure and align the sections. Each outer band of the great horseshoe bearing required a piece of plate steel four and one-half inches thick, five feet wide, and forty-seven feet long. Bethlehem Steel used a 12,000-ton forging press to form these sections to the desired curve. To align components prior to welding, and particularly while machining, the factory used surveyor's transits on the floor of the shop. All the welding, except for the circumferences of the hub and spokes of the declination bearings, was done by hand. Rein Kroon's design of the interior of the horseshoe had left enough room for the welders to get to every seam.

The foundry at South Philadelphia was one of the tallest industrial buildings in the world, but even with low-slung rail cars to move the telescope sections, the tracks leading into the building had to be low-ered five feet so the sections of the mounting would clear the doorway. And even the large furnace Westinghouse built at the South Philadel-phia Works would only anneal the fabrications in sections. The cage that formed the top of the telescope tube, twenty-two feet in diameter and twelve feet high, would just fit into the furnace. The biggest machine tools at the South Philadelphia plant couldn't do the finish machining on the horseshoe, the largest journal bearing in the world. The complete horseshoe bearing would weigh 400,000 pounds, and even the huge floor mill at the Westinghouse East Pittsburgh Plant had to be extended with supplementary rollers and tracks to handle the final machining.

The engineering calculations, in South Philadelphia and in Pasadena, went on as the work was in progress. When the final draw-ings and calculations were finished for the horseshoe, it was obvious that even with the superb internal framework Rein Kroon had designed for the bearing, the horseshoe would sag out of round as it turned. The ◡ turned on its side, would become a C and the top horn of the C would sag from its own weight. Manipulating a large struc-ture to mill a shape other than a circular arc or a straight line was a challenge. The 400,000-pound horseshoe was the largest and heaviest single piece of equipment ever handled by Westinghouse, and possibly the largest single unit ever machined. The floor mill at the East Pitts-burgh plant, which had previously been used to machine the thirty-foot gates for the Boulder Dam, was the only mill large enough to machine the surface.

The solution to milling the shape was proposed by one of the engi-neers, whose name, like the names of so many who worked on the telescope, is lost. His suggestion has the clean simplicity of Mark Serrurier's trusses or Rein Kroon's bicycle-spoke mounts for the declination bearings. If the ends of the horseshoe were "pinched" and the center pushed out *before* it was machined, he suggested, when the tension on the horseshoe was relaxed, the shape it assumed would be exactly correct. The calculations of how much to pinch were done by Rein Kroon. To Kroon the calculations on a slide rule seemed simple and obvious; to others, including the machinists who ran the big floor mill, the numbers seemed mysterious black magic. The calculated forces—450,000 pounds pushing out in the middle and 260,000 pounds squeezing the horns in—were optimized for observations of stars that would be within forty-five degrees of the zenith. The difference in dimensions from the new procedure to boring the horseshoe unstressed was a few hundredths of an inch—the difference between the telescope maintaining a proper alignment or sagging.

The Westinghouse workmen welded a brace across the middle of the temporarily assembled horseshoe, and a huge turnbuckle across the horns to apply the needed force. The engineers tried a test run and discovered that the horseshoe expanded as it turned on the boring mill. By late afternoon sections of the horseshoe had expanded as much as 13/1000 of an inch, more than two and one-half times the permitted tolerance of 0.005 inches. And the expansion wasn't even. Part of the horseshoe expanded by only 7/1000 inch in the course of the day.

An engineer watching the test figured out that the problem was sunlight through the overhead skylights heating the horseshoe unevenly. The Westinghouse engineers filled reams of paper trying to chart the expansion of the steel so they could adjust the grinding wheels of the mill to compensate. When the figures weren't reliable, they painted the skylights with dark blue paint. That helped, but to keep the temperature even enough they finally had to build a forty-six-foot-diameter sunbonnet a few inches over the horseshoe. The milling went on for weeks. Stewart Way, a Westinghouse research engineer, studied the surface of the bearing with a microscope to search out ridges and valleys until the entire surface was within the 0.005-inch tolerance. When they pulled the 318,000-pound structure off the boring mill and removed the bracing, the horseshoe "sprung" open a few hundredths of an inch to the shape that would compensate for sag. They would know if it worked when the telescope was assembled on the mountain.

Westinghouse completed the tube for the telescope first. In press releases Westinghouse liked to point out that turbines of that size—fifty feet long and twenty-two feet in diameter—were routine stuff for the South Philadelphia Works. But this product was different. In April 1937 Westinghouse sent out invitations and press releases to announce a ceremony marking the completion of the tube. William Ladley, an employee of Westinghouse for forty-eight years, was to have the honor of tightening the last bolt. The guest of honor would be none other than Albert Einstein.

For the great day, April 30, 1937, rows of chairs were set up on the factory floor, and a dais was erected at one side of the tube, high enough so speakers could reach out and touch the immense structure. Ormondroyd's celluloid model of the telescope shared the front of the dais with a microphone. An audience of dignitaries and reporters were joined by newsreel cameras to hear A. W. Robertson, the chairman of Westinghouse, trumpet the potential of the great telescope: "Sometimes one almost thinks the Good Book was right when it said that man was made to be only a little lower than the angels." The *New York Times* reported that Ladley "skillfully inserted the bolt and tightened it."

Einstein, with his familiar halo of hair and an old-fashioned stiff collar, looked uncomfortable on the dais. His remarks that day have been lost. Afterward, at a reception, he met with several of the Westinghouse engineers. Einstein was clearly awed by the sheer size of the structure. "What happens if somebody makes a mistake in manufacturing?" he asked Rein Kroon.

Kroon, awed by the presence of the great scientist, shrugged. The answer seemed so obvious. "We build it over again."

Einstein winked. "My work is so much simpler. When I make a mistake I just tear up the paper I wrote on."

Hale had said from the beginning that the need for the big telescope was so acute that it should be built as rapidly as possible, though not by sacrificing the capabilities of the instrument. He had made that speech during Sandy McDowell's initial interview in Max Mason's office. McDowell, who liked nothing better than to manage a complex project against a demanding schedule, had taken Hale's words literally. Even before he arrived in Pasadena, he had begun to organize the telescope project as if he were preparing a warship for battle. John Anderson remained the executive director of the project, but his interest was optics, and with the arrival of the mirror, he had his hands full. The construction of the observatory and the telescope was McDowell's bailiwick, and he flung himself into it, pressing the project as if to make up the lost years of preparation. He had been there only a few days when he reported, "I have been quite upset at the slowness of things getting under way."

Because he had not been there for the long, frustrating days of mirror work at GE, the disappointment of the first mirror at Corning, or even the long process of designing the mounting, McDowell tended to be impatient with the astronomers and engineers. As an outsider he believed he could see beyond the parochialism of astronomers who thought their machine unique. He had supervised the construction of large gun turrets and sighting devices for the Bureau of Ships; if the telescope was larger and required higher precision, to McDowell the changes were incremental. From his years of supervising underwater research for the navy in New London during the war, he believed he knew the style and foibles of scientists. He was a great advocate of research and had publicly advocated a much-expanded research program for the navy. But he also believed that scientists needed the guidance of a man like himself to get a project finished.

McDowell got on well with the engineers and sales managers at Westinghouse. They were accustomed to production schedules, flowcharts, organizational diagrams, and other management aids that were second nature in the navy or in industry. He got along with Byron Hill, not the easiest of men, if the two allowed each other room. Between McDowell's acknowledgments of every sentence, which triggered the voice-controlled transmitter at the Pasadena end of the link, and the static from cars on the street outside, communications between Pasadena and Palomar were fairly imprecise during the first months. Ben Traxler later built one-hundred-watt amplifiers for both ends of the link that made radio communications reliable. But by then Hill had already implemented many of his own ideas regarding the site work.

All power, telephone, and propane lines at Palomar were to be installed underground. The original specifications didn't take the highly acidic soil into account. Within weeks of installation, the outer coverings of the lines rotted and the wires reacted electrolytically with one another and with the DC phone lines. The autodialer in the powerhouse would signal the trouble by suddenly rotating without stopping, phone communication between sites on the mountain would fail, and Hill would rant about the "fool engineers" as he led another party to dig up the ground searching for the new break. Hill replaced the cables with heavy-duty neoprene-sheathed cable, but the gophers liked the shield. He ultimately switched to lead-covered cables, installed with a Byron Hill–designed machine that extruded concrete around the cable.

Work on the mountain had to follow a lockstep flowchart, because each stage of work depended on the previous one. The million-gallon water tank was a priority, because Hill needed water to mix concrete for the footings of the observatory. When the footings were in place, Consolidated Steel, from Los Angeles, began work on the base frame for the telescope and the dome supports. Much of the work on the mountain was routine, complicated only by the remoteness of the site. San Diego County work crews were building the new road, but it was slow going, especially in bad weather. Some of the road was usable for deliveries at night, when the road crews weren't working, but much of the heavy steel still arrived up the old Nate Harrison Grade. One load of heavy construction steel, overhanging the back of a tractor-trailer rig, tipped the trailer backward on a steep grade. A crew of workmen with the Caterpillar tractor and some chain righted the truck, turned it around, and backed it up the slope. The "zombies" prided themselves on resourcefulness.

The dome of the observatory, another construction job for Consolidated Steel, had been the subject of a good deal of engineering study in Pasadena. Russell Porter had drawn the original dimensions for the observatory. The relatively fast focal ratio of the telescope meant that the height of the dome could be the same as the diameter, and the finished dimensions, quite accidentally, are almost identical to the dimensions of the Pantheon in Rome. Porter had strong ideas about the design of the observatory. He wasn't a writer, preferring to articulate his ideas in sketches, but for the observatory he made an exception: "Aside from the principle that a building should express the functions of the mechanism that it covers," he wrote in a memo to the Observatory Council, ". . . I have felt the importance of expressing extreme simplicity along with the appearance of permanency." He included large base moldings on the building, to make it appear that the building was solidly rooted. He had noted, as many visitors do when it is pointed out, that the dome of the one-hundred-inch telescope on Mount Wilson appears to stop a foot above the ground, which creates the odd and unpleasant sensation that it is floating. The only decorations Porter included on his design were a simple ribbing on the dome, and moldings around the entrance. "I wish to call attention to the absence of superfluous decorations. Any attempt—to me— to embellish the wall surfaces with flutings, panels, medallions, etcs. so prevalent at the present time, will be as obsolete fifty years hence as the hoop skirt and bustle are today."

As the telescope rotated, or slewed, to different points, the dome of the building had to rotate in perfect alignment, and with no measurable vibration. It also had to provide sufficient insulation to maintain the temperature inside during the day, so that precious hours wouldn't be wasted each evening waiting for the telescope to reach thermal equilibrium. The engineering of the dome was assigned to Mark Serrurier. Serrurier elected a welded monocoque construction for strength and light weight, and turned to his own professor at Caltech, Romeo Martel, and Theodor von Karmann of the Aeronautics Department, for ideas on the design of what would constitute the largest welded structure ever built.

Von Karmann, then much sought after by the aircraft industry for his engineering work, noticed the similarity of the shape of the dome to the end of a zeppelin, and turned to the Goodyear Zeppelin Company, a forerunner of the blimp company, for information on their own designs. Von Karmann added some calculations, Martel added additional ideas, and Serrurier came up with a design for a model to be fabricated by the machine shop. The model consisted of a copper hemisphere floated in mercury, with measured forces applied by levers, to test the strength of the dome structure and the shutter doors that opened on one side against the potential loading of earthquakes, wind, and snow.

The lower portion of the dome was designed with double walls of reinforced concrete, thirty feet high, built on the load-bearing steel framework. A twelve-inch airspace between the walls would allow heated air to rise and escape from a row of vents. The steel dome above the walls was also designed with vents at the top and aluminum-foil-filled insulation panels that fit over the gridwork to form a four-foot airspace. Ventilation equipment could be used to purge heated air from the spaces between the layers to speed the cooling process in the early evening.

Consolidated Steel, working from Serrurier's drawings, raced to get the dome supports in place before winter. Scaffolding, then steel, rose quickly, as a circle of columns went up. Byron Hill watched the first columns go up and protested. They should start with an absolutely level circle at the bottom, he argued, then machine the columns to length before they put them up, so the tracks for the dome would be exactly level. The Consolidated Steel workers pointed to the plans. Serrurier had specified push-pull bolts at the top to level the support for the dome. Consolidated's job was to get the steel up. Someone else could level the top.

That job fell to Hill. He had to set up a transit at the top, in midwinter. They could only align and level the bolts at night, because direct sunlight would heat the steel unevenly. One storm blew the 2-x-12-inch scaffolding planks down the mountainside. Hill cursed the desk engineers.

Consolidated was supposed to bend and clamp the rails for the cars that would carry the dome to tolerances of one-sixteenth of an inch before they were welded. They couldn't make the tolerances, so the rails emerged with rough welds at each joint. It would take months of hand-grinding, by workmen sitting in the cold at the top of the steel structure, to smooth the welds. When the welds were down to tolerances, the astronomers then insisted that the rails be ground all the way around until they were absolutely smooth and level. "What sort of tolerance do you mean?" Byron Hill asked. He expected the figure of one-one-thousandth of an inch, which had become their standard. For the rails, it wasn't good enough. Any bump in the rail would affect the motion of the dome, which would in turn be picked up by the telescope. He was told to grind the rails so there was *no* measurable vibration when the car was moved. Jerry Dowd, at Mount Wilson, had worked out experiments to test the vibration and slippage of a dome truck on the rails.

Johnny Kimple from the Mount Wilson shops was sent over to take charge of the grinding operation. A traveler built on an arm from a central pillar carried the belt-driven grinding machine. The crews ground the rails every day for six months before they were smooth enough. That same winter of 1936–37, the plans called for pouring the floor of the observatory. Work crews collected firewood for a makeshift furnace and coil to heat water for the concrete.

Gradually, steadily, the building went up. Railroad-style carriages rode on the rails to carry the load of the dome. The sections of the dome, huge curved pie slices of sandwiched steel plates with lightweight girder sections between them, were fabricated in Los Angeles, trucked up the mountain, and hoisted into place. Consolidated Steel had built a crane in the center of the building to hoist the sections. Workmen aloft welded the sections together, creating the monocoque that Serrurier, von Karmann, and Martel had designed. Byron Hill and his workmen used the steel company crane and leftover materials to fabricate the walkway around the base of the dome. When it was finished, for a first trial, they hooked up the Caterpillar tractor to rotate the dome. The grinding had paid off. Workmen came from all over the mountain to stand inside the building and watch the dome; it turned so smoothly that it was hard to believe that it wasn't the dome that was standing still and the building turning.

Sandy McDowell, eager to find portions of the project where his own experience was relevant, dived into the dome project. The engineers and astronomers had gently dismissed his suggestions for guidance mechanisms and spotting scopes as far from the mark. Turning big structures, like a battleship gun turret, was an area he knew. He took the plans for the dome rotation to Westinghouse and GE to ask them to bid on motors to drive the dome. They studied the plans, with heavy, solid truck tires driven by DC motors turning the dome by friction, and turned him down. The draftsmen had put the motors under the gears. The motor engineers said that oil would drip onto the motors and short them. Hill agreed. He suggested that they turn the motors on their sides and use a right-angle gearbox. But McDowell liked the clean look of the motors mounted under the truck. The drawings had already been entered in a contest and won a prize.

McDowell got the firm that built the gears for the trucks to supply DC motors with magnetic brakes. They looked like the drawings, clean, compact, a proper modern design for the greatest telescope in the world. When they were installed, the seals on the gearboxes leaked, oil dripped onto the motors, and caused arcing and charring. Using grease instead of oil on the bearings helped, but the prizewinning design was never quite right.

John Anderson, Francis Pease, Russell Porter, and even Mark Serrurier and the other engineers at Caltech didn't take to McDowell's careful organizational charts with names and titles in boxes connected by lines of authority and reporting. They had worked together for years, in some cases for decades, without a formal structure. Hale had set up committees, and the more important of the committees held regular meetings, less as a forum to decide questions than as a means of keeping one another up to date on the progress of various aspects of the project. Serrurier would report on the latest tests of the tube structure while Pease reported on his work with bearings and mirror mounts. Most communication, before McDowell arrived, had been oral, with scant weekly or monthly minutes for the committee meetings.

At a distance Sandy McDowell was a good employer. He was comfortable around men like Byron Hill, who acknowledged his authority and who spoke in the practical terms of engineering. But in Pasadena he was seen as a usurper, grabbing work and authority from his equals. John Anderson, still executive officer of the project, came out McDowell's equal on the charts of authority McDowell sent out. The other scientists didn't fare as well. Sinclair Smith, an experienced astronomer who was working full-time on the control system for the telescope, was treated as an errand boy; McDowell would send him off to the East Coast to look up contacts or to check on the progress of work at Westinghouse and Babcock & Wilcox. Francis Pease, who had as much experience with large telescopes as anyone in the world, but little respect for McDowell's style or knowledge, was sentenced to an oblivion as the design work was gradually taken away from him and given to the engineers.

George Hale and later Max Mason tried to smooth over the abrasions. When Mason arrived in Pasadena and took over the chairmanship of the Observatory Council, he found that he had to do "some organizing. It is necessary, I believe, in order that the peculiar abilities of some of the members of the group shall be used at their full capacity." The real problem was that McDowell and the scientists didn't understand one another.

As a group the Caltech scientists and engineers were self-confident, almost arrogant, convinced that they could do the job that others had called impossible. Part of their confidence was California, a place where everything, even science, seemed to happen faster. In Berkeley, Ernest Lawrence was building ever-bigger cyclotrons, probing deeper into the atom, raising and spending state funds and foundation grants at a phenomenal rate. Caltech scientists, after only a decade and a half of formal existence of the school, were pushing rapidly in physics, chemistry, geology, biology, and engineering. California, the land of the highest mountain and the lowest desert, the biggest telescopes and cyclotrons, the fastest-growing cities—was a land where anything was possible.

The Pasadena scientists and engineers were undaunted by either expertise or experience. Francis Pease and Sinclair Smith had put in their time on large telescopes, and they understood what an observer or an astrophysicist wanted from a big telescope. If they weren't trained engineers or experienced at solving engineering problems, they had the background in physics that let them understand problems. And because everyone in the group shared that background—the Caltech-trained engineers like Mark Serrurier had strong training in physics and math—the problems and solutions to the design of the telescope received a thorough grilling.

Sandy McDowell knew whom to see at every big company, knew how to get through to the president of the company, the head of production, or the head of sales. But while he could negotiate a contract with Westinghouse or Babcock & Wilcox and could list specifications for what they were to produce, at bottom he lacked the common background in science that had been so important in the conception and design of the telescope. Men like Rein Kroon or George McCauley, men who could understand and explain problems in the language of physics, were quickly welcomed into the group designing and building the telescope. McDowell remained an outsider.

To McDowell the scientists—Pease, Anderson, Adams, Smith— were unrelenting theorists, pie-in-the-sky dreamers who would go on designing a telescope forever, working out theoretical problems, reinventing the wheel as they conceived solutions to problems that in only slightly different form had been solved elsewhere. When they tried to explain that the telescope was a unique challenge, that no machine had ever been built to that scale and those demands for precision, he would point out that a triple gun turret on a battleship was *almost* as large, that the targeting requirements for a naval gun were only a few orders of magnitude less exacting. Their insistence on exploring, designing, and engineering every step of the project from scratch— doing basic research on subjects as varied as oil bearings, the wind resistance of dome sections, and the chemistry of glass—was for McDowell a sure route to a quagmire of indecision that would never see the telescope built.

McDowell believed he had gotten the scientists off dead center by forcing the end of design and the beginning of construction. By 1937 Babcock & Wilcox was near completion of the complex welded mirror cell, Westinghouse was well along on the tube and yoke, the astrophysics machine shop working on the demanding job of grinding the gears that would set and drive the telescope, and the work on the peak was moving along so well that visitors had begun to come up in droves to admire the buildings and sites. Through it all McDowell seemed always to see problems of management, of breaking up logjams, attending to details, keeping the project moving. A memo ordering the use of spring faucets in the showers and washhouse on the mountain to conserve water would receive the same attention and priority as a discussion of the control system, bearings, corrector lenses, or even the state of the primary mirror of the telescope. For McDowell the telescope was a project. He had done others before and would do others after.

Men who were building a unique machine, an instrument perfect enough to explore the secrets of the universe, didn't like that attitude.

26

Fine Points

There was no one moment in the optics shop when the big disk was suddenly all right. By January 1937 George McCauley had identified the cause of the fractures as contamination from the refractory bricks in the melting tank and offered his best-guess explanation that the contamination had been introduced by drainage down the sides of the tank as the level of glass in the tank fell. If he was right, the first Pyrex poured into the mold, from the top of the melting tank, should have been pure, and the checks and fractures would begin to disappear as the surface, the material poured last, was ground off.

At the end of January 1937 Amory Houghton and McCauley traveled to Pasadena for another secret meeting with Anderson, McDowell, Mason, Millikan and Weaver from the Rockefeller Foundation. McCauley described the various experimental procedures he had followed and his plans for producing a new disk if one was needed. Houghton thought his offer to produce a disk at Corning's cost generous; Mason and Millikan thought Corning should at least share the expense. Houghton, secure in McCauley's assurances that the disk they had already produced would prove satisfactory, said he would discuss the possibility of Corning sharing the expense with his board of trustees.

McCauley's prediction was as much hope as science, but as the flat grinding approached the final level at the edge of the disk, the checks began to diminish. A few troublesome fractures remained on a cord between two of the pouring points, but the frequency of new fractures decreased, and the gravity of the remaining fractures was less troublesome than the early deep ones. In midsummer 1937, the correspondence about the disk was all still in letters marked PERSONAL, but John Anderson and Max Mason had let their tentative negotiations for a new disk lapse.

Despite the good signs, the initial grinding had gone *much* slower than anyone predicted. The disk had been in the shop for almost a year and Marcus Brown and his crew still hadn't begun the real task, grinding a precise concave shape that would transform the glass disk into a telescope mirror.

"How long will it take to finish the mirror?" everyone asked— reporters, visitors, even officials from the Rockefeller Foundation. For those who read or heard about the telescope episodically, their attention piqued only when press conferences, public announcements, magazine features, Robert Benchley's jibes, or spectacles like the railroad journey of the disk from Corning to Pasadena brought the telescope back to public attention, it seemed as though the project had already gone on for as long as anyone could remember. In 1928 George Hale's predictions that the telescope might take as long as six or seven years to build had seemed pessimistic. By 1937, almost a decade into the project, the actual shaping work on the mirror hadn't even begun. Figuring the mirror of the one-hundred-inch telescope had taken almost six years, and that was working with plate glass, a familiar material. After the preliminary grinding of the faces of the Pyrex disk had taken a large crew of men most of a year, and used up five tons of carborundum to remove two and one-half tons of glass, no one on the project needed more evidence to realize that even the most pessimistic estimates of a finishing date were wildly optimistic.

Though it seemed an eternal job that would leave everyone in the optics shop permanently deaf, the surface grinding finally halted in the spring of 1937. There were still visible checks in the disk, but it looked as though they would be ground away when the disk reached a concave shape. The next work, grinding the perimeter of the disk, the edges of the central hole, and reaming out the pockets for the support system, took three months. Special equipment had been built for these tasks, using hollow iron cylinders fed with water and carborundum as the grinding tools.

To visitors the work in the optics shop looked and sounded the same from one day to the next. The disk was on a rotating table in a huge machine. Men hovered over and around it, tending the slowly rotating tools that ground away the glass. But to the optician, there was a great difference between the "mechanical" work of preparing thirty-six pockets, each of precise dimensions and spacing, or trueing the edge of the disk—all work done to tolerances that the best of machine shops can achieve with care—and the ultimate optical work, figuring the surface shape of the disk to tolerances unknown outside the optics lab. The edge grinding and the preparation of the pockets was boring and tedious. Everyone in the shop looked forward to the start of making a mirror.

It was almost the anniversary of the disk's arrival when Marcus Brown ordered a thorough cleaning of the shop, the sort of search-and-scrub that made the normal daily cleanings seem casual. The turntable of the grinding machine was covered with two layers of one-inch sponge rubber. To assure even seating of the disk, the sheets of rubber had been tested with a fixed load; only sheets with measured and tested compression within a tight range were used on the table.

Corning had cast a forty-inch Pyrex disk to fill the center hole in the mirror disk during the grinding of the mirror. Trueing the plug to a perfect circle was a relatively simple task, but Anderson had to think awhile to come up with a scheme to lower the 1,400-pound plug into the hole in the disk, keep it there during the grinding and polishing of the mirror, and be sure of removing it without harming the disk. His solution, described regularly to visitors, was a staple story for the optics shop, guaranteed to evoke "Why didn't I think of that?" smiles.

Anderson designed a wooden lifting clamp that was attached to the top half of the plug, leaving the lower ribbed section, approximately fifteen inches deep, projecting. The overhead crane could then lift the plug and swing it over the disk. To lower the plug into the hole in the main disk, Anderson had Brownie's workmen place a large cake of ice, tall enough to support the plug, in the hole in the disk. The crane lowered the plug in place onto the ice, the lifting frame was removed, and as the ice slowly melted, the plug slipped into its exact fit in the disk. For optical lab workers, used to watching minuscule progress after a day or a week of polishing a mirror, watching ice melt wasn't boring. A room full of workmen exhaled all at once as the plug settled into place without harming the mirror. Brownie used plaster of paris and waterproof cement on the seam to hold the center plug in place.

Brownie was almost ready to start shaping the disk. During grinding and polishing the disk would rest on its back, supported by foam rubber cushions on the machined turntable. For testing, the disk would be raised to a vertical position, both to permit the long light path that the test procedure required and because a vertical orientation would more closely simulate the most extreme loads on the disk in use. In the vertical position the disk would require the internal support system suggested years before by Elihu Thomson, and refined in years of sketches by John Anderson, drawings by the Caltech engineers and draftsmen, and model making in the astrophysics machine shop.

The internal supports were amazing machines. Counterweighted levers were engineered so that in any position, the weights would exert the correct forces on the back of the disk to compensate for the sags or other deformations in the overall shape of the disk. The forces required are so small that the fundamental problem in designing the support system was what an engineer calls stiction, the initial friction of a system that is mostly *not* in motion. Bearings and lubricants were tried and rejected. The design grew more complex, until there were dozens of parts in each support. The thirty-six supports would work together, automatically translating the motions of the counterweights to the precise combination of tiny pressures on the disk that would compensate for its changed position. The blueprints of the devices got so complicated that only a few men in the machine shop understood how to put the devices together. Among those who didn't have to decipher the drawings, the supports were black art. Rumors started that only one or two men on the Caltech campus understood the supports.

The mirror supports were installed in their pockets in the back of the disk before the shaping of the disk began. Initially the mechanisms were disconnected, lest the delicate bearings and cams be subjected to the motions and pressures of the largest grinding tools. Later, when the disk was ready for testing, the supports were reconnected. The forces exerted by the support system were so small that the only real test of the supports was the optical tests of the mirror. Like the great oil bearings, the corrector lens, the readout devices (servos) to tell where the telescope was pointing, and dozens of other technologies that had been developed and refined for this telescope, the supports were innovations, designed for an instrument so precise that the technologies could ultimately be tested only with the most precise and exacting test of all—the light of a star.

In the summer of 1937 Brownie and his crew began shaping the mirror. Day after day the same small crew of workmen in white cotton surgical suits and canvas shoes worked the big machine. Occasionally they traded jobs. Usually regularity was more important: The same man did the same job each day. One man stood on a scaffold under the machine, greasing the main gears for the turntable to make certain they didn't gall. Another man continuously washed the edges of the disk with a hose, sloshing away the excess grinding slurry. A man stood by the power switch in case of an emergency. The seventeen-and-one-half-foot-diameter table of the grinding machine turned slowly. The disk was so large that at a rotating speed of one-half turn per minute, the outer edge moved by the grinding tool at only twenty-six feet per minute.

Hour after hour, day after day, week after week, month after month, the disk turned, while the grinding tool above turned in its own serpentine lissajous figures. Occasionally the routine was interrupted so they could change the glass blocks on the face of the tools. Every few months they would change the grinding tool, from the one-third- and half-size tools to the full-size tool, a disk as large as the mirror itself. A sharp-eyed visitor might notice a slight change in the configuration of the machine. Most visitors would watch for a few minutes, amazed that men could work hour after hour, day after day, doing the same job, in the same windowless room, with the same droning machines, and without seeing any progress in their work.

The concave shape Brownie and his men were grinding into the disk would be approximately three and three-quarter inches deep at the center of a two-hundred-inch-diameter circle. It would take months before the curve was apparent to the naked eye. The men in the room stopped guessing how long it would take to grind and polish the mirror to the approximately one-millionth-of-an-inch precision the final figure would require.

Astrophysics could hardly wait for the telescope. In the years since the project began, remote galaxies, the "island universes" of the Washington debate, and the expansion of the universe had become the focus of much astronomical research. Hubble had finished a book on the nebulae, detailing his morphological scheme for classifying nebulae, the famous tuning-fork diagram that looked as if it demonstrated an evolution of types of galaxies. Hubble's work got onto the covers of the newsmagazines. His discoveries not only attracted attention to astronomy, astrophysics, and the telescope project but drew students to Caltech, even though Hubble did not accept students.

While Hubble's observations were great fodder for the magazines, the evidence he and Milton Humason found was disturbing. Hubble's observations that other nebulae (the term *galaxy* was not used in the Mount Wilson offices until after Hubble died, in 1953) were receding from one another seemed incontrovertible, but the consequences of Hubble's and Humason's evidence troubled the cosmologists and other critics of the new astronomy.

From the Hubble-Humason evidence, it seemed that the remote galaxies Hubble had photographed and recorded on his spectrographs were all much smaller than the Milky Way. M33 in Trapezium was, from Hubble's data, one twentieth the size of the Milky Way. Our nearest neighbor, the Andromeda galaxy, was one fifth the size of the Milky Way. Why? Failing a good reason, the cosmologist likes to believe that the universe is uniform, regular. It was possible that our own galaxy was uniquely large, but the differences in size of the galaxies, without an explanation, were troubling.

Hubble had derived the distance of the Andromeda and Trapezium galaxies from the period of Cepheid stars—the same celestial yardstick Shapley had used for globular clusters. If Hubble's distance figures were correct, the intrinsic luminosity (absolute magnitude) of the globular clusters he had photographed in the region of the Andromeda nebula were much too faint compared to those he had calibrated in the Milky Way. Cosmologists like to believe that objects of the same type have fairly uniform luminosity, anywhere in the universe. If the globular clusters around Andromeda were the same as those in the Milky Way, Andromeda must be twice as far away as had been thought previously—a scale that didn't fit Hubble's measurements. No one had a better scale for the universe, but Hubble's nonetheless seemed fishy.

Even more disturbing was the apparent contradiction between Hubble's derived age for the universe and the age geologists had derived for the earth. From the few galaxies for which he had calculated distances, and from his calculated *rate* of expansion of the universe—the exact number changed, as he and Humason refined their observations—Hubble was able to run the expansion of the universe backward to the beginning, the not-yet-named "big bang," deriving an age of the universe. It was heady, mind-boggling science, using the red shifts on tiny spectra of distant galaxies to derive a geometry of the universe and then using the distances to the nearest of those galaxies to convert the geometry to a time scale since the Creation. Even if they understood only fragments of what Hubble was doing, journalists loved it.

When Hubble did his calculations, the age he derived for the universe was under 2 billion years. Geologists, working from rock samples with various dating techniques, believed that the earth was closer to 4–5 billion years old, twice as old as Hubble's universe. Hubble and Humason accumulated more red-shift data, refined their figures for the rate of expansion, rechecked their calculations. No matter what they did, the age they came up with for the universe was *less* than the age geologists had derived for the earth. How, the skeptics asked, could the earth be older than the universe?

A few cosmologists offered theories to explain the contradiction. "After all," de Sitter wrote, "the 'universe' is an hypothesis, like the atom, and must be allowed the freedom to have properties and to do things which would be contradictory and impossible for a finite material structure." Hubble wanted no part of the sleight-of-hand explanations:

> We face a rather serious dilemma. Some there are who stoutly maintain that the earth may well be older than the expansion of the universe. Others suggest that in those crowded, jostling yesterdays, the rhythm of events was faster than the rhythm of the spacious universe today; evolution then proceeded apace, and, into the faint surviving traces, we now misread the evidence of a great antiquity. Our knowledge is too meager to estimate the value of such speculations, but they sound like special pleading, like forced solutions of the difficulty.

Hubble's answer to the dilemma was more observations, looking ever further into space, cataloguing and measuring the red shifts for more galaxies, until somehow the discrepancies could be resolved. "The next step will be to follow the reconnaissance with a survey—to repeat carefully the explorations with an eye to accuracy and completeness." With the completion of the two-hundred-inch telescope and the Schmidt camera, which would survey likely areas in preparation for the deep probes by the telescope, the answers would emerge. Hubble planned to map the universe. The new telescope would be his tool.

★   ★   ★

304 ✳ THE PERFECT MACHINE

Hubble wasn't the only astronomer with plans for the big telescope.

In 1931, the year Hubble met Albert Einstein, Walter Baade
arrived at Mount Wilson from Hamburg, bringing not only Schmidt's
idea for the wide-angle deep-space camera, but a remarkable skill at
observing with large telescopes. Baade had spent 1926 at the Mount
Wilson Observatory as a Rockefeller fellow, and he knew exactly what
he wanted to do on the big telescopes. He would leave the discovery of
ever-more-distant nebulae to Hubble; he wanted to know more about
the nebulae. In photographs, even with the biggest telescopes, the
nucleus of even a close nebula like Andromeda was a glowing mass. A
few, like the nebula in Draco, gave off a bright line spectrum, as if they
were a glowing gas. Others, like Andromeda, emitted what appeared to
be a continuous, or "white," spectrum, as if the nucleus were a glowing liquid or solid, or perhaps a collection of stars packed tightly
together. No one had succeeded in resolving the nucleus into individual stars.

Baade was a small, lively man, with a limp in one leg, a reluctance
to publish until he was absolutely sure of his material, and a generosity of spirit that stood out in an increasingly competitive field of
astronomy. He didn't like the midnight snacks that the observatory
provided, so he brought his own cheese and sausages, which he would
share with the night assistants. Baade was a dark-time man, who
observed when the moon was down. The light-sky men, those who got
time when the moon was up, used their spectroscope gratings to analyze stars. Baade hoped for a lucky black night of clear weather and
still skies, the superb seeing that would let him stretch the resolving
power of the telescope to an even fainter image. "Those spectroscopists," he said. "They don't know how to eat, how to drink, how to
love."

For a few years in the mid-1930s, Baade collaborated with Fritz
Zwicky, who was using the eighteen-inch Schmidt camera on Palomar
to find supernovas. It seemed a natural partnership: They both had
German as a native language (Zwicky, though born in Bulgaria, was
Swiss), and they shared an interest in the edges of cosmology, questions like the sequence of evolution of stars that would lead to supernovas and other extraordinary celestial happenings. Their agreement
was that when Zwicky found supernovas with the Schmidt camera,
Baade would follow up with a study of the light curves with the large
telescopes on Mount Wilson. The collaboration worked for a while,
and they did some promising work together, including pioneering
investigations of the concept of neutron stars. But Zwicky, who got
along well with secretaries and night assistants, couldn't work with a
coequal for long. He accused Baade of reneging on his part of the
research and stealing credit for Zwicky's own work. For good measure
Zwicky threw in an accusation that Baade was soft on Hitler. Before

long Zwicky's verbal threats became so intimidating that Walter Baade refused to be alone with his former colleague. Baade later told other astronomers he was physically afraid of Zwicky.

At Mount Wilson Baade was safe from Zwicky. Working as many dark-time nights as he could get telescope time, Baade searched for the secrets of the distant galaxies. Harlow Shapley had argued that the nuclei of the galaxies were only a glowing gas, but in one photograph taken with the sixty-inch telescope in 1920, Baade thought he could almost see individual stars in the nucleus of M33. He and Joel Stebbins, who had pioneered work with photoelectric devices to measure the intensity of light, refined the accuracy of measuring star brightness, and Baade began a long program of observing M31 in Andromeda. He was an extraordinarily careful observer, and before long he had photographs of the outer region of the nucleus that registered star images with the smallest angular diameters yet recorded. Yet, no matter what emulsions, corrector lenses, or tricks he tried on the one-hundred-inch telescope, the resolution of stars in the nucleus of Andromeda eluded him. It was, everyone agreed, a task for a bigger telescope. Baade joined the queue of astronomers eagerly awaiting the two-hundred-inch telescope.

The dream of resolving the nucleus of Andromeda was only the first step of Baade's ultimate goal. Hubble had searched for the geometry of the universe. Baade wanted to understand stellar evolution. He wanted to break down the populations of stars in the distant galaxies, to document and understand their evolution, and ultimately to be able to age-date the stars and the universe. Elsewhere chemists and nuclear physicists were exploring the complex reactions at the core of stars, the processes of nucleosynthesis that created the energy of the stars and the elements of the universe. Hans Bethe, a physicist, was exploring a theory of stellar energy, showing how almost all the energy generated by the most brilliant stars stems from a fusion reaction in which hydrogen is the fuel and carbon the catalyst. The theoreticians seemed to be outstripping the observers. In the Monastery at Mount Wilson and in the laboratories and seminar rooms at Pasadena, talk about the progress on the two-hundred-inch telescope was daily fare.

The anticipation of the new telescope was much accentuated when Max Mason sent around a memorandum to the astronomers asking their thoughts on the idea of setting up facilities at the new observatory to permit an astronomer to be accompanied by his wife. There had been a long-standing debate on whether the astronomers' residence at Mount Wilson was called the Monastery because of three early astronomers who had worked there: Monk, Abbot, and St. John; or because of George Hale's rules that the accommodations remain strictly bachelor quarters. The newly married Harlow Shapley had put up with the then-difficult commute from Pasadena instead of accepting free accommodations in the Monastery, and astronomers making long runs on the telescopes had routinely griped about the too aptly named Monastery.

Fritz Zwicky, taking advantage of the lack of rules at Palomar, had taken his wife for his runs on the "little" Schmidt camera and said he thought it had been a good idea. Walter Baade, agreeing for a change with Zwicky, urged breaking "occasionally from the tyrranic [sic] rule of the monastery. . . . For individuals like myself it would be a decided improvement over the present system as being practiced on Mount Wilson." Not everyone agreed. Milton Humason, who had put in his time on the mountain as a mule driver and staff worker before he began observing, noted that if wives were there, observers would not make maximum use of the telescopes. He and Walter Adams opposed having more than one cottage available for use by an astronomer bringing his wife. No one raised the question of facilities for unmarried women, or a woman observer bringing her husband for an observing run. Women observers were still virtually unknown at Mount Wilson.*

Mason decided to provide one cottage for observers, along with the residence facility. The name "Monastery," brought over from Mount Wilson, stuck even before Russell Porter had finished his designs for the building or the site, nestled in a wooded grove at some distance from the big dome, was chosen. The residence was solidly built, with metal studs, metal lath, and plaster walls. George Hale put in his voice on the furnishings, urging that the simplest furniture would be the most appropriate. "On Mount Wilson we never used window drapes, though rolling window shades are necessary."

The original design called for a copper roof, but the redheaded woodpeckers on the mountain found the copper inviting, punched holes to store their acorns, and later returned to harvest insect larvae from the acorns. The result was leaks in the roof, which ultimately had to be replaced with composite shingles. The woodpeckers also dropped acorns down the toilet vent pipes, so that the waste pipes were soon clogged with oak roots.

The eighteen-inch Schmidt telescope, a few hundred yards down the slope from where the dome for the two-hundred was going up, was the only working telescope at Palomar, but astronomers from Mount Wilson and other observatories visited, marveling at the size of the new dome and the caissons that would support the mounting of the

---

* As late as 1955, when Margaret Burbidge, a superb observer, applied for a position at Mount Wilson, she was told that there were no toilet facilities for women on the mountain. With aplomb, she replied that she would use the bushes. Her husband, a theorist, was hired as a postdoctoral fellow at Caltech, which granted him privileges at Mount Wilson and Palomar. Astronomers looked the other way when she showed up at Mount Wilson to use her husband's allotted time on the one-hundred-inch telescope. He read books in the darkroom while she observed.

telescope. They had talked about the Big One for years, studied the drawings, discussed myriad details. Seeing the dome going up, visiting a site crawling with workmen, and the reality of a residence for astronomers, increased the anticipation.

Palomar was still a wild, undeveloped spot. The climb up the mountain was difficult, and the dome site at the top, with the trees and shrubs cleared, created a barren plateau. But under the night sky, far from city lights, it was hard not to feel the closeness to the cosmos.

In 1937 Byron Hill went on a camping vacation, and Mark Serrurier was asked to temporarily take over the supervisor's task at the top of the mountain. While he was there he asked a woman he had been dating, Naomi, to visit him for a weekend. Serrurier waited until a dark night, when the carpet of stars was spread overhead, to ask Naomi to marry him. She felt so overwhelmed by the sky, the stars, and the power of the place that she deferred an answer until the next day. It wasn't until she was halfway down the mountain that Naomi felt far enough away from the magic of the peak to say yes.

As long as astronomers asked new questions, there would always be dreams of newer and bigger telescopes. This one felt different. It wasn't just the size and already-growing fame of the two-hundred-inch telescope: The questions of astrophysics were so ripe, and the answers to the eternal cosmic riddles seemed so close, tantalizingly just beyond the grasp of the one-hundred-inch, that it was hard not to feel incredible impatience for the new telescope.

27

Passing the Torch

Sinclair Smith, a bright young astronomer, came to Mount Wilson in the early 1920s. A year of study in England "settled him down," and he was soon a regular member of the staff, investigating the physical constitution of nebulae and star clusters. By 1931 he and Fritz Zwicky were independently studying clusters of galaxies. Smith scrutinized all the available data on the Virgo cluster, measuring the differences of velocities of galaxies in the cluster, and concluded that the galaxies were moving too fast to stick together; a group with that little mass should fly apart. Zwicky coined the name "dark matter" for the missing mass. Cosmologists today are still searching for the dark matter to balance their equations.

Smith was recruited to the two-hundred-inch telescope project to work on electronics, particularly instrumentation and the drive system for the telescope, the combination of motors, gears, and black-box magic that keeps the telescope accurately tracking objects as the earth turns through the evening. At first glance the task seems easy. The rotation of the earth makes objects appear to move across the night sky. Move the telescope to compensate for the earth's motion and the objects appear to stand still. The equatorial mounting of the telescope, with its polar axis exactly parallel to the axis of the earth, made the basic motion a simple rotation. The basic task of the drive mechanism was to turn the telescope at the precise speed of the earth's rotation, but in the opposite direction so that the sky appeared stationary.

But turning the telescope the equivalent of one full rotation in twenty-four hours to match the rotation of the earth isn't enough. The bigger the telescope, the more exacting the requirements for a drive mechanism. Atmospheric refraction, a minuscule offset of images as the azimuth of the telescope was raised or lowered from the horizon to the zenith, is magnified in a large telescope. The big telescope would be sensitive to eccentricities in the bearings, gears, or mounting structure. A bearing surface as true and smooth as the machine shop could produce, the surfaces honed to the precision of a watch, would still introduce periodic errors, "bumps" in the motion of the telescope. Over the course of a long exposure, an uncorrected bump would ruin an image or degrade the quality of a spectrum by moving the light from the object off the slit of the spectrograph.

The sixty- and one-hundred-inch telescopes, after decades of use, still had balky quirks in their drive and control systems. Baade used to test budding observers on the sixty-inch telescope at Mount Wilson by seeing if they recognized and compensated for the periodic error in the gear train that had the telescope creep ahead of the stars it was tracking every eighty seconds. By listening for the relays, Baade could tell if the observer was pushing the East and West buttons to compensate. Good observers, like Milton Humason and Walter Baade, were magicians with the machines, with more tricks, bumps, and grinds than a burlesque queen in their repertoire of techniques to get the telescopes to behave. The tricks distracted from the primary task facing the observer, and as the size of an instrument reached the scale of the two-hundred, the idea of *horsing* a five-hundred-ton machine into position became absurd.

Smith wasn't an electrical engineer by training, but he was a veteran of many nights on telescopes, who understood what astronomers would want from a drive system. At a conference the Pasadena astronomers decided that the drive system should not have any periodic error of more than $\frac{1}{10}$ of a second of arc (a second of arc is $\frac{1}{60}$ of $\frac{1}{60}$ of a degree, or 1/1,296,000 of a circle) for periods of five seconds or more. They also wanted automatic and manual controls of the dome and telescope, with repeater stations so the telescope could be controlled at the various foci.

Their ideal was for the night assistant to dial in coordinates for the precise area of the sky the astronomer had selected, and for the control mechanism of the telescope to do the rest. The astronomers' dream list included simple control panels at each observing station, with indicators of right ascension and declination, buttons for guiding and slewing the telescope, switches for adjusting the focus, and a telephone for communicating with the night assistant. Instead of horsing with the telescope or contorting himself into strange positions, the astronomer would use as much as possible of his precious time on the telescope for actual observing. Smith understood the wish list.

Robert McMath, who had designed an operating and control system for telescopes in Michigan, and later for telescopes at the Lick Observatory and the new eighty-two-inch telescope of the McDonald Observatory in Texas, was invited to join the project as a consultant. "In the very nature of things," he wrote,

A project like the 200-inch telescope forces the engineer to extrapolate. . . . In this case, we are extrapolating many important items, such as the 200-inch mirror, the pedestal truss, the oil pad bearings, no polar axis defining bearings, the gimbal declination axis connections, the declination axis radial and thrust bearings, the polar axis torque tube, etc. Doubtless most of these items will prove satisfactory in service. Unfortunately, just one of them can spoil the job. . . . I again urge concentration of your available personnel on the problem of building the simplest possible telescope, considered as a whole.

McMath particularly thought the plans for the electric drive system unnecessarily complex. The system he had designed took care of most corrections, except that it required that the observer manually change the rate of drive from time to time, based on what he observed in the guide scope eyepiece. His much simpler scheme, he pointed out, added "one-half of one percent more burden on the observer." McMath's control systems were good, but for this telescope the astronomers wanted even more.

It wasn't just the astronomers and engineers who had ideas for the control systems. In 1934 Max Mason had recommended that Hale and his colleagues study the "automatic curve-following mechanisms" which were being designed at MIT under Vannevar Bush. Hale had written to Bush, who thought his analog-computer mechanisms might be more accurate than any manual control of the telescope. The idea was to use photoelectric cells to track a guiding star and to trigger signals to control screws that would move the plate holder at the heart of the telescope in response to the apparent motion of the guide star. It was a superb idea, decades ahead of its time, but calculation showed that the field of view of the two-hundred-inch telescope was so narrow that the best of the guide stars would be magnitude 10 or 12, too faint for the photocells then available.

Sandy McDowell, with his long years of experience supervising the construction and installation of naval gun turrets, also considered himself an expert on control mechanisms. He had worked for years with Hannibal Ford, whose small company on Long Island had developed pioneering servo control systems for big naval guns. McDowell wanted to use the Ford work in the telescope. The accuracy required by the drive mechanism for the telescope was orders of magnitude more demanding than the needs of naval guns—someone once calculated for publicity purposes that an error the size of a quarter at three miles would have been unacceptable—but to McDowell it was only an incremental difference. Ignoring the work of the Ford company, he said, was reinventing the wheel.

On an early trip east, McDowell took Sinclair Smith with him to meet Hannibal Ford. Smith had spent his working life at Mount Wilson and had little experience dealing with large corporations or specialized defense contractors. McDowell insisted that no work take place on the drive controls until the Observatory Council had a proposal and estimate from Ford. Smith was given the job of conveying specifications to Hannibal Ford. Hannibal Ford and his company were not familiar with astronomical telescopes or equatorial mountings, so Smith faced a formidable task in explaining the requirements of pointing a telescope accurately to engineers accustomed to the much simpler task of moving a gun turret. The proposal and estimate from Ford were repeatedly delayed.

While McDowell waited for the Ford proposal, Smith worked on his own ideas for a control system. He worked alone much of the time, and colleagues didn't notice that he was often pale, easily tired, and sometimes in apparent pain. When the Ford proposal seemed stalled, McDowell wanted Smith to go east again to "consult" with Hannibal Ford. Smith was too ill to make the trip. By September he was in the hospital. Mark Serrurier and others on the project visited him. Serrurier arranged to give Smitty a blood transfusion. Smith didn't talk about it, even with close friends, but a doctor had told him he had cancer. He worried that he might not see the drive system completed.

By October 1937 Smith was out of the hospital, in remission. The Ford proposal and estimate finally arrived, and Smith had the job of studying the drawings and estimate. McDowell was eager to sign a contract with Hannibal Ford. Smith reported to the construction committee that the Ford proposal didn't really meet the needs of the telescope. He estimated that a better drive system could be built locally for one-fourth of Ford's estimate. The committee of Caltech engineers and astronomers, overruling McDowell, voted to have Smith develop his own plans and estimate. They gave him three months, until January 1, 1938.

He met the schedule, and by mid-January, the committee voted to accept Smith's plans. He pushed ahead on working drawings. As soon as he had a sketch finished, it would go to the draftsmen and off to the machine shop. He reported to the construction committee that the controls would be finished in two months. He was often short of breath and pale, but he pushed on, converting ideas to working drawings and control systems. He told no one that the doctors had given him only a few months to live.

A young Caltech graduate in electrical engineering, Bruce Rule, was recruited to work with him, but the control system was so complex that only Smith understood its full workings. The heart of the system was a corrector unit that would make compensating adjustments in the alignment of the telescope. The "errors" that Smith had isolated were minuscule, far smaller than had ever been worried about on a telescope. He had designed corrections for atmospheric refraction, flexure of the mounting of the telescope, misalignment of the polar axis, and a sinusoidal (cyclical, like a sine wave) skewing of the yoke. He had isolated the remaining uncompensated errors—a nonsinusoidal element in the skew of the yoke, a potential eccentricity of the telescope bearings, and a slight rotation of the field of view when the telescope was pointed near the pole. The first of those, he concluded, could be corrected by the machining of the horseshoe.

Smith was working frantically on the last remaining problems when he was again hospitalized. After two months in the hospital, Sinclair Smith died on May 18, 1938. He was thirty-nine years old. Max Mason got permission to pay six months' salary to his widow. John Anderson took time off to write an article on Sinclair Smith for the astronomy journals.

Smith's death was a terrible loss. He was a bright young astronomer at the prime of his career. He had temporarily suspended his promising astrophysics research to work on the control system for the telescope. He had lost the race to finish and document his work by weeks.

Three months before Smitty's untimely death, at the beginning of February, Francis Pease was hospitalized for cancer surgery. It had seemed a routine operation, but surgery before antibiotics—sulfanilamide and penicillin were not yet generally available—was never routine. On February 11 Pease died from complications of peritonitis and septicemia. Like Smith, he had put astronomy research aside to work on the telescope. The project had begun with his drawings and models, and Pease had worked to the end to refine the design of the telescope. His work had increasingly been shunted aside by Sandy McDowell, who favored the work of engineers he had known from his navy days and personally disliked Pease. To Pease's credit, he recognized when alternatives were better than his own designs. He had championed roller bearings, but when Rein Kroon showed that oil bearings would work for the telescope, it was Pease who moved for the adaptation of the radical new design.

Francis Pease's life had spanned two generations of big telescopes. He had designed the one-hundred-inch telescope almost alone. Others contributed details, but the design and the drawings were his. Twenty years later the new telescope belonged to an era of Big Science, of projects so complex that it was no longer possible for one man to understand all that was involved.

George Hale, who had been confined to his dark room for months, hadn't been able to attend Pease's funeral. His nervous affliction had been compounded by a new symptom, attacks of violent vertigo coming without warning and severe enough to confine him to bed. With this, as with his other medical conditions, Hale was not candid, even with trusted friends. From his terse descriptions, the symptoms sound like Ménière's syndrome, a disorder of the inner ear. It was not treatable, and the unpredictability of the attacks left Hale confined to his house, frequently unable even to go to his beloved solar laboratory.

A few days after Pease's funeral, Hale felt well enough to be wheeled outside. He looked up at the sky and said, "It is a beautiful day. The sun is shining and they are working on Palomar." It was his last word on the telescope. He died a few days later, on February 21, 1938.

More than any other man, George Hale's name had been synonymous with big telescopes in the United States. He was known everywhere for his research in solar astronomy, for the Yerkes and Mount Wilson Observatories and telescopes, as a cofounder of the California Institute of Technology, a longtime supporter and officer of the National Academy of Sciences and the National Research Council, and the founder of journals of astrophysics. For years, as the public had read the continuing saga of the design and building of the great two-hundred-inch telescope, they were reminded that George Hale had conceived the idea and found the people, the funding, the companies, and the institutions that would contribute to the cooperative project. It was his web of academic, government, and business friends who constituted the old-boy network that had made science and technology on a national scale possible; his prestige that had persuaded the Rockefeller Foundation to commit the largest grant ever made for a science project; his ideas that had pushed the technology beyond limits; and his leadership that had kept the project from foundering when the demands of the telescope grew too large.

Even as his health deteriorated, in lucid periods Hale kept his fingers on thousands of details of the telescope. He would fire off memos on salaries for starting engineers and draftsmen, or budgeting five hundred dollars for improvements to the old road up the mountain. As late as the fall of 1935, when vertigo attacks had joined the demons to plague his days, Hale still planned trips east to inspect the machining work on the mirror cell at the Baldwin Locomotive Works in Eddystone, Pennsylvania, to see Robert McMath's recent work on telescope drives in Michigan, and to consult with Vladimir Zworykin at RCA on the latest work on photocells.

In 1936 Harlow Shapley invited Hale to attend a symposium in his honor at Harvard. Hale was too ill to go. With typical modesty, he urged that the honor be conferred on someone else. "Old and battered fossils retain a certain antiquarian interest," he wrote. "But in the midst of recent revolutionary advances, they are rapidly outclassed." At the gathering, in the library of the Harvard College Observatory, Shapley announced that he had planned the symposium with two thoughts in mind: to recognize "Hale's remarkable contributions to science and to the techniques and equipment of science" and to call the attention of younger astronomers "to the great debt we all owe to one man for the commendable position of astronomy in America at the present time."

In its obituary the *New York Times* urged that the two-hundred-inch telescope be dedicated to George Hale. The suggestion was echoed privately in the coming months. A few years later Millikan raised funds to commission a bust of Hale from a Danish sculptor named Jensen who just "happened along" in Pasadena. The bust, replaced many years later by a new bust by Marian Breckenridge, a friend of the Hale family, was installed in the entrance to the dome of the two-hundred-inch telescope.

In an era of simpler science, the deaths of three men as central as Smith, Pease, and Hale might have ended the project. But the two-hundred-inch telescope had gathered a momentum of its own. By the spring of 1938 hundreds of men were working on the telescope, in factories in Philadelphia, glass foundries in Corning, labs and shops in Pasadena, offices in New York, and on a lonely mountaintop. The telescope had already touched the lives of thousands of men and women all across America—mechanics, engineers, supervisors, professors, workmen, and researchers. The great telescope, an achievement of American science and technology in the midst of the most terrible depression anyone could remember, had become a part of the American consciousness, a symbol of pride and achievement. Railroad engineers with thousands of miles of service would tell of the greatest honor of their railroad lives—the time they had driven the shortest train of their career, only two cars, at a speed of twenty-five miles an hour, carrying the "great eye." Glassworkers who had worked an entire lifetime at Corning, who had watched hundreds of thousands of bottles, casseroles, and dishes leave the factory and seen their products become part of every American household, would remember most of all the role they played in casting the most famous piece of glass in the world. In 1938, while the first components of the telescope had just begun to arrive on Palomar Mountain, the perfect machine had become part of American folklore.

The unveiling of the telescope tube, in the august presence of Professor Einstein, was Westinghouse's last great publicity venture on the project. As the teams of arc-welders finished the other sections, assembly after assembly went through the boring mills and the annealing ovens. Some of the fabrications, especially the three sections of the great horseshoe, were larger than the tube, among the largest structures ever machined. But the Caltech engineers had refused to pay the costs of modifying the factory and test-assembling the sections, so the full majesty of Westinghouse work couldn't be demonstrated to the reporters and public. Even without photographs, the numbers were impressive. The largest journal bearing ever constructed had a diameter of forty-six feet and a face width of fifty-four inches; it weighed 375,000 pounds. But to the press and public the horseshoe looked even less like a telescope than did the bare frame of the tube. A lucky photographer caught glimpses of the huge sections on railcars, on their way to the docks in Philadelphia. The gargantuan components of the horseshoe spanned two tracks on the siding where they waited.

McDowell was eager to get the components to Palomar before winter weather set in. He used his navy pull to get access to the Philadelphia Naval Yard and to get local rail traffic rescheduled to make way for the huge sections of the telescope mounting. The only crane large enough for the job was commandeered from the yard to load the tube and the other parts of the telescope mounting as deck cargo on the *American-Robin*, the *Pacific*, and the *Pennsylvanian*. The assembled telescope tube was the largest single deck cargo ever shipped, but a journey through the Panama Canal could not generate the popular appeal of the slow train carrying the disk across the country.

The ships began arriving in San Diego in mid-October. The mounting components were the first major cargo to go up the new road to the summit. Snow had begun to fall as the sections of the horseshoe went up on a massive trailer, pulled by two heavy tractors and pushed by a third. The trucks moved so slowly that men walked in front and alongside the cargo.

In its press releases Westinghouse said of the telescope mounting, "On the site a fine job of rigging will be necessary to get the telescope parts into the dome and to erect it." It was a gracious understatement. For Byron Hill and the workmen on the mountain, the arrival of each piece was like opening another box at Christmas. Mark Serrurier had written long memorandums explaining exactly how the pieces were to be assembled, but as so often happens with Christmas toys, the assembly didn't go quite as easily as the instructions suggested.

The landowners on the mountain welcomed the completion of the new road, their joy prompted less by the improved access than because the closing of the WPA camp meant an end to the Saturday-night rowdiness. For years Captain Bolin's store had sold out of Don Leon wine each payday. The resulting hilarity from the WPA camp and from the "zombies" on the mountaintop had led to some hijinks that Byron Hill, and Colonel Brett before him, had to smooth over with apologies and reassurances.

The closing of the WPA camp also meant that there was no longer a resident physician on the remote mountain. One workman had first-aid training and his wife was a nurse, and there was a well-equipped dispensary, but one man's death from a heart attack during the earlier work and the potential danger of the work with the heavy telescope components prompted McDowell to appoint a resident physician. The work camp hired a cook who had worked at the Agua Caliente racetrack. The rude mountain was turning into a research facility.

Cottages went up for the resident staff. Hale had specified that the residences should be simple, but Byron Hill, knowing that what was built on the mountain would have to be fixed on the mountain, added his personal touch to the designs. Hill liked concrete. It didn't rot, woodpeckers didn't eat it, and squirrels didn't bury their winter cache in it. He provided each residence with a six-inch-thick concrete wood-shed. The walls were steel lath with stucco, the floors concrete, the roofs copper foil. The mountain wasn't a ski resort. The cottages would be there as long as the telescope.

The question of power for the observatory had come up early in the planning. The telescope and its instrumentation would require steady, uninterrupted, regulated power. Variations in voltage that would cause no more than a dimming of lights in a home would be crippling for a telescope and its instruments. Mount Wilson had at one time generated four-hundred-volt DC on the mountain, then later purchased electrical power from a local utility. The purchased power suffered frequent outages that shut down telescopes and lab equipment. A generating plant on the mountain entailed the risk of vibrations that could be transmitted to the telescope, noise that would be distracting to astronomers and residents, and poorly regulated power from the smaller equipment available for a local generating facility. E. M. Irwin, a Caltech engineer assigned the task of researching the question, concluded that "the desirability of the two systems is about equal."

Enough astronomers had lost a night of observing to blackouts on Mount Wilson for the independence of a generating plant on the mountain to win out. Two diesel generators from the Enterprise Engine Company in San Francisco, a primary unit and a backup, were installed in a powerhouse, with tunable spring mounts to isolate the vibrations and heavy insulation to mask the noise. The units Irwin selected, and the installation on the mountain, were quiet enough to not disrupt work on the telescopes, although the rumble of the big diesels was hard to miss from the recreation room next door.

With the dome finished, the footings in place, the powerhouse installed and running, and huge machined sections arriving by truck up the new road, Byron Hill and his men set about building a telescope.

The Caltech design engineers who had produced thousands of detailed blueprints of the assembly of the telescope tube, yoke, and mounting, had designed an overhead crane for the observatory, a fifty-ton unit built into the dome, to unload the cargoes. The crane ran up and down a track from the edge to the top of the dome. With rotation of the dome, the crane could service any area inside. A second, five-ton crane supplemented the main crane. In addition to the blueprints and Mark Serrurier's memos, Hill had Russell Porter's drawings of finished assemblies. For machinists, who often understand a machine better by taking it apart and remembering how it goes together, the vivid three-dimensional images in the Porter drawings were sometimes more useful than the file cabinets of blueprints. Porter's charcoal drawings, with beautiful cutaways, translated design sketches and engineering drawings into reality. The subtle shading of his drawings, much of it done with smears of a thumb to represent the grit of machinery, conveyed better than photographs or the most complete set of engineering drawings the feel and scale of the telescope.

The Caltech engineers had calculated the size of the hatch in the dome the way a mover can calculate whether a sofa or piano will fit through a doorway. The opening was mathematically large enough for the largest components—the sections of the horseshoe and the lower section of the yoke mounting, which held the two tubes that connected to the horseshoe. A draftsman with a slide rule could demonstrate that with the right twists and turns the hatch would accommodate everything that had been shipped.

As each component arrived, Hill and his crew swung into action immediately, eager to get the cargo unloaded and into the observatory. Hill, an efficient man and proud of his record, would do anything, including working all night long, to avoid demurrage charges from the trucking companies.

Most of the unloading went well. Hill would take the controls of the crane himself—a tricky operation because the motion involved rotating the dome and raising the crane on its tracks as well as the hoist itself—to lift the assemblies off the trailers and through the hatch. For one piece, the bottom section of the yoke, the engineers at Caltech had designed a special lifting harness to bring the assembly off the truck and up through the hatch. The harness didn't arrive in time for the unloading, so Hill and his crew did it with lifting hooks and slings they put together on the spot. The unit had been trucked lying on its back and had to be tipped onto its side to fit through the hatch. Tipping a structure while it is hanging from rigging is a tricky operation, because the center of gravity of the item shifts as it turns. The fit was tight. The next day, when the yoke had been squeezed through the hatch and the exhausted crew had gone to bed, Hill said he finally understood what women went through at childbirth.

The only damage the structures suffered in the trip from the Philadelphia factory to the observatory floor was some minor denting, nicks, gouges, and defacing with green paint by the stevedores in transit. One piece was dropped from the rigging and made a dent in the concrete floor of the observatory. It would have no effect on the telescope, but Frank Fredericks raged for days at anyone within earshot. There weren't supposed to be dents in the floor of a perfect machine.

The shipment of the big Westinghouse components got enough publicity to attract carloads of tourists up the new road. There were no facilities at the top, not even a paved parking lot. The trees and brush had been cleared from the area around the observatory, leaving a barren moonscape; there were no pathways or signs; and only the intrepid could find their way to the door of the dome. Still tourists showed up, eager to see the great machine.

The area that had been designed as a gallery for visitors, just inside the main entrance, wasn't enclosed yet, except for a hastily erected chicken-wire partition. Visitors crowded into the area, watching the workmen and asking enough questions to distract them from the complex assembly procedure. Finally, after he had been interrupted with questions so many times that he couldn't concentrate, Ben Traxler put a neatly printed sign on a pipe standard outside the chicken-wire enclosure:

DON'T TALK TO THE PRISONERS
ASK THE GUARD

It took Byron Hill a few days to figure out why the tourists had suddenly quieted down. Hill was capable of laughing at a dry joke, but he could imagine the reaction when rumors reached Pasadena that the telescope was being built by chain gangs. He tore the sign down.

Frank Fredericks had been at the Westinghouse plant in South Philadelphia when the alignment and bolt-holes for the components of the mount were bored. He came to Palomar to supervise the assembly. Some of the holes had to be rebored because of dents in the structures through shipping. Hill argued against relying totally on the alignment pins. He was a Caltech engineering graduate but, with a mechanic's sense of a machine, skeptical about paper specifications.

Once the three sections of the great horseshoe were aligned and bolted, the mounting began to take shape. The balance of the telescope depended on the weight of the primary mirror at one end of the tube, so for testing purposes Hill cast a dummy blank of eighteen tons of concrete. The final balancing depended on adjustments, but the design process, sometimes jumping from astronomers to draftsmen, had skipped the stage of rote mechanical engineering, when the mechanical assemblies would be checked for ease of adjustment, lubrication, and servicing. The adjustment studs and screw assemblies were an afterthought; the counterweights were on four-inch Acme screws with unlubricated nuts. The south bearing, a ball suspended on oil pressure pads, was brilliantly engineered, but the only access to align the bearing was through manholes. By the time the telescope was assembled, Hill's knees were gone from too many trips down the manholes, and he had tasted enough dripping oil for a lifetime.

Llewellyn Carlson, a salesman for Mobil industrial products, had pitched Mobil's experience in developing special oils to get the contract to develop the oil for the pressure bearings. The oil had to maintain a film thickness of three one-thousandths of an inch, in a temperature range from 20 to 100°F. Mobil made a publicity fuss about the special work of the General Petroleum Corp. engineers in the Vernon Lab and the Research and Development Department of the Socony Vacuum Laboratories in New Jersey in developing Mobiloil 95 for use in the telescope. Later they marketed Flying Horse Telescope Oil for other observatories.

Mobil and Westinghouse, in their publicity releases, bragged that the structures were designed and machined so carefully, and the oil bearings worked so perfectly, that 1/160,000 horsepower would be sufficient to move the telescope at the sidereal tracking rate of one revolution per day. As an example of the ease of movement, the engineers claimed that a milk bottle on top of one end of the perfectly balanced great horseshoe would be enough force to move five hundred thousand pounds of telescope. The skeptical "zombies" and Byron Hill couldn't wait to try that experiment.

The drive and control system was almost but not quite complete when Sinclair Smith died. His use of servo motors and generators was almost a decade ahead of its time, and not in general use for complex control systems until late in World War II. Smith's design for the drive and control system placed every control for the telescope at duplicate consoles for the night assistants. Wherever the astronomer was working, the prime-focus cage at one end of the telescope tube, the Cassegrain focus at the other, or the Coudé room below and south of the south pedestal, there was a console nearby. The consoles themselves were meant to be foolproof, even at the end of a long session in a dark observatory.

The first task of the drive and control mechanism was to point the telescope accurately, rotating the telescope around the right ascension axis, parallel to the axis of the earth, in sidereal (star) time, so the objects in the focus of the telescope would appear to stand still. The telescope would be moved by huge worm gears. The machinists in the astrophysics machine shop had already polished the gears for a year, achieving the precision of a chronograph on the largest gears ever cut. But no amount of polishing could eliminate every trace of backlash, and no bearings could be built without some measurable play. To achieve pinpoint exposures on the faintest possible objects, at the optical resolution that was anticipated for the telescope, the drive mechanism had to compensate for those mechanical aberrations with minuscule corrections in the motion of the telescope. That was the first task.

As the telescope turned, the dome would also have to turn, its motion synchronized so that the shuttered opening was in line with the field of view of the telescope. It seems simple until you realize that the dome turns in azimuth, rotating on a vertical axis perpendicular to the earth, while the telescope is turning in right ascension, on a polar axis (tilted thirty-four degrees from the horizontal at Palomar). Curve the index finger and thumb of one hand into a C and hold it parallel to the floor. Hold one finger of the other hand, angled at approximately thirty-five degrees from the floor inside the C and slowly twist and lift it. Now try to keep the opening of the C aligned with the end of the tilted and turning finger. That is the task of the dome drive mechanism.

Wind blowing into the shutter opening of the dome can disrupt the motion of the telescope and change the thermal equilibrium of the optics. A heavy canvas windscreen was designed to fit the shutter opening. On windy nights the screen would be raised high enough to block the wind yet left low enough not to block the telescope's field of view. The motion of the windscreen is in altitude, perpendicular to the ground; the motion of the telescope it is protecting is in both altitude and azimuth, rising and falling on its equatorial axis. You would need three hands and Houdini's coordination to imitate the motions with your fingers.

Finally, one major design goal of the telescope had been that various focal points would be switchable, by the night assistant alone, without the loss of a day or two of engineering time. The ability to shift from prime-focus work to the Coudé room in the course of an evening meant that if the weather or seeing were not good enough for deep-space work, the observer could switch and spend the evening on spectrographic work on nearby stars. The tube, prime-focus cage, and auxiliary mirror placement had all been designed to permit the relatively quick switches. The control system needed to include motorized controls for the auxiliary mirrors that would swing the mirrors into place and remotely lock them in precise alignment.

Sinclair Smith had worked at a furious pace to finish the drive controls before his death. He left behind drawings and schematics, some complete, some in sketches. The system in its entirety was so complex that Sandy McDowell turned to Vannevar Bush at MIT for help. Bush, famous for his analog computer systems, recommended his best student, Edward Poitras, as the man to take over the work. McDowell, still eager for Hannibal Ford to build the drive system, suggested that the remaining work on the control system be shifted to Ford's company, with Poitras, who was then with the Lombard Governor Corporation in Ashland, Massachusetts, serving as eastern liaison. McDowell assured Ford he could get security releases from the Navy Bureau of Ordnance and individual approval from the resident naval inspector for any work Ford did.

Ed Poitras was intrigued by the idea of working on the telescope, but before he agreed, he went to talk with his mentor Vannevar Bush. He found Bush, on a steamy hot July day, at his farm in New Hampshire, perched on his tractor, mowing a field of hay. Bush took a break, got a pitcher of lemonade, and the two men talked for hours about the problems of controlling a telescope. When the day was finished, Bush's hay wasn't mowed, but the project had a new man to take over the work on the drive and control system.

It is difficult to determine where Smith's work stopped and

Poitras's began. The drive system is a complex maze of gears, servos, motors, and controls, with hundreds of miles of wiring tying the various units to one another and to the consoles for the night assistants. The heart of the system Smith and Poitras created is a tiny phantom telescope. Models that trace the movement of large systems were a fascination of early-twentieth-century technology. The control room for the Panama Canal, a technical marvel of an earlier age, included a working model that duplicated each operation of the locks: Tiny aluminum fender chains rise and fall with the movement of the controls, aluminum pointers representing the gates swing over the blue marble that represented the lock chambers, upright indicators show the positions of the rising valve stems, and indexes show the level of water in the chambers.

The phantom telescope at Palomar is less than 1/200 the size of the real telescope, small enough to fit under the counter of the console at the head of the stairway to the Coudé room. There are no optics in the phantom telescope, but the tube, yoke, dome slit, and windscreen of the phantom mimic every movement of the huge instrument above it. At the edges of the dome slit on the phantom sensitive microswitches open and close at a contact of the phantom tube. A small recorder next to the phantom, with a typewriter ribbon, could be engaged to record automatically the right ascension, declination, sidereal time, and the guide rates on both axes, giving the observer an automated log of the observing session.

The clean, simple appearance of the consoles the night assistant would use to control the telescope belies the complexity of the systems they control. The stark black consoles, stripped of ornament, are both a reflection of late-1930s and 1940s design and an effort to minimize mistakes in the control of the telescope. Telescope domes are cold at night, and even the most rested of observers and night assistants can begin to make mistakes after eight or ten hours of concentration at subzero temperatures. Astronomers who had rare seeing conditions wasted because of a human error in the wee hours, or who had used up valuable telescope time battling a balky instrument, saw their dreams realized with the two-hundred-inch telescope. The controls were set up with simple spinner dials and readouts. When it was time to turn to another object, the night assistant would spin the controls to the new right ascension and declination, turn on the fast-slewing drive if necessary, then engage the slow drive until a second set of indicators showed that the telescope was pointed at the object. From there the drive systems would take over.

With no further intervention the telescope would track the object, the dome would track the horizontal movements of the telescope, and the windscreen would track the vertical movements of the telescope. For fine control, to compensate for instantaneous seeing effects in the atmosphere, the observer would still have the familiar paddle with control buttons. The entire drive and control system requires some sixty-five motors and more than four hundred miles of wiring. To maintain the electrical connections as the dome turns required more than four miles of dome slip rings. Emergency trip buttons guard against the telescope moving too far in any direction, a drop of oil pressure on the bearings, excessive acceleration, or electrical failures. Much of the wiring is concealed, making the telescope look simple and sleek. The balky and sometimes mysterious controls of the Mount Wilson telescopes, which had experienced observers like Milton Humason or Walter Baade doing bumps and grinds against tubes and clock drives to get the telescope to behave, were gone.

By 1939 enough of the mounting and control system had been assembled that the instrument inside the dome began to look like the telescope in Russell Porter's drawings. Astronomers who came to visit, even Walter Adams, who had been director of the largest working telescopes in the world for many years, were astonished at the size of the machine. The proportions of the dome, more classical than the taller domes at Mount Wilson, minimized its sheer size. Inside, the immense battleship-gray machine made the Mount Wilson telescopes seem toylike. Sometimes Byron Hill would make the machine perform for visitors, firing up the oil bearing pumps and drive and control systems.

The great machine looked and worked just like a telescope, if you didn't know that in place of a primary mirror it had a disk of concrete.

28

Testing

The first optical test of the mirror was in September 1938.

It took most of a year to grind a spherical figure into the mirror. As the shape progressed, the coarse grades of carborundum were replaced with successively finer grades of carborundum and emery. When the thirty-six-inch spherometer, an instrument to rough measure the curvature in the glass, indicated that the concave shape was a sphere of the proper radius, the grinding, the first stage in making a mirror, was done. Brownie ordered another complete scrub of the optical lab. The skirt that surrounded the disk to facilitate washing was removed; the disk was moved off the grinding machine to its storage easel; and the entire lab was hosed, scrubbed, wiped, vacuumed, rescrubbed, revacuumed, and rewiped to remove every trace of ground glass or metal fragments. Men spent days on their hands and knees searching for a single errant grit of carborundum or glass. Before the disk went back on the machine, the walls were dressed with cedar oil to make them sticky enough to capture stray dust that escaped the filters in the air-conditioning system.

Grinding used progressively finer carborundum to remove glass from the disk. Polishing, which would carry the surface from a rough sphere to a fine one, and then slightly deeper to a paraboloid, would be done with rouge, mixed into a slurry with water. For polishing, the five-ton full-size tool was mounted on the grinding machine. The glass facing blocks on the tool were covered with pitch pressed to the shape of the curve in the disk. Optics shop workers who hadn't ever polished a mirror discovered that the polishing, though less noisy, was even more tedious than the grinding. To prevent the tool from galling on the glass, one man and sometimes two stood on a bridge over the disk with squeegees on long poles. As the polishing tool slowly turned, they made sure the rouge stayed in suspension and that there was always a slurry of rouge and water under the polishing tool. Even an instant of dry pitch against the glass disk would leave scars. Hour after hour they would slosh the slurry under the polishing tool. They would then lift a hose, suck on the end to start the siphoning, and wash away the accumulated mud of fine abrasive and glass. Hour after hour a man on the scaffolding under the machine lubricated the main drive gear by hand.

The full-size tool consumed fifty pounds of polishing rouge per hour, most of it washing away over the edges of the disk. The rouge was expensive. When he saw the trucks bringing barrels of rouge to the lab, John Anderson—who had learned pencil counting from George Hale—decided that they would polish with smaller tools and use the full-size tool only when they absolutely needed it.

The routine was deadening. At least the rough grinding had showed some minuscule, visible progress. Workers who had been on the project during the grinding quit in frustration at the monotony of the polishing. Marcus Brown was fair, but he wasn't an easy boss. Shaping the mirror was a compulsion for Brownie. He had no time for arguments. "I know what I want done," he would say. "Do it!" By 1938 the depression was easing. There were other jobs.

In September 1938 the disk was raised to a vertical position for the first optical test. Even for the largest mirror in the world, the testing procedure was straightforward. J. B. L. Foucault, who devised a pendulum experiment to demonstrate the rotation of the earth, had also developed a simple and reliable test for measuring whether a mirror had a true spherical shape. Foucault discovered that if a point source of light is shined at the center of curvature of a *perfect* spherical mirror, all the light is reflected back to a point. A knife edge moved so that it cuts off the light at the focus will cause the mirror to darken evenly all over. A slight movement of the knife edge across the point of focused light will cause no moving shadow in the mirror.

For testing the two-hundred-inch mirror, a Foucault test station, with a micrometer-adjustable eyepiece, knife edge, and point source of light was set up at the far end of the room, 120 feet away. The point source focused the light through a pinhole 1/1000 inch in diameter onto the mirror. Brownie or Anderson would then move the knife edge until it just cut off the light. If the mirror was perfect, the entire diameter of the disk would suddenly go black. But no mirror is perfect. The secret to the testing procedure was to read the significance of the patterns of light and shadow created by the knife edge, to identify areas that required additional polishing. Good opticians read the shadows the way a sailor reads dark spots on the water for wind, seeing significance in patterns that escape the eye of the amateur.

Brown and Anderson tested the mirror on Saturdays. The optical lab had been designed as much for testing the mirror as for polishing: The big five-hundred- and one-thousand-watt lights in the ceiling were set behind panels of heat-absorbing glass, and the vents in the walls and attic were oversize. The accuracy of the tests depended on the optical homogeneity of the air in the room. The air-conditioning and ventilation equipment would all be shut down one half hour before testing began, to allow the air in the room to steady itself. The heavy cork insulation on the walls and ceiling were sufficient to maintain the interior temperature, in summer or winter, while tests were in progress.

On the first test of the two-hundred-inch mirror, the Foucault patterns weren't hard to discern. Brownie and his crew had achieved a fair spherical surface in their preliminary polishing. Anderson reported that there was still a trace left of the worst of the fractures in the glass, which showed up as a "very fine dark line." They could live with only one fracture. "Brown and I are jubilant over the results."

Some zones of the disk were too high or too low, which wasn't surprising for a relatively preliminary stage of the shaping of the mirror. The mirror also seemed to suffer from astigmatism. The curvature in the vertical plane was approximately one millimeter shorter than the curvature in the horizontal plane. The linear astigmatism was approximately 0.05 inches, which didn't seem that bad on a disk almost seventeen feet in diameter that didn't yet have what Brownie or Anderson would consider an *optically* smooth spherical surface.

Anderson and Brownie rotated the mirror ninety degrees and tested it again. The astigmatism was still there, and still in the vertical plane. That was troubling. If the astigmatism was in the shape of the mirror, it should have rotated when the mirror was turned. Anderson repeated the tests, checking that the error wasn't introduced by the test procedure itself. The astigmatism remained.

There were enough surface errors—high and low spots revealed by the knife-edge tests—to keep Brownie and his crew busy with the polishing tools and rouge while Anderson worried about the astigmatism. More months of fine polishing improved the accuracy of the overall shape of the mirror from approximately 0.01 inch to 0.001 inch. But the vertical astigmatism was still there. As the accuracy of the surface permitted finer measurements, Anderson discovered that when the disk was tested with its axis horizontal, then rotated by one hundred eighty degrees, so that what had been the top was now the bottom, the astigmatism was not exactly the same, but differed by as much as 0.01 inch in the value of a focal length in the two directions—an amount too large to be accidental.

The results were enough to send Anderson back to his slide rule.

Physicists who choose optics as their field crave precision and predictability. An astrophysicist sometimes has to live with vague answers, theories based on the paltry fragments of evidence he or she can squeeze from observations at the limits of a telescope's reach. In optics the materials are predictable. The qualities of glass, even new glasses like Pyrex, can be measured. The response of the glass to tension, compression, and heat can be measured and extrapolated. The optician relies on that predictability to bring an optical surface to the incredible level of precision—on the order of one millionth of an inch—that a large telescope mirror requires.

John Anderson was a precise man. He wore a neatly tied bow tie and wire-rimmed glasses. His hair was carefully parted in the middle. He was a worrier. He had worried about an earthquake when the disk arrived at the optics shop. When Brownie began the surface grinding, Anderson worried that they would not get through the contaminations to good, clear glass. Now the behavior of the mirror disk troubled him even more, because he wasn't sure he would explain it. The surface grinding seemed to have removed the checks and fractures. The glass that remained was consistent. Tests of samples from the disk, and George McCauley's tests of glass from the batch of Pyrex that had been used to cast it, didn't differ from the parameters they had used in designing the disk. They had gotten the glass they expected. But something was wrong.

Anderson suspected the support mechanisms in the back of the disk. The supports were supposed to compensate for any tendency of the disk to change shape as it moved from horizontal to vertical. The designs of the complex devices had been gone over and over; the parts had been machined to close tolerances in the astrophysics machine shop; and in tests the supports had performed exactly as the calculations predicted. In the disk they didn't seem to be doing the job.

He had the supports removed and retested. The thirty-six supports were precision machines, an assembly of levers, counterweights, gears, and ball bearings like a fine wristwatch, but large enough that it took two men to carry each of them. They were machined to the precision of a watch. But in a watch, where parts move continuously, and the largest force is the turning of the hands, simple jeweled bearings are sufficient to keep the parts moving freely. The lever arms of the support mechanisms moved only short distances and infrequently; the loads they had to push were on the order of 850 pounds.

When the supports were retested, the mechanics found that some of the support mechanisms required three or four times as much force to overcome the stiction as the freer-moving supports. The mechanics went to work on the supports, stripping them down, testing batches of bearings, repolishing shafts. When the mechanics finished, some of the components were brought back to the optics shop and polished with rouge by the opticians, honing the surfaces to an optical smoothness. When the rebuilding was finished, every support mechanism tested within the same narrow parameters.

Even as he had the mechanics rebuild the support mechanisms, Anderson realized that the repairs wouldn't fully fix the problem. It took him a long time to figure out what was really wrong.

The design of the support mechanisms was clever. The levers of the mechanisms were designed to push up, against the tops of the pockets in the back of the disk, to counteract the force of gravity pulling down on the disk. The supports were four inches behind the actual surface of the disk. The distance from the support to the front of the disk acted as a lever, pushing the upper part of the surface in front of each support forward, and pulling the lower part back to create a vertical S. In the language of the optician, "The deformed condition will consist in the addition of a very weak convex cylinder to the upper half and a similar concave cylinder to the lower half." With thirty-six supports, the thirty-six areas of the mirror should each have been equally deformed: thirty-six barely detectable Ss on the surface of the disk.

But there were no local Ss, or if there were, they were too small to measure. Instead the tests of the mirror showed a vertical concavity of the entire surface. It was as if each support were producing only the "concave cylinder" on the lower half of each area in front of the pockets. Together, those minuscule concave sections added up to a deepening of the shape of the disk in that plane—astigmatism.

Anderson stared at sketches on the blackboard. He ran numbers through his slide rule. He studied Russell Porter's drawings of the support mechanisms. He combed engineering and optics texts, trying to find an analogy to the behavior of the disk. There were no data on the behavior of masses of glass as large as the disk. No one had ever analyzed the interaction of support mechanisms and a thin, ribbed disk. And no one had ever tried to figure and test an optical surface as large and demanding as the two-hundred-inch disk.

The problem, Anderson finally realized, was that despite the interlocked webbing around the pockets, the front of the disk was stiffer than the honeycombed back. As a result the effect of the support levers pushing upward in the pockets was asymmetrical: A smaller area would become convex from the upward and outward pressure of the lever arm, and a larger area below would become concave. The Ss had a tiny upper half and a giant lower half. Instead of the concave and convex distortions canceling one another, the net local effect was concave.

Anderson's solution to the astigmatism was to design twelve "squeeze" correctors that would press on the rear portion of the edge of the disk. The edge correctors also worked with lever arms and counterweights. When the mirror was horizontal (the telescope was pointing to the zenith), they would have no effect; when the mirror was vertical (pointing to the horizon), they would exert the maximum squeeze on the edges of the disk. The compensating squeeze was calculated to compensate exactly the tendency to vertical astigmatism. The machine shop fabricated the supports, and they were installed in the mirror cell with the rebuilt pocket supports. The troublesome vertical astigmatism all but disappeared. Brownie and his crew of twenty-one men went back to work.

By 1938 the trajectory of the project had leveled off. The years when design work was in partial suspension, when a crew of opticians waited in the optics shop, busying themselves with experimental disks while they waited for news first from GE, then Corning, when work on the mountain was tentative, lest they build an observatory without a telescope—were long gone. Everywhere work was in full swing. On the mountain work crews worked on the dome and mounting for the two-hundred-inch and the forty-eight-inch Schmidt telescopes, auxiliary buildings, the residence for astronomers, cottages for staff, and the powerhouse and utility building. Sandy McDowell was taking daily reports from Palomar and firing off his memos in eight copies.

Even after three years, McDowell still didn't get on with the scientists in the project. The objections didn't arise because he was an outsider. Max Mason had stepped in from the Rockefeller Foundation without ruffling feathers. As chairman of the Observatory Council, he stayed out of the day-to-day work on the telescope, but he understood the technical problems and the working style of men like John Anderson and Sinclair Smith. Men like George McCauley at Corning, Rein Kroon from Westinghouse, or John Strong from Johns Hopkins, who had come to work on experiments on the coating of the mirror, were accepted almost immediately. They spoke the language of scientists and played by the rules of science—documenting their work, commanding the facts and figures to make their points, and producing solutions that could stand the test of the physicists.

McDowell had never learned the rules and style of science. Used to navy command and his own old-boy network of the Bureau of Ships and the companies that contracted for them, he had never gotten over the feeling that what the group really needed was a "kick in the pants." He never fully grasped how different the telescope was from the machines he knew, like a battleship turret gun, or why the criterion of "good enough" that produced the best balance of cost and function in expendable military hardware wasn't an appropriate criterion for a unique scientific instrument that was expected to function effectively for a century.

McDowell wrote reports on the progress of the project for journals and gave talks wherever he could. Even for an audience like the readers of *Scientific American* he would write that the control system had to be able to hold a rifle on a quarter at a distance of three miles, the sort of analogy that might be appropriate to naval gunnery but that missed the real challenge of a telescope control system that had to compensate for minuscule atmospheric refraction and periodic errors that were not a factor in holding a gun on a target. Early in the project Sinclair Smith investigated the possibility of using photosensitive devices as detectors for the telescope. He worked with Joel Stebbins, an astronomer who had pioneered work in this area, and Vladimir Zworykin, director of research at RCA, to find out the state of the art in detection devices, including those used for early television cameras. McDowell's conclusions from this research were: "Because of the possible public demand for seeing results of the 200-inch telescope, and particularly because of the desire not to allow the public inside the dome at night, it seemed that a television projection of planets, moon, etc., as seen through the 200-inch telescope, would be of interest." It was the kind of suggestion that was guaranteed to make astronomers cringe.

The astronomers began to worry about the direction of the project. Especially at Mount Wilson, where the telescopes had been designed by working astronomers, they worried that the calculations for the two-hundred-inch telescope were by engineers and physicists rather than people with experience on big telescopes. "The tendency," Walter Adams wrote, confidentially, "is toward an instrument equipped with a multitude of experimental devices instead of a simple and rugged telescope." To counter the trend he urged that Hubble be appointed astronomical director of the Caltech project.

Others on the project tried to avoid McDowell, but except for the optics shop, where "only God and Brownie" were allowed and John Anderson had the absolute say, McDowell's authority extended to every corner of the project. No one was exempt from his memos. In confidential letters Max Mason admitted that some "improperly considered and ill-advised moves by McD[owell] . . . rather thoroughly shocked the observatory crowd." To mend the damage an Engineering Advisory Committee was established that had to approve in detail all of McDowell's decisions. Nobody liked the arrangement.

The final blow came late in 1938. David Woodbury, who had been following the project since 1935, had visited Corning and spent three weeks in Pasadena interviewing anyone who would talk to him. Woodbury was hoping to have his book appear when the telescope came into use. John Anderson told him that the telescope "*may* possibly be ready for use by the end of 1939, but this is on the assumption that everything goes perfectly from now on. It probably will not." Undeterred, Woodbury submitted an article from his manuscript on "The Glass Giant of Palomar" to *Reader's Digest,* got what he considered "a large price for it," and decided to go ahead and publish the book. At the request of his publisher, Woodbury sent copies of the manuscript to Walter Adams, John Anderson, George McCauley, and others, hoping that Caltech would endorse his book and that he could get a jacket blurb from one or more of the major figures on the project.

Woodbury had written a lively though sometimes fanciful book. He had workers at Corning swimming around the flooded basement of the factory, sparks flying as the water shorted the annealing circuits, a state of total confusion in Pasadena before McDowell rode into town to save the telescope, Brown as an untrained man who took over the glass grinding with the sheer force of his ambition, a landing strip at the observatory where astronomers would arrive by small plane for their observation runs, and all manner of marvelous, charming stories about the individuals that were captivating to readers but not quite true. He apparently never heard an anecdote he didn't believe. Woodbury hadn't so much made up his material from whole cloth as selectively created his story, exaggerating the role of a few individuals as suited his narrative fancy, omitting facts that were inconvenient, and muddling the scientific and technological details of the telescope.

Woodbury didn't get the endorsements he sought. Adams wrote that full corrections would require a rewrite of "a large portion of [his] text," and questioned whether Woodbury needed to write "an interesting romance rather than a bit of history of the Mount Wilson Observatory. Certainly it could not be used for purposes of reference in the future." George McCauley sent Woodbury a list of "statements [from the manuscript] that are in opposition to facts, notwithstanding their value as fiction," and pointed out that news articles, which seemed to have been Woodbury's chief source, were notoriously inaccurate and certainly not a useful source for an "accurate historical account" of the building of the telescope. "You will pardon me," McCauley concluded his letter, "if I seem to wish for facts along with my fiction."

Even Harlow Shapley, at Harvard, was unhappy over Woodbury's manuscript, because it left the impression that astronomers were jealous of the small group in Pasadena who would have the opportunity to use the great instrument. Shapley, who had privately argued that the two-hundred-inch telescope wasn't really necessary and probably wouldn't yield the expected results, still maintained cordial relations with a few of the astronomers at Mount Wilson and Caltech. Once the telescope was going to be a reality, he didn't want to burn his own bridges.

Copies of the manuscript were passed around, and before long every scientist who had worked on the project "was disturbed, to put it very mildly" by what Woodbury had written. Some thought the manuscript so completely bad in general spirit and organization that it was pointless to make detailed objections. Others argued that since it would be published anyway, it was better to correct as much as possible. Together the scientists objected to about 50 percent of the book as misstatements of fact.

Scientists are notoriously reluctant to see personal details and human disputes in a narrative about their scientific work. Some of the mistaken details that Adams, McCauley, Anderson and others pointed to were picky, the sort of mistake that could be excused as a reporter's license, especially in a manuscript pitched at a *Reader's Digest* audience. Other mistakes suggested that Woodbury didn't really understand glassmaking, the process of figuring a telescope mirror, the function of the various systems in a large telescope, or the significance of recent developments in astronomy.

Woodbury's sources were obvious to those who worked on the project. "Many of the stories in the book evidently originated with Dowd, Brown, Porter, and McDowell," Anderson wrote. Not surprisingly these four emerged as heroes of Woodbury's story. As Woodbury told it Jerry Dowd, a Mount Wilson electrician who returned from retirement, was the one man who could make sense of the complexities of the wiring. Marcus Brown was the untrained, dedicated workman whose passion and commitment made the grinding of the mirror possible. Russell Porter was the amateur designer and outsider whose fresh, brilliant notions cut through the squabbles of those who were mired in old ideas. And Sandy McDowell, a man destined for flag rank in the navy, sacrificed his career to work on the telescope for the good of science.

Woodbury hadn't taken the time to confirm his sources. Jerry Dowd, a superb Mount Wilson electrician, worked in the electrical workshop on the Caltech campus. He did almost no work on the Palomar project except for some early wiring on the "little" Schmidt telescope. Marcus Brown liked to present himself as self-educated, but he had been trained to grind mirrors in a demanding apprenticeship at the Mount Wilson optical laboratory. Russell Porter had been isolated from the others in Pasadena because of his age, his impaired hearing, and his lack of scientific credentials and experience with engineering on the scale of the Palomar project; he was bitter enough at his isolation and the lack of credit given his designs to exaggerate his own contributions.

But the portion of the Woodbury manuscript that aroused the strongest reactions in Pasadena was the portrayal of McDowell and his role in the project. "The whole account of McDowell is exaggerated and much of it is contrary to fact," Max Mason wrote. "The tale of confusion in Pasadena and how he set it right is nonsense. . . . The picture of a man like Henry Robinson of the Observatory Council being stunned by McDowell's energy is ridiculous."

From the beginning the scientists and many of the engineers had resented McDowell's condescension and his insistence that *his* procedures, *his* friends, and *his* contacts in industry were the only course for the telescope. The version of events McDowell fed to David Woodbury was grossly unfair to men like George Hale, John Anderson, Francis Pease, and others who had borne the bulk of the responsibility for the design and construction of the telescope. Even if he hadn't exaggerated his own role and made many in Pasadena look like bumblers, McDowell's stories were, in the minds of scientists, "conduct unbecoming." They had struggled to avoid sensationalism, to keep public information about the telescope accurate. After the manuscript circulated, and it was clear that McDowell had been the chief source of what many thought a badly distorted book, Max Mason quietly asked McDowell to leave the project at the end of 1938. After the firing Mason reported that "the somewhat complicated personnel situation now seems to be entirely straightened out."

McDowell put a good face on it, explaining to anyone who would listen that while the telescope wasn't *finished*, "so far as I can see, all the work that has come under my jurisdiction will be completed within the estimates set up." In his final report he commented favorably on several of the engineers on the project. The only scientist about whom he had a kind word was Sinclair Smith, who was dead.

Forty years later a picture of Sandy McDowell, posing with the one-tenth-scale model of the telescope, was still hanging in the basement offices of the astrophysics building. No one on the Palomar staff was quite sure who it was.

29

Almost

The world was too busy to pay much attention to Palomar in the fall of 1938. Shooting wars of surprising passion were being waged in almost every corner of the globe. The Spanish Republic had been fighting the Franco rebels for more than a year, with German, Italian, and Russian "volunteers" and the International Brigades using Spain as a warmup for the expected death struggle of fascism and communism. The Japanese were campaigning in China, forcing the Chinese government to abandon its capital; Jewish and Arab groups were in open rebellion against the British government in Palestine; fascist movements were bidding for power in South America; Italy had invaded and annexed Ethiopia; Germany had annexed Austria and was in the process of dismembering Czechoslovakia. No one needed a telescope to see the coming world war.

Most Americans, urged on by Charles Lindbergh, the America First movement, and the isolationists in the Senate, did their best to ignore the stirrings of war that threatened every continent except their own. The United States had already turned down membership in the League of Nations; in 1935 the Senate rejected adherence to the World Court; and in the following years neutrality laws canonized the oft-quoted injunctions of George Washington's Farewell Address. A few critics pointed out that the American army and navy were not prepared for war, but their warnings went largely unheeded. After sending armies to fight the "war to end all wars," much of America had no enthusiasm for seeing more of its young men fall on foreign soil.

Still, it was impossible to ignore the whirlwind that seemed to have gripped much of the world. News from Europe was on the front pages every day. Commentators read the tea leaves of Hitler's speeches. Occasional frightening articles outlined the extent of Japanese military and naval might and territorial ambitions. By 1939 even the two great oceans didn't seem enough to keep America isolated. The only consolation was that a few far-seeing entrepreneurs suggested that war business might end the lingering depression.

Many Americans hoped that progress—the technology that had built the Empire State Building, the Golden Gate Bridge, and Boulder Dam—would be enough to bring the United States out of the doldrums. That same technology, the newspapers occasionally reported, was finishing the greatest scientific instrument ever built. By 1938 the articles about the telescope included photographs of the disk, the mount, or the observatory—concrete proof of America's unique mission: Let other nations fight their wars. America would build a better world.

For visitors to Caltech and Pasadena, a stop in the observers' gallery of the optics shop and a walk through the halls of the astrophysics building, where Russell Porter's drawings of the telescope hung, were the height of the tour. Visiting astronomers and observers at Mount Wilson were often treated to a visit to Palomar to see the telescope. Even men who had spent many hours on large telescopes had trouble believing the scale and precision of the machine they saw going up. When John Anderson took Vannevar Bush, Walter Adams, and their wives up in April 1940, the telescope was still 15,000 foot-pounds out of balance, but the oil bearings were working well enough to demonstrate that a milk bottle on one side of the great horseshoe would actually move 1 million pounds of telescope. To answer the most common question, Anderson said that the summer of 1942 was a likely completion date. The last step would be the installation of the mirror. Once that was ready and installed, he said, the telescope would be in operation within two days.

Even the most cynical of eastern skeptics had put a moratorium on voicing doubts about the telescope. Hubble, Baade, and others were already planning how they would use the new telescope. Baade had concluded that only the light-gathering power and resolution of the two-hundred-inch telescope would give him evidence to support his new and important theory of stellar populations. Hubble and Humason had used the special Ross lenses developed for the two-hundred-inch to stretch the range of the one-hundred-inch telescope beyond the limits anyone had anticipated. Hubble had plans for an extensive program with the two-hundred-inch telescope, beginning with a extensive sky survey by the forty-eight-inch Schmidt telescope. Pleased by the popular attention his research had attracted, and caught up in his own celebrity inside and outside the world of astronomy, Hubble believed that the first priority for the new telescope should be his work.

The astronomers at Mount Wilson and in the spacious library on Santa Barbara Street could talk and plan, but Caltech had the keys to the two-hundred-inch telescope. They also had the nucleus of an astrophysics faculty in Fritz Zwicky, spectroscopist Ira Bowen, and cosmologist Richard Tolman. There were some at Caltech who suggested that the institute should run the telescope alone. The lack of experience at building large telescopes certainly hadn't stopped the ambitious and confident engineers and scientists. Maybe the lack of experience administering a large observatory could also be overcome. The Caltech boosters overlooked the fact that Caltech had no funds to operate the observatory.

The terms of the original grant called for Caltech to raise an endowment suitable to pay the operating expenses of the observatory. Henry Robinson, the chairman of the Caltech board of trustees and the member most enthusiastic about a telescope, had pledged $3 million when the grant was first received, fulfilling the Rockefeller Foundation requirement and obviating the need to search for endowment funds. Robinson hadn't lost his enthusiasm, but the stock market crash had decimated his portfolio. Aside from an interest in the Bolsa Chica Club, which he had purchased almost by accident before oil was discovered on the property, he had nothing to offer the observatory. As of 1938 no portion of the endowment was in hand, and with the depression still vitally affecting business and philanthropy, there seemed little prospect of raising it.

The obvious partnership, which the public assumed already existed, was between Caltech and the Mount Wilson Observatories. Mount Wilson, a department of the Carnegie Institution of Washington, had a long record of experience with large telescopes; offices conveniently in Pasadena; and a substantial staff of opticians, engineers, electricians, and trained night assistants. The Mount Wilson staff astronomers, men like Walter Baade and Edwin Hubble, were experienced observers with unparalleled experience on large telescopes. The Mount Wilson optics lab had done work on the early quartz and glass disks, correcting lenses, and other instrumentation for Palomar; portions of the telescope instrumentation and optics had been tested on Mount Wilson telescopes; and Mount Wilson staff members, including Sinclair Smith, Francis Pease, and John Anderson, had for many years worked half-time on the Palomar telescope, with that portion of their salaries paid by Caltech from the telescope budget. Mount Wilson staff had been accustomed to attending colloquiums and symposia at Caltech and had worked with Caltech faculty in physics and geology.

However sensible the relationship between the two institutions might have seemed to an outsider, the Observatory Council and John Merriam, the president of the Carnegie Institution, had long memories. After the roadblocks Merriam had erected in 1928, when he tried to sabotage the grant, and again in 1934, when he appointed his special committee, the Observatory Council and the staff and trustees of the Rockefeller Foundation thought it best to soft-pedal any talk of cooperation, because of the "personal difficulties which sometimes surround the activities of J. C. Merriam."

Merriam occasionally tried to court the project. In 1936 an exhibit on the two-hundred-inch telescope was included in the annual Carnegie Institution Christmas display in Washington. Early the next year Walter Adams asked Max Mason if he would consider appointment as an associate of the Carnegie Institution. Mason said he would be honored, but John Merriam apparently changed his mind. The appointment never came through.

The mistrust went both ways. Merriam wasn't popular, even among the Carnegie Institution staff and trustees. But he wasn't the only one at the Carnegie Institution who felt that Hale, as an officer and an employee of the Carnegie Institution had been openly disloyal when he worked out the original plans for the telescope. It was no secret that Hale had used his personal friendships with some trustees and Elihu Root to sidetrack Merriam's initial obstruction of the grant.

Underlying the resentment of the grant and Hale's actions, there was also a deep-seated Carnegie dislike of Robert Millikan and Caltech. The Carnegie Board was made up of conservative men, mostly easterners with strong ties to the Ivy League and quietly contemptuous of Millikan's aggressive fund-raising, faculty-stealing, and penchant for publicity. After the stir his comments on religion had made in the 1920s, Millikan followed up in the 1930s with strong public positions against federal support of science, all the while arguing that increased scientific research—meaning the kind of research he had raised funds to support at Caltech, rather than economic tinkering—would create jobs and end the depression. Even Carnegie trustees who agreed with Millikan's opposition to the New Deal resented the publicity he drew to Caltech, often at the expense of Carnegie departments like the Mount Wilson Observatory.

Still, to many a marriage between the two institutions seemed inevitable. Frederick Keppel, a trustee of the Carnegie Institution, was at Max Mason's retirement party, given by John D. Rockefeller. He drew Mason aside to say that the Carnegie Institution was "vitally interested" in the two-hundred-inch telescope and wanted to discuss the possibilities "as soon as Merriam was out."

Merriam retired on schedule in 1938. His replacement was a surprise—Vannevar Bush from MIT. Although Bush had advised Sandy McDowell on a number of questions, found Edward Poitras to replace Sinclair Smith on the control-system work, and suggested people and firms to work on the project, no one at Caltech really knew him.

Bush took over the Carnegie Institution by storm, appointing special committees to review every area of operations and personally investigating large projects. He recruited the most famous of California engineers, former president Herbert Hoover, a trustee of the Carnegie Institution since 1921, to head a special subcommittee on astronomy. After leaving the White House, Hoover had become an active California booster. He visited Palomar before construction work on the observatory began, eating dinner with Colonel Brett in his cabin. Over the years Hoover followed the progress of the telescope, and even before Bush came to the Carnegie Institution, Hoover tried to raise money in Southern California to bring the Mount Wilson and Palomar Observatories together under a single organization, independent of both the Carnegie Institution and Caltech.

Hoover's plan was that the Carnegie Institution turn over the Mount Wilson Observatory and $6 million of endowment to the new organization. Caltech would turn over the Palomar telescope. Bush liked the plan, with one exception: He thought that the Mount Wilson Observatory and the Carnegie Institution expertise was a sufficient contribution *without* an additional $6 million for operating endowment. What Bush liked about Hoover's plan was the centralization of authority. Millikan and others at Caltech had suggested that a group of scientists from Caltech and a group of astronomers from Mount Wilson could work together at Palomar under a joint committee. Bush thought running an observatory by committee an unworkable idea.

Bush went to Warren Weaver, the head of science programs at the Rockefeller Foundation. The Rockefeller Foundation, he told Weaver, had made a serious mistake in letting Caltech go so far with the two-hundred-inch telescope. Caltech didn't have the funds to endow the operation of the telescope, and they didn't have the technical or astronomical staff to operate the facility. Lacking the resources to embark on their own astronomical program, they should stick to physics and leave astronomy to the Carnegie Institution. Two departments would not only be wasteful, but would inevitably lead to competition. Even if Caltech didn't try to start a separate astronomy department, if the Rockefeller Foundation was not effectively to lose the funds they had put into the project, Bush warned, they would have to be prepared to face the necessity of committing between $1 and $2 million in additional grants to support the maintenance and operation of the telescope, and be prepared "actively to enter the situation and manipulate the plans for the operation of the telescope."

Bush's warning struck sensitive chords. Any foundation officer dreads two possibilities: the need to commit more funds to "rescue" a project, and the need actively to enter the management or control of a funded project. Warren Weaver urged Raymond Fosdick, the new president of the Rockefeller Foundation, to stay out of the fray and not to make any move that could be construed as an offer of additional funds, even though "back of all this theorizing there remains one hard and disagreeable fact. It is true, and of inescapable significance, that the Rockefeller boards will have a very heavy investment in this Observatory. It is true that it would be intolerable and unthinkable that the Observatory not be properly utilized." For the present, at least, he was reluctant to consider the possibility that the Rockefeller Foundation might have to add $1 or $2 million to their commitment "rather than see the former investment invalidated."

Bush told Weaver that the only reasonable administration for the new telescope was for a member of the Mount Wilson staff to be appointed director of the Palomar Observatory, and for the astronomical staff to be Carnegie Institution employees. When Weaver asked whether the Carnegie staff didn't already have their hands full at Mount Wilson, Bush assured him that the present staff of Mount Wilson was fully capable of utilizing the facilities of both observatories. Under Bush's plan Caltech would own Palomar and furnish a minimal thirty thousand dollars per year for basic maintenance. Caltech scientists would be *allowed* to work at Palomar, with the understanding that their interests would be in physics rather than astronomy. All publications from the observatory would give full credit to Caltech, and an advisory board from both institutions would recommend research programs and allocation of observation time. Bush, convinced that publicity was the real objective of Caltech, even offered to structure any agreement to save face and provide maximum publicity for Caltech. The crucial point for Bush was that final authority would rest with the director of the observatory, who would be a Carnegie man.

As he explained his idea to a somewhat incredulous Warren Weaver, Bush argued that Caltech was overextended, that Millikan had lost his ability to raise funds and no longer commanded respect from other scientists, that Caltech was not capable of utilizing the Palomar Observatory without the Carnegie Institution, and that the Rockefeller Foundation had no choice but to intervene to protect its investment. At the same time he said he would not present his plan directly to Max Mason or the Observatory Council, because they would instinctively oppose placing the ultimate authority for the observatory in the hands of a Carnegie Institution officer and would only accept the idea if it were forced on them by the Rockefeller Foundation.

Bush was in a powerful position. The Carnegie Institution of Washington not only enjoyed the wealth of its own considerable endowment, but Bush could appeal to the Carnegie Corporation, a separate body, for additional funds. Caltech, which had promised to raise an endowment for the telescope, did not have the funds at hand, and Millikan, who usually had the knack of making Southern Californians think it was a privilege to give money to Caltech, had had no success in raising an endowment for the telescope. The long years of depression didn't help. Even if the economy had been better, endowment funds are less-glamorous causes than capital construction, and the observatory was so closely associated with George Hale and the Rockefeller Foundation that any potential donors knew they would have little chance of the immortality that Lick, Yerkes, or Hooker had achieved with the telescopes bearing their names.

The negotiations between Bush and Mason went on indirectly.

Max Mason had one advantage. As the former president of the Rocke-feller Foundation, he enjoyed enormous respect and support from the foundation. Warren Weaver sent copies of his diary records of meet-ings with Bush to Mason and made it clear that the Rockefeller Foun-dation would not intervene in support of Bush's plan.

When Bush and Mason finally met in May 1939, they were able to split many of their differences. Bush was willing to agree to a joint committee to run the observatory provided its chairman was a member of the Mount Wilson staff and had "reasonable power." Members of the Mount Wilson staff would have honorary appointments at Caltech, where they could conduct seminars and accept graduate students. Astronomy-minded members of the Caltech faculty would be named associates of the Carnegie Institution—an honor that hadn't yet been accorded to Mason himself. Bush thought that by consolidating basic services of the two observatories, the Palomar facility could be run from an endowment of $1.5 million. He tentatively agreed to include the sum in his requests for funds from the Carnegie Corporation.

Mason liked the plan. The day after he met with Bush, he met with Weaver and Raymond Fosdick at the Rockefeller Foundation to explain that the plan would require no additional commitment from the Rockefeller Foundation, even though "some individuals might interpret this arrangement to mean that the RF had given an observa-tory to CIT and that CIT was in effect giving it away to the Carnegie Institution."

In the offices of the Caltech astrophysics building, named Robin-son Hall after the death of Henry Robinson, Caltech physicists, astro-physicists, and cosmologists had already begun talking about estab-lishing their *own* program in astrophysics to take advantage of the new observational facilities. The Mount Wilson people were quick to proclaim their expertise and experience at running large telescopes, and to point out that Caltech did not have the funds to provide the operating expenses for the telescope. It was hard for many at Caltech to shake the feeling that the Carnegie Institution was trying to steal *their* telescope.

Bush and Millikan were both ambitious men, with great appetites for power and publicity. The control of the most expensive and most famous scientific instrument in the country was a plum that neither was willing to give up easily. Millikan had slowed down—"By actual clocking," the Pasadena wags reported, "it takes M[illikan] twenty-two minutes to ask whether the R[ockefeller] F[oundation] would object if CIT named the astrophysics building Robinson Hall"—but he was still a fierce defender of Caltech. After a year of talk Mason and Bush were close enough that "less than one sentence stood between them, and they could write out the partial sentence in a way which would be sat-isfactory to both if they could keep Rob Millikan out of the room while they wrote it."

In return for having a Mount Wilson man as director, the Carnegie Institution was willing to allocate up to three million dollars of endowment for the telescope, if not in one grant, in a series of maintenance grants and progressive endowments. Bush was convinced that in the end the Rockefeller Foundation would contribute toward the endowment of the telescope. When Mason offered to bet that they wouldn't, Bush backed down. Carnegie would pay to run the telescope that Caltech built.

At Corning, once a year, the accounting department would bring up the question of when it would be time to bill Caltech for the 10 percent "profit" on acceptance of the mirror. When the question came up in 1937 and 1938, the issue was shunted aside. In 1939 Eugene Sullivan, second only to Amory Houghton at the glassworks, decided that Corning would send a bill when the figuring was completed, "which probably will be sometime next year." Corning had been paid $329,347.27 by Caltech for the entire series of mirrors. The unbilled "profit" for their work on the mirrors was $29,129.14, 10 percent of the expenses other than depreciable major equipment.

McCauley had successfully cast the last of the disks for auxiliary mirrors, and work was underway in both the Caltech and Mount Wilson optics shops to grind and figure the difficult convex secondary mirrors. His analysis of the striae that had emerged in the two-hundred-inch disk had led to a decision that all future disks would be cast by a new process. Instead of ladling the molten glass and risking contamination from the dropping level in the melting tank, he would mine blocks of cullet from glass that had been melted in a tank built of unused refractory brick. An appropriate weight of this clear glass would then be "sagged," melted in place, in a mold. To ensure against striae, if there wasn't a single block of the correct size, smaller blocks of cullet would either be ground and polished on every face or chosen from cleanly fractured blocks, then fused together to achieve the required mass of glass before they were sagged into a mold.

In April 1938 McCauley got an opportunity to try the new process when Caltech ordered a seventy-two-inch mirror for the big Schmidt camera.* McCauley mined blocks of Pyrex from a fresh batch of cullet, including blocks of 3,200 and 2,372 pounds. The larger block was used to sag the Schmidt mirror. The specifications called for a five-and-one-

---

* Telescopes like the two-hundred-inch are measured by the diameter of their primary mirrors. Schmidt cameras are measured by the diameter of their correcting lenses, which are smaller than the primary mirror. The forty-eight-inch Schmidt camera has a mirror seventy-two inches in diameter. Comparing that mirror to the mirror of the sixty-inch telescope, which had been the biggest working telescope in the world twenty-five years before, is an interesting index of the growth of astronomical instruments.

half-inch hole in the center of the disk, which would have required a core in the center of the mold. Since that core would interfere with the placing of the huge block of glass, McCauley had the disk sagged without the hole and Corning technicians later ground a hole through the nine inches of glass. The disk was shipped in November 1938 and was soon on a grinding machine at the Mount Wilson optical labs. McCauley's new process was reliable, and another wave of orders poured in from observatories around the country.

Don Hendrix, chief optician at the Mount Wilson optics shop, was put in charge of figuring the mirror and corrector plate for the Schmidt telescope. For the thin corrector plate he ordered a carload of three-eighths-inch-thick plate glass from the Fuller Plate Glass Company, inspecting the sheets of glass one by one until he found one optically clear enough to be figured as the corrector. The fine ridge he ground into the corrector plate was so subtle, varying in contour less than five-thousandths of an inch over the surface of the fifty-inch-diameter plate, that it could not be seen with the naked eye or felt with the fingers. The spherical shape required for the primary mirror of the Schmidt camera required substantially less work than the paraboloid of the two-hundred-inch disk, so despite a delayed start, progress on the Schmidt was soon ahead of the bigger telescope.

When Ray Fosdick originally approved the expenditures for the Schmidt telescope, in May 1937, on the grounds that the original grant for the "construction of an observatory, including a 200-inch reflecting telescope with accessories" was broad enough to include the Schmidt, the estimate for the construction of the forty-eight-inch Schmidt telescope was fifty thousand dollars. A year later Max Mason admitted that their original estimate was low. With the machine shop busy on drive gears, machining tracks for the fifty-millimeter ball bearings on the polar axis, welding the huge cannonlike tube out of $\frac{5}{16}$-inch plate, machining interior supports of Invar to keep the distance from the photographic emulsion to the mirror fixed, and constructing equipment to prebend the glass emulsions, and mandrils to hold the emulsions in the required curved plane for exposures, the estimate for the cost of the telescope was now closer to one hundred fifty thousand dollars. Max Mason knew the Rockefeller Foundation well enough to know that they wouldn't object to the cost overrun. "If it were $250,000," he told Warren Weaver, "the group would unanimously endorse it."

The pieces of the observatory were rapidly coming together. Anderson believed that another year of polishing and tuning of the supports would lick the astigmatism problems. One more year after that to parabolize the mirror, deepening the curvature from the spherical figure by about $\frac{1}{200}$ of an inch, and the mirror would be ready to go up to the telescope. The drive and control mechanisms would be completed in the machine shop by early 1941. One right ascension drive gear was already on the mountain, and the machinists were almost finished with the fine cutting of the declination gear and the second right ascension gear. Once the gears and the rest of the drive and control system were installed, Byron Hill and Bruce Rule estimated that it would take approximately eight months to tune the control system with the dummy mirror so the telescope would be ready for the real one.

The plans all came together. Everything would be finished at the same time. The tentative date for taking the mirror up to the mountain was January 1, 1942. A few months more for final tuning, and the telescope would be ready for the astronomers.

30

Impossible Circumstances

No one wanted war. Opinion in the United States covered the spectrum from Lindbergh, Henry Ford, and the America Firsters, who saw German victories as the triumph of civilization over the "red hordes," to the president and some of his close advisers, who thought America's entry into the European conflict inevitable and were biding their time until events turned the opinion of the nation. Even without open conflict America began gearing up for war production. Industries that had been moribund in the depression were soon working at capacity to supply the needs of Lend-Lease and the increased buying and production for the U.S. Army and Navy.

Though no one wanted war, the talk in the United States was about war. The stories that had once monopolized the news—the debutante parties of Brenda Frazier, Wrong-Way Corrigan's comic flight to Ireland, or the superlatives of dams, skyscrapers, and telescopes—faded alongside the banner headlines and photographs of marching armies. Newspaper readers and radio listeners could say that the rape of Nanking or the fall of Paris didn't affect them, but when England hung on by virtue of a handful of fighter pilots, and no nation seemed ready to restrain the Japanese, even Americans who felt safely isolated by two oceans asked what would come next. Would the Japanese attack Singapore? The Dutch East Indies? The Philippines? How long would England hold out alone? Where would Hitler, now master of the entire European continent, turn next? Another assault on England? Or would he attempt what even Napoleon had failed and attack the Soviet Union?

Even on Palomar Mountain, connected to the world outside only by the daily radiotelephone calls to Pasadena and visitors, the war intruded. Five years before, in 1935, there had been no shortage of labor: men had only to hear a rumor of a job, and they would drive, hitchhike, or walk up the mountain road for the chance of work. For fifty cents an hour they were willing to live on an isolated mountaintop, sleep in a rude bunkhouse, put in a six-day week, and spend Saturday night with the boys and a bottle of cheap wine. By 1940 men weren't so desperate for work. The New Deal was no longer new. Nine and one-half million Americans, 17.2 percent of the total work force, were still unemployed, but it was a far cry from the depths of the depression, when some heavy industries, like steel, had slowed to the point where more than half of all workers were out of work, and companies like U.S. Steel had so parceled and split jobs that they could honestly declare that they had *no* full-time workers on their payrolls.

The outbreak of full-scale war in Europe and Asia completed the economic fix that the New Deal had begun. War orders lit the furnaces and started smoke up the idle stacks. In 1939 the aircraft plants in San Diego and the Los Angeles area were again hiring. By the end of the year Douglas Aircraft, bristling with $18 million of back orders, had ten thousand men on the payroll. Factories that built airplanes or tanks or trucks bought parts and raw materials from hundreds of other companies, bringing work to aluminum smelters in upstate New York, rubber factories in Ohio, and engine plants in Connecticut. Building trades picked up. Companies and contractors competed for electricians, construction workers, and mechanics—exactly the jobs that were needed on Palomar Mountain to finish the telescope.

Some workmen left the mountain, especially family men tired of commuting once a week to wives and children in Escondido, and men who had come to Palomar only because there were no other jobs. Working on the mountain suited some men. Whether they were fascinated by the machine they were building and proud of their role in bringing it to life, or enraptured by the beauty and mystery of the mountain, many stayed even when better-paying jobs came along. The mountain had its own rewards, the beauty of the trees and snow and air at five thousand feet, the evening skies, and the pride of working on a machine so huge and famous that the word "Palomar" evoked pride in those who built it and immediate recognition in others. Like the windowless optics shop, the mountain was a place men loved or hated.

Through 1940 and 1941 the work went on. In August 1941 one of John Anderson's Saturday tests determined that the mirror surface had achieved a satisfactory sphere with a radius of curvature of 1335.7 inches. On August 30 Marcus Brown and his men started the final stage of figuring—parabolization—polishing the shape to a slightly deeper curve. By the end of September the parabolization was 90 percent complete. What remained was the final figuring, bringing the entire surface to ultimate smoothness, so that the deviations from a perfect optical surface would be measured in fractions of a wavelength of light.

The original plans for figuring the mirror, drawn up when George

Hale first turned to Corning for glass blanks, included the 120-inch disk both as a trial run for the casting of the two-hundred-inch, and to use as an optical *flat* mirror for use in testing the big mirror. The machine shop had built a special grinding machine for the 120-inch mirror, and the dimensions of the optical shop had originally been chosen to permit room to test the telescope mirror with a flat mirror at the proper distance.

Figuring a large flat mirror is a difficult task. The normal procedure for grinding mirrors, rotating one disk over the other with a slurry of carborundum and water between the disks, produces a concave curve in the lower disk. Creating an optically flat surface on the lower disk requires special care. A flat used for testing had to be figured to tolerances as demanding as the big mirror, lest imperfections in the flat mirror confuse the tests. Brownie had estimated that it would take a year or more to figure the flat.

Anderson didn't want to wait. In 1941 he and Frank Ross, who had been developing corrector lenses for the prime focus of the telescope, asked whether it wasn't possible to test a big mirror accurately *without* using a large flat mirror. They fiddled and finally came up with a scheme that would use a much smaller half-silver flat mirror and a special lens to focus the light from a light source in a way that would precisely test the zones of the big mirror. It looked right on paper, so they had Brownie's men build the lens and the small flat mirror—relatively simple tasks. Their scheme—they never claimed much credit, because they weren't sure no one else had ever tried it—worked beautifully. In the hands of a skilled optician like John Anderson, the test showed patterns of dark and light in the eyepiece that told him which zones needed additional polishing.

All that remained was to polish down those zones, in tiny increments, until the mirror took the perfect paraboloid figure. Some weeks Brownie would polish with a small tool for only a few minutes each day, resting the mirror for hours to let the heat of the polishing dissipate before the mirror was tilted up in test position to measure the effects of the polishing. When asked, Anderson and Brownie would predict that a few more months of polishing could finish the mirror. The figure on the mirror had reached the point at which Anderson was doing much of the testing late on Saturday night, so traffic on the street outside would not disturb his measurements. Sometimes alone, often with Brownie there, Anderson would study the shadows and write notes. Some Saturday night, Brownie knew, Anderson would finish his tests and announce that the mirror was ready.

The point when an optician stops figuring a mirror is arbitrary. The small mirror of an amateur telescope might be figured to one-fourth of a wavelength of light, a precision of hundreds of thousandths of an inch. The surface of the two-hundred-inch mirror would be figured to one-fortieth of a wavelength of light, a precision of approximately one two-millionth of an inch. It was the most formidable optical task ever attempted. The closer they got, the more obsessed Marcus Brown and the men in the windowless room became. They were polishing smaller and smaller areas, narrowing the zones of imperfections, bringing the entire surface closer and closer to the elusive goal of a perfect paraboloid.

In the machine shop next door, the machinists on the big-gear-grinding machines worked toward a similarly elusive goal. The smoothness of the motions of the telescope, and the freedom from backlash that would enable the control and drive systems to keep the telescope precisely guiding on faint stars, depended on the machined precision of those gears. The design goal was for the drive mechanism not to vary by more than one arc second per hour at the normal drive rate of one revolution per day, and for short period errors to be limited to one-tenth second of arc per five seconds of motion.

There were three big gears, one for the declination axis, tipping the telescope tube in the great yoke, and two for the right ascension axis, turning the yoke on the great horseshoe bearing. One right ascension gear was for slewing the telescope, moving at relatively rapid speeds when the observer wanted to point the instrument at a new area of the sky; the second gear was for the slow-motion drive that would keep the telescope moving synchronously with the motion of the earth. By using a separate gear for the relatively rapid slewing motion, wear and tear on the main drive gear would be reduced.

Earle Buckingham of MIT had done much of the design work on the gears, working without compensation. Each gear was fourteen feet in diameter and weighed ten tons. The gears were cast by Westinghouse and rough-machined at their South Philadelphia Plant, then shipped west. Special gear-cutting machinery had been designed and built in the astrophysics machine shop.

These were the largest high-precision gears ever made. One second of arc equals 0.000445 inch on the pitch circle of the gears, which meant that the 720 teeth on each gear had to be machined to a tolerance of one ten-thousandth of an inch, crude stuff by optical standards, but demanding for mechanical work on that scale. The process, from the rough-cutting to the fine-polishing of the gears, took two and one-half years. To limit the expansion and contraction of the gears, the work was done in a subroom inside the machine shop, where air-conditioning could maintain the temperature at an even seventy-five degrees. The mechanics who ran the machines needed patience to match that of the opticians in the building next door. They examined the mating surfaces of the gears with microscopes. Soon only special measuring gauges could record the fine change in the surfaces.

Even in the sealed optics and mechanical shops, where men worked at the relentless pursuit of perfect surfaces, the world of war was closing in. In 1939 and 1940 a group of scientists, remembering the lack of preparation for World War I, had tried to organize a national research effort to harness the strength of American science. Vannevar Bush, from his office in Washington, had taken the lead, drafting an executive order that would establish a National Defense Research Council (NDRC) in 1940. His idea was that the need for close cooperation between the scientific community and the military could best be assured by having a scientific organization with its own funds that reported directly to the president. Franklin Roosevelt signed the order on June 15, 1940, a day after the fall of Paris, and Bush soon had James Conant, president of Harvard; Karl Compton, president of MIT; Frank Jewett, president of the National Academy of Sciences; and Conway Coe, the commissioner of patents, working with him. Vannevar Bush wasted no time: the NDRC promptly recruited scientists on university campuses and private industry for projects as prosaic as periscopes and as dramatic as cornering supplies of uranium ore from the Belgian Congo in case experiments then going on at Columbia University and the University of Chicago proved successful in unleashing the power of the atom. The United States wasn't officially at war, but on campuses and in defense plants, the scientists and engineers were already fighting.

Caltech, which had played a role in World War I research in its earlier incarnation as the Throop Institute, was quick to rally and offer its services. By mid-1940 Caltech facilities were engaged in contract work for military-related research and production projects. Max Mason, who had worked on sonar research at the Sound Laboratory at New London during World War I, supported the effort, and by mid-1941, fully half of the optical shop had been turned over to war-related work. At one end of the room Brownie and a crew worked on the big polishing machine, isolating smaller and smaller zones for fine corrections to the surface. The rest of the shop was gradually given over to opticians working on roof prisms for periscopes and optical range finders, mirrors, and corrector plates for Schmidt-type aerial reconnaissance cameras, and optical parts for aircraft research in wind tunnels. In the astrophysics machine shop next door, machinists who had been finishing components of the control and drive system and the mounting of the big Schmidt camera put the work aside to work on fire-control computers for naval and antiaircraft guns, and navigation computers.

The optics and machine shops had prided themselves on their quick response to problems. Rein Kroon, coming from the more structured world of Westinghouse, had been astonished that the machine shop could immediately produce the models he needed without paperwork or special authorization. The same quick response made them a productive prototype shop for military research and development. Engineers and scientists on military contracts could take new ideas to the shops and get quick test models. The staff of the shop expanded from the normal twenty-four to seventy. By late 1941 the defense-related activities were so extensive that the accountants at Caltech couldn't distinguish charges for the telescope project from charges for NDRC projects. Specialized machines that had been designed and constructed for figuring auxiliary mirrors for the telescope were now grinding mirrors for aerial reconnaissance cameras. High-precision machine tools that had been purchased to bore and hone components of the telescope mountings were being used twenty-four hours a day for defense work.

The Rockefeller Foundation could have insisted that the equipment, which had been purchased or built with funds it had awarded for the telescope, be bought or leased by the NDRC. Robert Millikan, eager for the contract work, didn't want to be bothered with the details of distinguishing one project from another, and Max Mason, who had authority over the various shops as head of the telescope project, knew the Rockefeller Foundation would not insist on the letter of its grant agreements. Many Americans still hoped and prayed that the United States, and especially American boys, would stay out of the fray, but there was no question of withholding science and scientific facilities, including those funded for peaceful activities, from the national preparedness effort. The old boys who had gotten together to conceive, plan, and fund the telescope were the same old boys who threw their energies and their institutions behind the NDRC in an effort to bring the United States from the neutrality and deliberate unpreparedness of the 1930s to readiness for the total warfare of the 1940s.

The late summer and fall of 1941 felt odd. The weather was freakish: drought in New York and New England, hurricanes in East Texas, floods in Arizona, early snow in Montana and Utah. Factories were belching smoke and the breadlines had disappeared, Ted Williams was batting over .400 and Joe DiMaggio's incredible hitting streak reached fifty-seven games, but it was hard to ignore a pervasive uneasiness, an ominous feeling that a storm loomed on the horizon.

If the attack on Pearl Harbor in December was a surprise, the entry of the United States into the war wasn't. Senators and press commentators had long accused the president of plotting with Winston Churchill to get the United States into combat. And even as Americans still hoped that American boys wouldn't be in the fighting, by the beginning of December 1941, the spread of the war in Europe, Asia, and Africa had already turned into a global conflict.

The Japanese attack and the unexpected declaration of war by Germany meant little immediate change in America's involvement. But the day President Roosevelt said would "live in infamy" brought on an enormous transformation in attitudes. Those who had been involved only on the periphery willingly worked harder when they knew the products they were making would help win America's war. For the young the dreariness of the workaday world was replaced by the glamour of war. Lines formed at recruiting offices, as young men who had dreaded the search for a job now turned down a choice of jobs to enlist.

Many in Los Angeles feared a Japanese invasion or, at the very least, bombing and sabotage. Caltech, with its wind tunnel in the Guggenheim Aeronautics Laboratory and war-related research and production in various labs, seemed particularly vulnerable. Millikan appointed a special committee for campus security and recruited Caltech seniors as special campus guards, armed with ax handles (rifles and pistols were rejected as too dangerous) to patrol the buildings. The students heard banging noises, which they thoroughly investigated in the best Caltech engineering fashion, only to conclude that the suspicious tapping was routine steam noises. Fritz Zwicky, though not an official member of Millikan's security committee, offered his services and some special inventions for the emergency, including an inexpensive gas mask made of flour sacks, bicarbonate of soda, and a rubber "raspberry" from the noisemakers used to produce "Bronx cheers" at baseball games. Zwicky tested the gas mask in his bathtub, using chlorine gas. The next morning, for the first time anyone could remember, he was subdued and had a nasty cough.

"Something was wrong," he explained in his strong Swiss accent. "Maybe a leak. Maybe the seams need to be sealed. Maybe it just doesn't work."

Zwicky shifted to experiments on gluing cellophane to windowpanes to prevent shattering from a bomb blast. Zwicky, who liked to show off his strength by doing one-arm pushups on the floor of the faculty dining room in the Atheneum, threw a deflated football against glass panes to test the distribution of shards. He tested for an appropriate color of cellophane by taking panes to the foothills north of the campus with a flashlight and asking students on campus balconies to record the light intensities. Like the gas mask trials, these experiments led nowhere and did little to endear Zwicky to his colleagues.

While Zwicky experimented and the students guarded buildings with ax handles, Max Mason secured military exemptions for the workers in the optics and machine shops. Much of the work in the shops was already war related, and with the initial demand for fighting men easily met by enlistments, there was no pressure for the army or navy to take men from critical jobs.

Still, the attitude in the shops changed. Patriotic posters went up on the walls next to the pinups. Idle chitchat turned to the war and to the maps in the daily newspapers, showing the spreading amoebalike movement of the Japanese and German empires. The nightly radio broadcasts brought reports of German victories in Russia and the consolidation of Japanese conquests across the Pacific. Daily someone would report a rumor that the Japanese had shelled the U.S. mainland or that Japanese submarines were landing spies on the California coast. The unfinished drive gears were still the largest precision gears ever cut, but in the face of the call to build fifty thousand airplanes in a year, the tedium of two years of machining a single gear lost its allure.

The can-do confidence that hadn't hesitated at the impossible task of building a perfect machine would take time off to fight an impossible war.

Instead of the usual estimated budget he submitted each year, in January 1942 Max Mason asked the Rockefeller Foundation to make an appropriation of fifty thousand dollars to cover the current year without an exact accounting for the funds. The Caltech accounting office, he reported, was overwhelmed with the switchover to defense research in the various laboratories and shops.

The astrophysics machine shop, with an augmented staff, was working twenty-four hours a day on defense projects. All remaining work on the control and drive mechanism for the two-hundred-inch telescope, the final cutting and polishing of the gear trains, and the mechanical work on the mounting and drive mechanisms of the "big" Schmidt telescope had been suspended, some of it within months of completion.

In the optics shop all equipment except the big machine for the two-hundred-inch disk was committed to defense work. Crews working two and sometimes three shifts polished roof prisms for navy range finders and mirrors for wind tunnels. Machines that weren't useful for defense work, like the grinding machine for the 120-inch disk, were shunted to the sides of the room.

Marcus Brown and a small crew of men persisted on the two-hundred-inch disk. Brownie had been fighting trouble with his legs—the "bone was gone," he explained. He had been working on the big mirror for six years. War work, both in the shop and outside, had lured many men away, including some of the most experienced. The project was far enough along that Brownie thought he could persist with a small crew. If more men left, he would train others. He would keep polishing, getting ready for the Saturday tests. The war work might mean that instead of six months, the figuring would take a year. He had already been working in the Calfornia Street optics shop for almost ten years. What was another six months?

John Anderson had been ill on and off, but he regained his health by late 1941. Using his new testing technique, in January 1942 he pronounced the mirror to be within one wavelength of a true parabolic shape. A wavelength of light is a very small distance, approximately 0.000055 cm (5500Å). The only meaningful unit of measurement is the angstrom, a ten-millionth of a millimeter. It had taken more than six years to get the mirror that close. Brownie and his crew had used tons of polishing rouge. As the surface got closer to the elusive perfect shape, they mixed talcum powder with the rouge to cut down the abrasiveness. The mix gave them better control, but the polishing went even slower. The changes in the mirror from one week to the next were imperceptible except under the critical Saturday tests.

Up on the mountain, the domes for the two big telescopes were almost complete. The wiring and telescope mounting hadn't been installed in the dome of the "big" Schmidt camera, and the gear train and a few components of the control system were still missing from the two-hundred-inch telescope. With no further components expected from the machine shop, work on the telescope wound down. Some men worked on landscape and construction details, replacing telephone cables that the squirrels ate as fast as they were laid, rebuilding the roof of the Monastery after the woodpeckers had punctured the copper to store nuts, or planting trees to ameliorate the barrenness of the mountaintop.

Except for the figuring of the big mirror, the telescope project was at a standstill.

Every pundit offered daily predictions, but no one knew how long the war would last. Production plans were for the *duration*. What would happen to the telescope and to the crew of workmen, engineers, and scientists who had worked together for more than a decade on the project? Max Mason, who had been through a period of wartime research before, wanted to hold together as many of the workmen and engineers who had worked on the telescope as he could, by getting them involved on a war research project. Even better would be a project that also employed some of the Caltech scientists. Trained and experienced men, accustomed to working with one another, could prove an efficient war research group. In the back of his mind, Mason never forgot that the pace of work on the telescope could resume that much more easily after the war if the team stayed together.

Some key men were recruited for the war effort elsewhere. Hubble went to the Aberdeen Proving Ground in Maryland to head up ballistics research. Fritz Zwicky, whose sometimes-undisciplined ideas had included a water-burning torpedo, a terrajet that would drive through the earth, capturing an asteroid to mine its mineral resources, and creating an atmosphere on the moon, went off to do rocket research for newly created Aerojet General in Asuza.*

John Anderson took on optics projects for the navy, army, and the

---

* Zwicky's penchant for the peculiar and the lack of self-censorship in his ideas were just the qualification for no-holds-barred weapons research. He maintained his affiliation with Aerojet General after the war. He later claimed that a project of his in 1957 had achieved the launch of a small pellet of aluminum into interplanetary space—a first that paled alongside the news of Sputnik that year.

NRDC. By the middle of the summer of 1942, with American men already engaged in battle in the Pacific and the Atlantic theaters, even the stalwart crew working on the two-hundred-inch disk was unwilling to do anything but war work. Reluctantly Marcus Brown shut down the polishing machine. A few months later there was hope of reviving the polishing. Max Mason even got assurances from the regional draft coordinator that a team of five could be excused from active military duty.

But the fury of war swept up even those five. The priceless disk was covered with heavy timbers for protection. Brownie moved over to the other end of the shop to work on prisms. From where he worked, he could see the idle polishing machine and the timber-covered disk. Sometimes he and Anderson would discuss how long the final figuring would require. At least another year, they agreed. Perhaps two. How long could the war last? they asked each other.

Max Mason, experienced at directed war research from his work on submarine sonar at New London, organized a new naval research project at Caltech, directed toward rocketry and underwater ballistics, with laboratory facilities at Morris Dam and in the basement of Robinson Hall, the astrophysics building. Their goal was to build a rocket-powered antisubmarine bomb. The project expanded quickly, recruiting engineers like Bruce Rule and much of the staff at Palomar. Byron Hill, Ben Traxler, and others packed up on Palomar and moved to Morris Dam, where they built and tested the rockets and torpedoes. Bruce Rule lost part of a finger in one test, when he fired a miniature nonexplosive torpedo from a compressed-air gun in the Robinson Lab. With the typical reticence of an engineer, Rule explained the incident tersely: "It misfired."

On Palomar, with the work crew gone, the buildings were closed. Harley Marshall stayed behind as a caretaker, and once a month, a small crew came up to the mountain to "exercise" the machinery, checking equipment for corrosion, starting up the oil pumps on the two-hundred-inch telescope, turning the dome, and slewing the drives from one limit to the other to avoid the ravages of inactivity. When they left Marshall was again alone with the machines.

Marshall had been in charge of public relations and visitors before the war. Navy pilots in training at Southern California bases sometimes used Palomar as a navigation point, so Marshall would see planes overhead, but there were no more visitors to the observatory and no press releases. The world had forgotten the perfect machine.

The Mount Wilson astronomers, like Hubble, were also recruited for war projects, with one exception. Walter Baade had taken out citizenship papers soon after his arrival at Mount Wilson in 1931, but he had lost them when moving from one house to another and with cosmic absentmindedness, had done nothing about the matter. When war was declared the German astronomer was officially an "enemy alien." He elected to remain at Mount Wilson for the war years. With the other astronomers occupied on war research, Baade had the big telescopes to himself. The scares about a Japanese invasion led to nightly blackouts in the Los Angeles Basin, eliminating the light pollution that had begun to limit observing at Mount Wilson. Baade found himself with an astronomer's dream: virtually unlimited time on the biggest working telescopes in the world, with unparalleled observing conditions.

Baade had been studying the Andromeda galaxy for a dozen years. He had photographed Andromeda thousands of times, identifying novas, supernovas, and Cepheid variables in the outer reaches of the galaxy. But no matter how careful his observations, and how superb the seeing conditions, he had not succeeded in resolving stars in the central region of Andromeda. The core of the galaxy remained a mysterious glowing blob on his plates.

The only observer on Mount Wilson, with a skeleton maintenance crew, Baade resumed his pursuit of Andromeda. He chose nights when the combination of a blackout in the basin below and the superb seeing of Mount Wilson provided optimum conditions. He opened the dome of the observatory early in the afternoon so the temperature of the air and equipment inside would stabilize with the ambient air temperature, and made sure the slit of the dome pointed away from the sun so he wouldn't inadvertently repeat the disaster of the first light with the one-hundred-inch telescope when the sun had heated the mirror. He developed a technique of guiding on the off-axis image of a faint star magnified 2,800 times. Baade had learned to read the striations in the fuzzy coma image of the guide star for signs of turbulence in the atmosphere and minuscule changes in the focus of the mirror of the telescope during the four-hour exposure.

Night after night he tried. Each noon, when he showed up for breakfast at the Monastery, everyone on the mountain would know from his mood whether the previous night had produced a good plate. Through the best seeing months of late summer and early fall 1942, Baade got closer and closer to his elusive goal. He estimated that some plates showed "incipient resolution," that he could *almost* discern individual stars. All he needed, he estimated, was a tiny increment in limiting magnitude, not more than 0.3–0.5 magnitude, to resolve the stars of the distant nucleus. He had pressed the equipment to the limit. With the largest working telescope in the world, nearly ideal observing conditions, and every extraordinary precaution he could conjure, the stars still eluded him. The task, like so many other unresolved problems of observational astronomy and cosmology, would have to wait for the completion of the two-hundred-inch telescope.

It wasn't the only problem that cried out for a bigger telescope. Rudolph Minkowski, another Mount Wilson astronomer, had been studying the Crab Nebula, a supernova that Chinese astronomers had first observed in 1054. Minkowski was eager to get spectra of the two faint stars at the center of the nebula, which he suspected were remnants of the supernova. But even with Baade's careful observations, in the very best seeing, using a special three-inch Schmidt camera at the Cassegrain focus of the one-hundred-inch telescope, it was impossible to achieve sufficient contrast between the central star and the surrounding nebulosity for satisfactory spectra of the stars. By 1942, after dozens of tries, Baade and Minkowski had succeeded in getting only one satisfactory spectrogram. The Crab Nebula was left as another problem for the rapidly filling agenda of the two-hundred-inch telescope.

In the summer of 1943 Baade tried again to resolve the central region of Andromeda. He had previously used emulsions that were sensitive in the blue end of the spectrum, because the evidence from spectrograms of the stars in the arms of the Andromeda Nebula showed a predominance of high temperature blue-white stars there. After his failed efforts of the previous summer, and on the basis of studies he and Hubble had made of the Sculptor and Fornax stellar systems, Baade suspected that the stars in the central region of Andromeda belonged to an entirely different stellar population, that they were cooler and less luminous red stars. He also reasoned that red-sensitive photographic plates might cut down on the sky fog from the background of light scattered by interplanetary dust and scattered starlight.

He set his campaign for late summer, when Andromeda would be highest in the sky at midexposure, and when Mount Wilson has the best seeing. He waited for perfect nights, with optimal seeing and optimal figure in the sensitive mirror of the one-hundred-inch telescope, and exposed his plates from 11:00 P.M. to 3:00 A.M., the four-hour window around the time when Andromeda was on the meridian (at the highest point in the sky). Before he exposed the plates, he bathed them in a dilute ammonia bath to increase their sensitivity. Then he would patiently guide the telescope for the full four hours.

Finally he got his perfect night. The plate was labeled

> M32—the brighter, round companion of Andromeda, 103E plate (ammoniated) behind a Schott RG 2 filter, $\lambda\lambda$ 6300–6700—exp. $3^{\rm h}$ $30^{\rm m}$, Aug. 25, 1943.*

The next day the shy, normally reticent Baade was all smiles at breakfast in the Monastery. "After shooting was over," he recalled, "it

---

* M32 is the Messier catalog number, 103E is the Eastman Kodak emulsion type. The Schott RG 2 filter cut out the green and red auroral emissions but transmitted the red spectrum between wavelengths 6300 and 6700 angstrom.

was quite clear that all the precautions had actually been necessary; I had just managed to get under the wire, with nothing to spare." To the naked eye, his plates show only the dark (the plate is a negative) disk of M32, the unresolved blob he had captured on hundreds of plates before. Then Baade, with his wonderfully trained eyes, showed that under a magnifying eyepiece the surface of the disk was powdered with thousands of pinpoint stars.

The margin of visibility was so slim that when the *Astrophysical Journal* received Baade's article announcing the achievement, they would not trust reproductions of the plates. Instead individual enlargements were printed directly from the negative and bound into each issue.

Baade's achievement has been called the greatest "scoop" in the history of big-telescope astronomy. The consequences went beyond the extraordinary observational feat, as Baade used the evidence he gathered to distinguish two different populations of stars, a major step that would have broad consequences in the determination of cosmic distances. That those observations required years of effort by the finest observer on large telescopes, and the peculiar circumstances of a wartime blackout, proved how much astronomy needed the two-hundred-inch telescope that languished, half finished, in the laboratories in Pasadena and on a remote mountain.

31

Endless, Damnable War

War transformed America. For a generation that had grown up knowing only the depression, the nation suddenly seemed whole again, united and mobilized in a cause that few opposed.

The boom hit everywhere. Corning production lines shifted overnight from baby bottles and Pyrex baking dishes to lenses for gun sights. The Westinghouse plant that a few years before had been eager to build a telescope mount to keep its workers and machines busy was working full shifts on ship turbines. As men went to war, women left home for the factories. Rosie the Riveter showed up in cartoons, radio editorials, and patriotic speeches. Prudish critics were shocked when some women *liked* work outside the home, and when others, freed from the protection of the home, found entertainments and amusements outside work. Servicemen got "Dear John" letters or came home to surprises. The moral crisis the naysayers had predicted as far back as the 1920s, with the rise of hemlines and the sight of women smoking in public, had arrived with a vengeance.

The war also transformed the scientific and technological underpinnings of America's industrial might. Before the war an undertaking as complex as the two-hundred-inch telescope, a national engineering and design effort that drew on the resources, facilities, skills, and wisdom of dozens of companies, large and small, and hundreds of scientists and engineers from universities, private industry, government, foundations, and the military—had been a unique enterprise. There had been no formal institutional structure to bring together that vast spectrum of individuals and institutions except the regular if informal connections of the white, well-to-do, educated, Protestant, urban men George Hale drew into the enterprise. In the early days of the war effort, many of the same old boys whose companies, institutions, and agencies had designed and built the two-hundred-inch telescope showed up in positions of prestige and power in the war effort. But modern war on two fronts brought a demand for coordinated research and production that soon far outstripped the manpower, organization, and resources the old boys could command from their clubs.

Projects like the two-hundred-inch telescope were dwarfed by the magnitude of the war effort. Vannevar Bush and James Conant, under the authority of the NDRC, swallowed up laboratories, faculties, scientists, and programs, channeling the efforts of thousands of researchers and vast manufacturing facilities into war research and development. When the resources of existing laboratories and production facilities weren't sufficient, they could call for the construction of vast new plants and secret labs. The NDRC/OSRD took over the administration of atomic energy projects from a committee headed by Lyman Briggs at the Bureau of Standards, and in 1942 handed it over to the U.S. Army and General Leslie Groves, who formed the Manhattan Engineering District to administer a far-flung empire of facilities in New York, Chicago, Berkeley, Hanford, Oak Ridge, and Los Alamos. The facilities in Hanford and Oak Ridge, and the secret laboratories in Los Alamos, dwarfed every previous scientific venture. The full cost of the two-hundred-inch telescope, the most expensive scientific instrument ever started before the war, would have been a footnote to the budget of the Manhattan Project.

The complexity of managing and operating such diverse facilities, coordinating the productive efforts of entire divisions of companies like Du Pont, laboratories with hundreds of engineers and scientists, and physical plants with thousands of employees, created a new scale for big science. The world of George Hale had rejected navy captain Sandy McDowell's flowcharts as too bureaucratic and authoritarian, preferring the easy collegiality of informal meetings in their basement offices. The new science and technology would need men like Gen. Leslie Groves, veteran of building the Pentagon, who had the managerial and bureaucratic skills, and the authority, to clear every obstacle. Some Manhattan Project scientists resented Groves as much as the Palomar and Caltech scientists had resented McDowell. They couldn't fire him. The needs of wartime research had created a science too big for the scientists to run alone.

Compared to the decade of war that some countries had endured, America's war was mercifully short. Yet for scientists eager to do astronomy, it seemed an endless hiatus. Young men—there were still almost no women in astronomy and astrophysics—took leaves from graduate school or postdoctoral fellowships to work on rockets at Morris Dam, ballistics at the Aberdeen Proving Ground, or fission in Los Alamos, postponing or even derailing their careers. Senior scientists lost the momentum of research projects that were suspended for the duration. Those who had counted on using the big telescope at Mount Wilson or who planned research projects that could only be completed on the unfinished two-hundred-inch telescope waited and dreamed.

There had been no question of continuing the work on the telescope during the war. But even as the unfinished telescope languished, the politics of the telescope wouldn't sleep.

In 1940, more than a year before Pearl Harbor, Vannevar Bush pushed aside the proposals for joint operation of Palomar and Mount Wilson by saying that under the "present emergency" the details would be difficult to work out. He was wary of Robert Millikan's insistence that the institution be headed by a committee instead of an individual. Mount Wilson had always been run by a single director—first George Hale, then Walter Adams. And, like other Millikan watchers, Bush knew that for all his insistence on committee and joint decisions, Millikan was not a man ever to allow himself to be overruled. The negotiations to pair Carnegie money with Caltech's telescope trickled on through 1941 as the two proud institutions worked out the details of what constituted "joint operation." What names would be on the letterhead? How would the citations on scientific articles based on work at the facilities read? Behind these seemingly frivolous issues lay Mount Wilson and Carnegie Institution fears of Millikan's "acquisitiveness" and his publicity machine, and Caltech fears that Mount Wilson and the Carnegie Institution were somehow *usurping* their telescope.

War ended the bickering. It also changed the rules of the game. Bush, more than anyone, understood how the war would change the future of science. Government-funded war research had demonstrated the potential of science on a scale no one had dared imagine in the years before the war. Even without a peacetime equivalent to the Manhattan Project, it was clear to Vannevar Bush that government funding for future big science was inevitable, and that institutions like Caltech would be prime recipients. Before the war his bargaining point had been that tiny Caltech was overextended and couldn't afford to run the telescope. He knew Caltech would be a different sort of place after the war.

By the spring of 1945, as the Allied forces marched across Western Europe and the Marines hop-skipped up the chain of islands toward the Japanese heartland, even pessimists began to think of the end of the war. Bush, who was aware of the secret weapons the United States was developing in Los Alamos, knew that the war would not last out the year. Within a year or two of the conclusion of the war, the two-hundred-inch telescope would be finished. Walter Adams was ready to retire as director of the Mount Wilson Observatory. The joint facilities would soon need a director.

Bush put great store in the role of a director. The Manhattan Project, with the unlikely choice of J. Robert Oppenheimer as director, was flourishing. Oppenheimer was far from a household name outside the world of physics, but at the Manhattan Project the legendary quickness and breadth of his mind attracted first-rate men to the project, and his grasp of the project kept the wildest of the scientists from going too far afield. The joint directorship of the joint Mount Wilson and Palomar Observatories, Bush concluded, needed a personality of comparable fame and prestige in astronomy.

He first quietly floated the name of Harlow Shapley. As director of the Harvard College Observatory since 1921, Shapley was something of a dean of American astronomy, at least in the minds of easterners. He worked at a famed circular desk, keeping his many interests distinct on different sections of the desk. Shapley had enjoyed good relations with journalists, from the irascible H. L. Mencken to reporters from the influential *New York Times*, and as a regular participant in programs on the radio, in the newspapers, and at the Harvard College Observatory, he had built a public following, unusual for an astronomer. In the world of big science that Bush knew would follow the war, a widely known director who enjoyed easy rapport with the press and the public would be a strong asset.

Still, Bush wasn't sure of Shapley. When he solicited views, he found that many thought Shapley lacked "generosity . . . his own reputation and advancement have often been too keenly in his mind rather than the welfare of his organization or of his colleagues." Walter Adams shared Bush's qualms. Shapley's position in astronomy, Adams pointed out, was due to his early work at Mount Wilson, rather than to anything he had done since. "Hubble and Baade do not rate most of the Harvard work at all highly, and they should be excellent judges." The strongest argument against Shapley was that he had never been enthusiastic about the two-hundred-inch telescope; indeed, he had done his best behind the scenes to derail the project.

The only other astronomer as well known, among the public and scientists, was Edwin Hubble. In 1936, when it seemed that the engineers were taking over the project, Walter Adams had confidentially urged that Hubble be appointed astronomical director of the Caltech two-hundred-inch-telescope project. From a public relations perspective Hubble was a natural choice. In the news weeklies and science supplements, he was the discoverer of galaxies, the measurer of the universe. His name was mentioned in conjunction with Einstein—science journalists liked to write that Hubble had *proved* what Einstein predicted—and Hubble had done little to discourage the idea that the real purpose of the big telescope was to support his own work. Hubble enjoyed fame; he liked to hobnob and be photographed with celebrities, rarely turned down a chance to be interviewed by a reporter or to appear in a newsreel, and made a point of appearing at Mount Wilson, in his tweeds and with his acquired English accent and pipe prominent, whenever a distinguished visitor scheduled a tour of the observatory. Robert Millikan, who had never been known to turn down publicity for Caltech, made no secret of his own preference for Hubble as the director of the joint observatories, where Hubble's adeptness with the press might help Caltech as well as Mount Wilson.

In the agreement for joint operation of the observatories, Millikan had ceded the right to appoint the director of the joint observatories to Bush, probably because he assumed that Hubble was such an obvious choice that Bush had no options. But Millikan hadn't checked first with the astronomers and physicists. Hubble was not popular among his colleagues. They found him pompous, arrogant, self-centered, and narrow in his perspectives toward astronomy. His wartime assignment to the Aberdeen Ballistics Lab, far away from the other astronomers who stayed in Southern California, was welcomed by many.

The harsh opinions of Hubble weren't strictly personal. Many astronomers did not share Hubble's enthusiasm for his proposed survey of galactic expansion, which he considered the primary mission for the two-hundred-inch telescope. In the years before the war, astrophysicists had begun exploring stellar evolution and nucleosynthesis, the processes and stages by which stars transform themselves. Hubble's interest had been purely cosmological; he had sought evidence for a geometry of space. A new generation of astronomers and astrophysicists wanted to go beyond his goals, to explore the sequences of stars within globular clusters and galaxies and thence to understand the entire evolution of the universe. The key to unlocking those secrets would be the superior light-grasp of the two-hundred-inch telescope. If time on the new telescope were monopolized for Hubble's measurements of red shifts, other avenues of research would be closed or at least hobbled.

Bush had spent enough time around academics to ignore much of the backbiting between astronomers and physicists. The one man whose opinion he trusted was Walter Adams. Adams was a quintessential New Englander: tall, slender, reserved, with a strong Protestant ethic of work and propriety. He worked hard as both an administrator and astronomer and admired those who worked hard and observed the proprieties and manners of the observatory and science. He thought Hubble a bad choice as his successor. Hubble, he explained to Bush, was not interested in administration and did not have a "friendly interest" in other members of the observatory, from the scientific staff to unskilled workmen. Adams's delicate language did not hide his feelings, and Bush had administered enough science and visited enough observatories to realize that while the position could tolerate a prima donna, the close life of a mountaintop observatory didn't leave room for a snob.

By midsummer Bush had decided against Hubble. He was confident that Hubble would not leave if not appointed, but to make sure he recommended that a new position of chairman of the Committee on Scientific Programs be established for Hubble, with a salary comparable to that paid the director of the institute.

Bush's final choice of director was a surprise to many. Turning down other well-known astronomers, he picked Ira Bowen, a spectroscopist and longtime member of the physics faculty at Caltech. Some astronomers quietly protested. Bowen was a physicist, a student of Robert Millikan. He had done some superb work identifying unknown lines in the spectra of gaseous nebulae—an achievement Walter Adams identified as one of the most brilliant astronomical discoveries of the early twentieth century—but Bowen had done his work on a laboratory spectrograph, not at an observatory.

Still, Bowen was a superb physicist, well respected by colleagues, and free from the posturing and politics of Hubble or Shapley. Though he had spent limited time in observatories and on telescopes, he was an excellent optician, well equipped to supervise the completion of the two-hundred-inch mirror and the other optics for Palomar. The easterners and the Californians at Caltech thought of Bowen as an unmistakable midwesterner. He was serious and often could not understand a joke. He had little sense of fashion, in his dress or manner. For years he wore a hideous necktie made of some synthetic material he claimed was stain-proof, until it literally fell apart; he was apparently unaware of the quiet grins of others when he bragged that the marvelous tie, no matter how many times he wore it, was as good as new. He was penurious with faculty salaries and supplies. When one astronomer needed to use a paper recorder for an experiment, Bowen asked how much paper the experiment would use. The astronomer said, "One thousand feet per week." Bowen finally approved the experiment, "but only if you use every inch of the paper." Waggish astronomers would ask, "Do you know the difference between Ike Bowen and a brick wall? If you push long enough on a brick wall you can move it."

Walter Adams had been an admirer of Bowen's research, and he eagerly supported the appointment. Adams, at the compulsory retirement age of sixty-five, was no longer officially on the Mount Wilson staff after Bowen's appointment. Eager not to cramp Bowen's style, he moved his own office to Hale's old solar laboratory when Bowen took over.

Not every astronomer welcomed Bowen's appointment. Hubble tried to put a good face on his disappointment, writing that he welcomed the prospect of research without the responsibilities of administration. "On the other hand," he reminded Bowen, "the appointment of a physicist as director of the astronomical center of the world will not be welcomed by astronomers if it involves actual control and direction of research. . . . I will expect to have a free hand . . . in the field you call cosmology. Some of us have given much thought to the big problems in that field, and have rather definite notions on the methods of attack."

Hubble was used to a prewar world of astronomy, where the director of an observatory could function as a benevolent dictator on his mountain. The war, the growth of big science, and the complex institutional relationships that were needed for facilities like the Mount Wilson and Palomar Observatories changed all that. Astronomy, a profession of gentlemen, had yielded to astro*physics,* in which thirty was old. As Paul Dirac had once written in an end-of-term sketch:

> *Age is of course a fever chill*
> *that every physicist must fear.*
> *He's better dead than living still,*
> *When once he's past his thirtieth year.*

Jesse Greenstein, who would later come to Caltech as the first chairman of the astrophysics department, witnessed the transformation: "The world changed. The characters in the play changed. Instead of the gentlemen, we have as the ideal the brilliant, aggressive young genius interested in everything, careless of whose feet he steps on and very anxious to make the discovery of the week or the year or whatever. It's a loss, and it's a gain."

Twenty years before, Edwin Hubble had been the brilliant aggressive young genius. Now, a younger generation was waiting in the wings. One afternoon late in 1945, a group of astronomers and physicists, including Bowen, Tolman, Baade, and Adams, met at Hale's solar laboratory to make preliminary plans for the allocation of research time on the two-hundred-inch telescope. They reviewed the work in progress among astronomers at Mount Wilson and at other institutions and concluded that the world's biggest telescope, and especially the priceless "dark time" when the moon was absent and the telescope could be used for research on faint objects at the limits of the observable universe, was too valuable to be monopolized by Hubble's "rather definite notions" and methods. Hubble would get some observing time on the telescope, but nothing like the free hand he expected.

32

Starting Anew

The end of the war left a malaise at Caltech. Wartime research and production had been exciting. Now, at parties, men asked one another, "Well, now what are we going to do?" A few couldn't wait to purge the years of war work. Bruce Rule brought his wartime research papers home and burned them in the fireplace, asking his daughter formally to witness the deed.

One group on the Caltech campus knew exactly what they would do. Soon after the bombs fell on Hiroshima and Nagasaki, before the formal peace was signed, workmen started clearing the war work out of the optics lab and the astrophysics machine shop. Max Mason's plan of holding the essential work crew together on war projects had been successful. A few men had moved on to other projects, but a core stayed on. Men who had started as apprentices when the project began were now heads of much of the work staff. The chief electrician from Palomar and his assistant took new jobs after the war; Ben Traxler, who had started as a utility man and radio operator at Palomar and stayed with the scientists and engineers at Morris Dam during the war came back to Palomar as chief electrician. Mel Johnson, who had stuck it out in the optical shop, was Marcus Brown's chief assistant when the work on the mirror resumed.

Before they could clear away the timbers that had protected the disk and restart the gear polishing machines, there was an administrative nightmare to clean up, as Caltech and Pentagon accountants tried to figure out who owned what. By the time the war work was cleaned out of the shops, the Palomar blueprints, tools, jigs, and dies were retrieved from storage, and new men were trained for the highly specialized telescope work, six months had elapsed. Officially the OSRD, the Office of Scientific Research and Development, had purchased much of the specialized machinery in the shops from Caltech for the war effort, then sold them back to Caltech for their salvage values. The result of the transactions was that Caltech was able to apply $144,000 of the funds from military contracts to restoring facilities for the telescope project. The funds helped, but the war years had been expensive for the telescope; a skeleton staff had to be maintained at Palomar to guard and maintain the facilities, and the war had inflated salaries and the cost of equipment. By July 1945 only $4,000 of the original grant of $6 million remained to finish the telescope.

Max Mason wasted no time in applying to Warren Weaver, his former colleague at the Rockefeller Foundation, for an additional $250,000 grant to finish the telescope. It was such a reasonable request, eighteen years after the original grant, that the Rockefeller Foundation promptly approved the additional funds. By summer of 1946 sixty-five men were on the astrophysics payroll.

In the machine shop the trickiest single operation was finishing the machining of the drive gears. It had never been easy to find and train men with the patience to cut and polish gears to the needed tolerance. The war years, with their emphasis on producing machined goods to quick timetables, made for a rough transition to work with specifications so exacting it couldn't be rushed. Instead of three shifts, competing with one another to set output records, the shop was slowly returned to the old pattern, one shift of master machinists, working painstakingly on one-of-a-kind production. While men on the gear-cutting machine measured their progress with hourly examinations of the surfaces of the fourteen-foot-diameter gear with microscopes, machinists on the boring mill and milling machines put together the mount and tube for the forty-eight-inch Schmidt camera and the remaining mechanical and electrical components of the two-hundred-inch telescope.

Mark Serrurier, who had supervised much of the assembly work on the telescope, had left the telescope project to work with his father in the movie industry, where he later developed the Moviola and received an Academy Award for his work. Bruce Rule, who had worked on the electrical systems of the telescope before the war, took over the supervision of the mechanical and electrical components of the telescope. Byron Hill returned to Palomar Mountain to supervise the work on the site. Rule and Hill had worked together at Morris Dam during the war. Both were strong willed, and their years together on war work hadn't mellowed either man. Their "discussions" were legendary.

Next door to the machine shop, in the optical shop, Marcus Brown had been counting the days until he could return to his work on the telescope. Brown had done war work polishing prisms and mirrors. For some of the men in the shop, work was work, and the optical shop was a good job: They were exempted from the draft, got steady paychecks, worked indoors, and didn't complain. For Brownie polishing glass wasn't enough. The mirror was his mission. He wanted to finish it before his legs gave out.

In September 1945 Brownie and his crew—only a few men were veterans of the prewar telescope work—lifted off the timbers that had protected the mirror disk for three years. It took them three months to clean the entire optics shop with magnets, hoses, scrub brushes, and magnifying glasses. The polishing machine, unused for almost three years, had to be examined and lubricated. On December 17, 1945, Brownie pushed the button to restart the machine.

Most of the men were new, but the rhythm of figuring the mirror on the huge polishing machine returned quickly. For five days each week, Monday through Friday, Brownie and his crew would polish zones of the disk, bringing the surface closer to the elusive perfect figure. On Saturday John Anderson and Brownie would test the disk, studying the shadows of the knife-edge to find zones that would need more attention on the polishing machine. Day after day, week after week, the polishing went on. From the gallery it was as if the war had never come along.

But inside the shop, in the offices of Robinson Hall, even on Palomar Mountain—the mood of the men working on the telescope had changed. However much the machines and men looked the same in 1945 as they had before, the war had changed America too much to ignore. Some changes were obvious: Returning servicemen flooded the workplaces and universities, women returned home from wartime jobs, deflation replaced the raging wartime economy. Other changes were unanticipated: Wartime separations had taken a toll on marriages and families, as women who had enjoyed life outside the home and men who had taken advantage of overseas freedoms confronted one another. Often relationships that had survived long separations fell apart when postwar reality didn't match wartime dreams.

The changes weren't only in the family. Men and women who had been part of a crusade against evil were now working for a paycheck. The work—whether on an assembly line or in the Caltech optics shop—might be the same, but without The Cause, it felt different. Polishing a prism in the optics lab had been a small but essential part of a national effort. The same routine steps, on the same machine, after the war, was just a repetitive, boring job.

Even the excitement of working on the telescope had paled for some. Before the war the two-hundred-inch telescope had been the most exciting venture of American technology and science. Men had considered work on even a component of the project a privilege. But even the triumph of prewar technology paled alongside the achievement of the war. The United States, the fortress of democracy, had put together the greatest human economic effort ever organized. Military historians might argue that the battles between the German and Russian armies on the Eastern Front were bigger than any engagement the Americans had fought, but for men and women who had worked in factories, railroads, and ships, producing and transporting more airplanes, trucks, ships, guns, ammunition, gasoline, and uniforms than the world had ever imagined, the war had provided a unique experience of working for something that mattered. After that crusade the telescope paled in comparison.

The war also marked a watershed in the scale of science. Warren Weaver, from his position as science administrator of the Rockefeller Foundation, watched it happen. Before the war the Rockefeller Foundation had made some substantial grants to Ernest Lawrence, at the University of California, for work on his cyclotrons. The grants were relatively small compared to the $6 million commitment they had made for the telescope, but after Lawrence pyramided the funds with grants from the state of California, and especially after Lawrence's eighty-four-inch cyclotron became one of the fuel sources for the Manhattan Project and the University of California later took over as site manager for Los Alamos, nuclear physics research at Berkeley and Los Alamos dwarfed what had once been big science, including the two-hundred-inch telescope.

An inverse corollary of Gresham's law prevails in science: Good money attracts more money and good people. At war's end the big machines, publicity, and seemingly limitless funding at places like Berkeley and Los Alamos drew bright young scientists and senior researchers to particle physics and nuclear weapons research. Men like J. Robert Oppenheimer, who had explored astrophysics problems before the war, were now absorbed in the seemingly limitless enterprise of nuclear research and its potential consequences. "Doomsday" had become part of the vocabulary. When he heard that work on the telescope had resumed, Warren Weaver wrote Max Mason that "everyone agrees that it should be finished adequately and promptly. Indeed, with the way the nuclear physicists are carrying on, the astronomer had better get a good look at the universe while it (and we) are still here."

If they were no longer at the center stage of science, the work on the unfinished telescope went on. The Los Angeles newspapers carried an occasional feature on the telescope project that everyone vaguely remembered, but people in Southern California had been hearing about big telescopes for so long that Robert Benchley's humorous plaints—"They must just throw the glass away when it finally, after weeks of publicity, gets out there"—seemed accurate.

In the shops at Caltech, the once pathbreaking machining and polishing work began to seem routine. The first drive gear, machined to a precision never before achieved in a gear that large, had been a compelling task. The second right ascension gear and the declination gear, equally large and no less exacting in the machining requirements, were only routine. The first gear had proved it could be done.

Even in the optics shop, the polishing became tedious. Opticians are patient men, but the two-hundred-inch disk had been in the optics shop for ten years.

By early 1947 the surface was within one-millionth of an inch of a true parabola. John Anderson didn't like to use those measurements—for an optician fractions of a wavelength of light are a more useful scale—but he proudly admitted to visitors that the mirror had reached a trueness over its surface never before achieved on any optical device. As the surface came closer to the final figure, Brownie's precautions in the optics shop became fanatical. Workmen had to exchange their protective suits for a fresh one at the slightest suspicion. Twice a day a man would go over the floor with a magnetic sweeper. Some men did nothing all day but search the room for metal filings or specks of foreign matter. Brownie worried about the forced-air ventilation system. The system had been designed to work under positive pressure, so no foreign material could be brought into the shop. But what if the forced air picked up a grain of glass? A single speck under the polishing tool for an instant could set them back six months or more.

The Saturday tests crept up on Anderson's elusive goal. Anderson and Brownie worked together on the tests each week, taking turns at the eyepiece as they discussed the remaining work to be done. Anderson was pleased with the progress. The remaining high zones were gradually coming down. There was still work to be done on the disk. The outside edge had deliberately been left slightly high. Anderson's calculations for the mirror supports indicated that the edge of the disk, extending out beyond the outer ring of supports, would sag in the telescope. What seemed a high edge in the laboratory tests could be just right in the telescope. If the mirror didn't sag that much, they could polish the edge down at the observatory. The opposite error, if the edge were too low, would require repolishing the entire mirror, a formidable task. Anderson's only worry was the support system, "for which we have at present no really satisfactory test. As far as we can tell it is ok—but the final decision will have to wait until it is installed in the mounting at Palomar, sometime next year."

The work in the optics lab was nearing its end. The surface of the mirror was close to a perfect paraboloid. As the surface progressed, unspoken tensions developed between Anderson and Brown. Brownie wanted to keep polishing, to bring the entire surface closer to a perfect mirror. Some of the tests detected a barely perceptible "orange peel" effect on the surface of the mirror, which would reduce the pinpoint sharpness of images by scattering light. He wanted a few more months, perhaps another year. Anderson pointed out that the theoretical resolution of the disk, the ability to focus the light from a distant star into a fine point, was already greater than the best of atmospheric seeing would allow the astronomer to use.

Anderson and Bowen had already announced that after the mirror was moved to Palomar, Don Hendrix, chief optician at the Mount Wilson optical shop, would take over the remaining polishing. A superb optician, in 1947 he was just finishing the mirror and corrector plate for the forty-eight-inch Schmidt telescope, which he had started before the war. Hendrix had come back to the seventy-two-inch mirror in late 1945. With the famed two-hundred-inch mirror across town at the Caltech optics shop, with its visitors' gallery, the reporters paid little attention to Hendrix's mirror. The figuring went well, and the crises remained private. Once, when they tried to lift the mirror off the grinding machine for a test a vacuum had formed between the mirror and the table of the polishing machine, so that the table lifted with the mirror. Someone spotted it and Hendrix quietly pushed the DOWN button on the electric hoist. Everyone in the lab was shaking when the mirror and disk finally settled back in place. Another time the polishing tool galled while they were grinding one of the sockets at the back of the disk. Hendrix calmly sent one of the assistant opticians off on a motorcycle to get dry ice to pack around the tool to contract the metal enough to release it from the disk.

Hendrix's assistant on the mountain would be Mel Johnson, who had worked on the mirror longer than anyone except Brownie. Like Hendrix, Johnson was younger and healthier than Brownie. He also had Hendrix's calm, easygoing temperament, what the opticians called a Mount Wilson style. Neither was interested in astronomy. Hendrix and Johnson belonged to the same gun club. Weekends they went hunting, fishing, or target shooting. It relaxed them from the pressure of figuring glass.

Brownie had the Caltech temperament. In the shop, he would roll Bull Durham cigarettes and smoke with the men, and he had the patience for hours of polishing and testing. But on two subjects he could be hot tempered: the rival Mount Wilson optics shop, and *his* disk. He didn't like Hendrix, and he didn't want to give up the disk, not to Hendrix and not to the astronomers. Bowen and Anderson said that they had named Hendrix to finish the mirror because Brownie's legs were acting up, that his health couldn't take the working conditions for polishing the mirror on the mountaintop. He had put almost twenty years into the mirror, training himself in an apprenticeship at the Mount Wilson labs, working on trial mirrors and auxiliary mirrors in the new optics shop on California Street, and from 1936 on, devoting his life to the big mirror. What had been a job for others was his calling. He didn't want to let go.

The "mirror is at the really exciting stage now," Max Mason wrote in June 1947: "minute polishing, and tests every other day—working now with the millionths of an inch. Nobody knows how long—maybe a month—maybe bad luck and longer—maybe a lucky run, and sooner." Astronomers at Caltech and Mount Wilson, following the reports from the optics shop, counted the days before the mirror would be finished. The joke around Caltech was that the mirror had already focused its first image, a reflection of the pinup photographs on the wall of the optics shop. Everyone knew that there was much work to be done afterward, the final figuring on the mountain, final tuning of the support systems, the drive and control systems, and the corrector lenses and spectrographs. But the goal seemed near.

In the optics shop the men began talking about what they would do *after*. Some would stay on, finishing auxiliary mirrors for a few months before the optics shop was closed down. Brownie talked of going back to the farm he owned, which had grown up in sage in the twenty years he had been working on the telescope. He talked of other projects, maybe a jewelry-and-clock business. The talk didn't conceal his sadness.

It was October 1947 before Anderson was finally satisfied. Max Mason reported, "We are calling the mirror finished, as to figuring, and are happy over it." The Caltech publicity office issued its announcement on October 3. "The most daring optical job ever attempted by man was completed today—polishing of the giant 200-inch telescope mirror for the Palomar Mountain Observatory." The release quoted Mason: "Our last tests show that we have reached the goal toward which we have worked from the beginning—a parabolic (concave) surface accurate to within two millionths of an inch." The grinding and polishing of the disk had taken eleven years, more than 180,000 man-hours. More than five tons of glass had been ground off the original blank. The eleven and one-half years on the grinding and polishing machine had consumed thirty-one tons of grinding compounds and rouge, hundreds of white cotton surgical suits and pairs of canvas shoes. Twenty men at a time worked on the mirror. Over the years, dozens of men had worked in the shop. Only Marcus Brown had been there from the start.

33

Delicate Cargoes

The blueprints for the telescope filled cabinets in the basement of Robinson Hall. To a trained engineer each blueprint was clear; each assembly made sense. But no blueprint could make the entire assembly clear. The engineers, work crews on the mountain, even John Anderson, who had been with the project from the beginning, used Russell Porter's drawings to visualize how the parts would all go together. Whenever Anderson saw a visitor's eyes glaze over as he tried to explain some detail of the operation of the telescope, he would take him out into the hallways of Robinson Hall and find the right Russell Porter drawing, which made the complex seem simple.

Porter's drawings and the smooth welded construction of the telescope were deceptive. The gears, motors, and electrical components of the drive and control systems were concealed inside the horseshoe yoke, the two massive side tubes, and the base assembly of the mounting. The oil pressure bearings for the horseshoe were exposed, but the neat piping and the inconspicuous pressure pads made it seem simple. The south bearing, where a ball on three oil pads supported the weight of the lower end of the telescope, was concealed under the mounting, the access through manholes and crawlways.

As the pieces of the drive and control mechanism arrived, the fourteen-foot drive gears from the astrophysics machine shop at Caltech, the worm gears from a gear company in Connecticut, portions of the drive controller from Hannibal Ford's plant in Long Island, other portions from the Mount Wilson labs and the Caltech machine shop, Byron Hill and Bruce Rule and the crew on the mountain assembled and tested the telescope. Rule had designed the windscreen, a heavy collapsible canvas curtain, on a heavy lifting mechanism, that could automatically rise to block the bottom portion of open slot in the dome without obscuring the view of the telescope, to protect the instrument from wind. The control mechanisms for the windscreen were phenomenally complex. Rule ran the initial tests in the daytime and realized that no one had calculated the expansion coefficient of the canvas. When the sun shone on one side of the screen, it drooped.

With the forty-eight-inch telescope nearing completion at the same time as the two-hundred-inch, the engineers saved time and money by using the same technology, scaled up or down, for both. Sometimes the scaling worked. Bearings that had been machined for the two-hundred-inch telescope, then rejected in favor of a different design, were adapted to the forty-eight. Sometimes the men assembling the telescopes found that the promised economies didn't work out. Bruce Rule designed the windscreen assembly for both telescopes. On the forty-eight-inch telescope, a single fractional-horsepower motor could lift the windscreen. On the two-hundred-inch, the windscreen was so heavy that his plan called for two three-horsepower motors, tied together electrically so they would stay in phase as they lifted the heavy windscreen. The voltage drop when the two big motors engaged would make the relays that controlled the motors "chatter," which in turn would leave the windscreen jerking up or down by fits and starts. Rule, sure of his design, insisted the installation was wrong. The electricians on the mountain quietly installed capacitors and an auxiliary power supply to cure the chatter. On the mountain the men had learned that it was better to get the telescope working than to win arguments with the engineers.

When the mounting of the forty-eight-inch telescope was first installed, the telescope wobbled and settled out of alignment from the slightest shake. The foundation was modeled after and scaled down from the four-point mounting of the two-hundred-inch telescope. The weight of the two-hundred-inch telescope, divided between the north horseshoe bearing and the south bearing, was split among four foundation caissons. The smaller forty-eight-inch telescope, with a simpler fork mounting, put almost all of its weight on the front two supports. Byron Hill, a civil engineer and concrete man by training, wanted to take it apart and rebuild the foundation. Bowen, in charge of the observatory, with a budget and timetable to meet, ordered reinforcements for the front bearings of the mount. Sometimes compromises were necessary.

As the last parts of the control and drive mechanism were installed, Hill took on the tricky job of balancing the telescope. If the bearings had the low friction everyone promised, and if the telescope were perfectly balanced with minimal backlash in the gear assemblies, the tiny motors—the main drive motor is 1/12 horsepower, the size of a sewing-machine motor—would drive the huge telescope smoothly, without the spurts and jerks of a powerful electrical motor. As soon as the mirror was delivered from Pasadena, the opticians would want the telescope to work easily and reliably, so they could point it anywhere they needed for testing of the mirror and other optics.

Byron kept his notes on the balancing of the telescope in a tiny

handwritten notebook that even today sits in the drawer of the superintendent of the observatory. An engineer at Caltech had written up
the procedure for balancing the telescope, which made it sound simple.* But the engineer wasn't a mechanic. The counterweights were on
four-inch-diameter Acme screws with unlubricated nuts that depended
on the smooth motion of copper against graphite to turn freely. They
probably turned freely when they were first installed, but the years of
inactivity during the war had taken their toll. By 1947 the screws
screeched when they were adjusted. Hill cursed and drank dripping oil
trying to keep them lubricated. It was October 1947 when the telescope was balanced enough to take the dummy concrete mirror out to
await the real mirror.

By then as many as two thousand tourists a day were coming up
to Palomar to watch the progress on the celebrated telescope. Palomar
Mountain Stages offered tours from Oceanside, leaving at ten in the
morning and returning at four in the afternoon, at a cost of six dollars.
Others drove up in their cars, enjoying a Sunday afternoon drive after
years of war-rationed gasoline and tires.

The visitors found plenty of surprises. Gus Weber, one of the workmen on the mountain, had gotten into the habit of parking his pickup
truck inside the dome. Hill didn't like that, so he taught Weber a lesson
by waiting until Weber wasn't around and using the overhead hoist to
suspend the truck high over the two-hundred-inch telescope. Weber,
suspecting a practical joke, searched the mountain before he came
into the dome and saw the visitors staring up above the telescope. His
barrage of swearing was another treat for the visitors. A picture of
Weber's truck dangling over the telescope still hangs on the bulletin
board in the observers' room of the telescope.

Another evening Ben Traxler and Olin Wilson passed the time by
crafting a sign that they put on the old concrete mock-mirror disk,
which had been relegated to a spot alongside the walk outside the dome:

THIS FLYING SAUCER, DRAWN HERE BY THE GREAT
LIGHT-GATHERING POWER OF THE TWO-HUNDRED INCH TELESCOPE
HAS BROUGHT VISITORS FROM OTHER WORLDS
WHO ARE CURRENTLY GUESTS OF THE GOVERNMENT

After a woman visitor fled to the Forest Service office down the mountainside, demanding to be protected from the spacemen, Byron Hill
took the sign down.

---

* Bill McClellan, who for many years was the machinist with responsibility for
Palomar, and who is famed for successfully rising to Richard Feynman's challenge to build an electrical motor no larger than 1/64 inch on a side, has built
a model of the telescope to demonstrate the balancing procedure.

★　★　★

Reporters, who love superlatives, might have argued that the San Francisco Mint, moved in the midst of the depression, was a more valuable cargo than the mirror for the Palomar telescope. If value is counted by the digits on the insurance policy, the mint won. But the heavily guarded armored cars moved only money and specie from the mint. The two-hundred-inch mirror was sixteen years of work. Even without a war, it would take a decade or more to replace. No insurance payment could compensate for that.*

The move was planned for months. Irving Krick, a Caltech meteorologist, studied weather charts, using his own weather data and navy reports to track the behavior of each weather system that approached Southern California. The weather on Palomar Mountain could be volatile. The proximity of the ocean and the lack of nearby mountains allowed weather systems to move over the peak quickly. During the beginning of the rainy season in November, a day can dawn clear and bright, as if it were ushering in a spell of fine weather, only to be followed that evening by heavy cloud cover or a storm. Krick watched for a window long enough to move the mirror.

Byron Hill and a group of Caltech civil engineers surveyed the route from California Street in Pasadena, out to U.S. Highway 101, down to Escondido, and finally up the new highway San Diego County had built up the mountain. The route they chose added thirty miles to the most direct route but avoided suspicious bridges and underpasses and congested thoroughfares. The engineers tested every bridge, overpass, and culvert with strain gauges to make sure they could safely bear the load. Five bridges needed additional shoring. The Galivan Bridge on Highway 101, near the line between Orange and San Diego Counties, sagged more than the engineers liked. The high bridge couldn't be reinforced. The engineers decided the trailer would need special dollies attached for the bridge crossing to distribute the load over additional wheels.

The road up Palomar Mountain was new and in good condition, but it was built to San Diego County road-building specifications, not Byron Hill's. He didn't trust the culverts. From earlier disputes Hill had discovered that submitting a request to the San Diego County officials was like trying to clear a patch of brush near a wasps' nest. He skipped the permits and, using crews from the observatory staff, quietly reinforced the culverts to his own specifications.

Reporters, sensing an event, began asking about the move. George

---

* The insurance companies wanted $35,000 to insure the move of the mirror from Pasadena to Palomar, far more than Caltech could afford from the already stretched budget. In view of Caltech's record of safety with the mirror, the underwriter finally agreed to insure the journey for $1,800.

Hall, in the Caltech publicity office, fended them off. California had changed in the twenty years since the first announcement of the telescope. Drive-in restaurants with waitresses on roller skates had replaced the once innovative cafeterias, and oil rigs had begun to encroach on the miles of orange and lemon groves. But some things, like the fundamentalist preachers on the tent circuits, remained the same. After the announcement that work on the mirror was complete, anonymous callers and letters to Caltech and to the observatory threatened to destroy the mirror on its way to Palomar.

They might have been crank calls, but there were enough phone calls and letters, and a couple of appearances of men of the cloth at Palomar that Hill thought suspicious, to put everyone on alert. Hall asked the reporters not to write about the move yet. He told them the date wasn't set, the mirror wasn't quite ready, the telescope wasn't ready for the mirror, that they were waiting for the right weather—any excuse that came to mind. The reporters wouldn't go away. Local reporters had heard the talk in Pasadena, and newspapers like the *New York Times* had contacts like Harlow Shapley, who kept up with the latest reports on the mirror.

At the end of October, Hall sent out a press release asking all reporters and press agencies to cooperate by not announcing in advance any details of the move. Those who cooperated would be provided with opportunities for photographs, film coverage, or a place in a vehicle accompanying the move.

The reporters respected the news embargo, even on November 12, 1947, when the great doors in the optics shop opened for the first time since the disk had arrived eleven and one-half years before, and a huge sixteen-wheel trailer from Belyea & Sons was moved into the optics shop. Jack Belyea was famous for trucking impossible cargoes, from oil rigs in the Middle East to ships that he had moved from seaports to inland lakes.

Inside the shop Marcus Brown ordered a final wash for the mirror. To the men with the hoses and sponges, it seemed no different from the thousands of times they had washed the disk, often several times a day. This time, when the disk was clean, the rubber skirts that covered the edge to keep water out of the support mechanisms were stripped off for good. The disk was tipped up on edge, in the position normally used for Saturday testing. John Anderson and Ike Bowen showed up with a Caltech photographer, and the entire optical lab crew posed in front of the disk. The crew wore their cotton surgical suits. Anderson and Bowen wore three-piece suits. Brown and Anderson shook hands for the photographer. For some shots one of the workers sat up in the central hole of the disk, giving a better scale. There was an air of festivity, but Brownie didn't smile in the photographs. When the workmen went home at the end of the day, Brownie stayed late in the optics shop, as he often did.

The next morning Brownie told Mel Johnson to operate the crane to load the disk onto the trailer. Brownie would stay at the trailer, supervising the lowering of the priceless cargo. They spent much of the morning securing the slings to the disk, measuring and remeasuring the trailer and the heavy timber case, twenty feet square and eight feet high, that had been built for the disk. Lifting the disk in its heavy cell, moving the crane the length of the room, and lowering it horizontally onto two I-beams that had been welded to the trailer took most of a day. Sponge rubber was used to cushion the disk. The mirror cell was fastened to the trailer at three points, two fixed and one movable, to allow for flexure in the trailer. Brownie paced the floor on his bad legs, signaling each move to Johnson.

When the disk was finally on the trailer, workmen climbed onto the big crate to put a sheet of paper and a wooden cover, one inch thick, on top of the disk. One worker saw what looked like a scratch on the surface of the glass in the central hole in the disk. A scratch on the inside of the hole wouldn't have any effect on the mirror, but everyone in the shop knew that the central hole had been ground smooth and true before the temporary plug had been put in. The plug had later come out cleanly, and the hole had been carefully washed. From his perch the workman leaned down to examine the scratch.

It wasn't a scratch. He read the letters etched into the glass: *Marcus H. Brown 1947 A.D.* Brown said nothing about the signature. Craftsmen and artists had always signed their masterpieces.

With the protective wooden cover in place, a radio-crystal detector was mounted on top to measure vibrations the disk received, and connected to an indicator in the cab of the tractor pulling the trailer, where an assistant driver could monitor the vibration. The disk was then covered with aluminum foil insulation and a heavy square cover of two-inch planks. Belyea and Pacific Rigging had brought a banner they wanted displayed on the outside of the crate.

The mirror, cell, and wooden case weighed forty tons. The trailer was another twenty, making a total load of sixty tons, just within the limits of what one of the huge Belyea diesel tractors could pull on a level highway. The civil engineers had estimated that speeds of up to 15 miles per hour would be safe on the best stretches of road; for the final stretch up the road on Palomar Mountain, the speed would have to be held to 4 miles per hour. At those speeds they would need one long day for the 125 miles from Pasadena to Escondido, and another day for the 37 from Escondido to the Observatory. Krick, the meteorologist, had been told to aim for a window of two days of clear weather.

On Monday, November 17, the trailer was pulled through the archway outside the optical shop with winches. An engineer had calculated that the trailer and disk could make the sharp right-angle turn, but it took a whole morning with heavy steel slip panels to get the cargo through the arch, with inches to spare on either side. By 3:00 in the afternoon a Belyea diesel tractor was coupled to the trailer in the driveway. The area was roped off under heavy guard.

The California Highway Patrol had been asked to provide ten officers on motorcycles to escort the mirror. Towns along on the route were asked to have police on alert for crowd control. The Southern Pacific and Santa Fe railroads agreed to station extra signalmen at each railroad crossing on the route, to stop any ongoing trains if the mirror got stuck on a crossing. On the afternoon of November 17, Krick rechecked his weather maps and predicted that the next two days would be clear. The time of departure was set for 3:30 A.M. on November 18.

The reporters and film crews were already gathered outside the optics shop when Bruce Rule placed a final phone call to Byron Hill, who had gone down to Escondido to check the weather at that end. Hill said the sky was clear, and Rule gave the go-ahead signal. At 3:15, the Belyea driver started the engine of the big diesel tractor. Minutes later, amid a blaze of flash bulbs, the trailer eased out of the driveway. Highway patrolmen on motorcycles rode in front and along both sides of the trailer. Bruce Rule sat on the trailer with a portable chart recorder, testing the level of vibrations as the driver moved down California Street at five miles per hour.

It wasn't until they were outside Pasadena that the speed picked up to eight, then ten miles an hour. By then it was early morning. Farmers working close to the highway and workers held up by local police closing off the roads stopped to watch the strange procession. "What is it?" people asked. Those who kept up with the stories in the newspapers answered with the ubiquitous-if-wrong moniker the reporters had adamantly assigned to the mirror: "the Giant Eye." Those magic words were enough to silence crowds, who knew what they had seen was something to tell children and grandchildren.

The procession ignored stop signs and red lights. The lead motorcycle officers were far enough ahead of the truck to block intersections and order approaching motorists to pull over to the curb. Two patrol cars behind kept irate motorists from trying to pass. Every time the road or grade surface changed, Bruce Rule monitored the vibration gauges before he authorized an increase to fifteen miles per hour on Rosemead Boulevard, or ordered a cut to eight miles per hour on Los Alamitos Boulevard in Norwalk. As the day went on, the crowds got larger. Local police were called out in small towns, "in case." But there were no disturbances. Everywhere people watched in awe.

America had changed in the twelve years since the raw glass disk had crossed the country by rail. In 1936 the United States was in the midst of the depression. Americans were hungry for a focus, for reassurance of the capabilities and promise of science, technology, and industry. By 1947 Americans were accustomed to the prosperity of wartime production, jaded by atomic bombs, television, and the possibility of owning their own homes and cars. Yet even people who knew nothing about Palomar or telescopes knew from the escorts and the slow pace of the procession that this was a sight they would never again see.

It took six hours to reach the junction with Highway 101 near Santa Ana. On the smooth four-lane highway, the tractor-trailer could safely speed up to 20 miles per hour. At the Galivan Bridge, five miles north of San Juan Capistrano, where the highway spans the Santa Fe Tracks in a ravine 50 feet below, the procession stopped while the riggers from Pacific Rigging bolted dollies onto each side of the trailer. Everyone was itchy as they waited. The highway patrolmen parked their motorcycles and stood at the perimeters, guarding the disk. The additional sixteen wheels on each dolly spread the load of the disk over fifty-eight wheels. As the lead tractor and then the front wheels of the trailer inched onto the span, the bridge sagged. Caltech engineers monitored deflection gauges they had affixed to the bridge. When the entire load was on the bridge, the gauges showed a maximum deflection of three-eighths of an inch. The reporters wondered why Rule and the other engineers were smiling as the dollies were unbolted on the other side.

At 2:34 in the afternoon, in Carlsbad, a second heavy diesel tractor was hooked to the rear of the trailer in preparation for the grades of the coastal mountains. The speed dropped to only a few miles per hour as they approached the slopes. Toward sunset clouds scudded in from over the ocean. The drivers needed windshield wipers for the misty drizzle. It was dark enough for headlights when they reached Escondido at 5:02 P.M. and parked for the night. They had covered just over 126 miles at an average speed of just under 11 miles per hour. Everything was on schedule but the weather. The mist had turned to rain, and the thermometer was falling.

By morning, the rain was steady and cold. The weatherman said the visibility was 150 feet. Workmen remembered it as 50. There was nothing to do but set off. Despite the conditions, the drivers, now used to the load, covered the 20 miles to Rincon at an average speed of 6.4 miles per hour. There a third huge diesel tractor was hitched behind the trailer for the final climb up Palomar Mountain. The highway patrol motorcycles left at Rincon. The road up the mountain was wide enough only for the tractors and trailer.

Bill Marshall had grown up on the mountain. His father had worked in the office and as the wartime caretaker, and his mother was the schoolteacher. After years of waiting, the arrival of the mirror was too exciting to miss, so Bill took a day off from college in San Diego. Byron Hill assigned him to the crews that braced each bridge from Valley Center to the top of the mountain with twenty- to thirty-foot-long timbers. Moving the timbers was hard work, but there was an excitement in the air as they watched the truck majestically cross each bridge.

On the drive from the coast highway to Escondido, the driver of the tractor at the rear of the trailer had coordinated his speed and gear shifts by watching the puffs of exhaust from the stack of the lead tractor. Now the visibility was so poor the drivers of the two tractors at the back couldn't see the exhaust of the one in the front. They kept their throttles and gears synchronized with one another and with the lead tractor by listening to the sounds of the engine and transmission of the lead tractor.

They passed the Native American village of Pala. Caltech officials had been sent to explain how astronomers would use the telescope to solve the great questions of the origins of the sun, moon, and stars. The Pala had listened politely and nodded. They had their own explanation. Their myths told of a great bird that had flown into the sky, with a branch of flaming tule reeds in its beak. The Pala stood by the road, watching the strange procession toward the mountain.

The grades were steeper after Rincon. The temperature dropped at the higher elevations, hovering around freezing. The rain turned to freezing rain, then sleet, and the road surface froze in patches over the culverts and overpasses. There were some hurried discussions about waiting for a clearing. The drivers wanted to go on. The schedule had called for a speed of four miles an hour up the road on Palomar Mountain, but with the weather worsening, the drivers picked up their pace to double that. Caltech engineers and the reporters in the cars following the big tractors felt their tires slip and skid on the slick road surface. They wondered how the drivers of the big rig did it.

At the bottom of the mountain Byron Hill met the trailer. After thirteen years on the mountain, he knew the treachery of the weather. The temperature on the mountain could drop suddenly. Sleet and freezing rain could turn to snow, covering the road so fast that the centerline and edges would be invisible. The switchback turns of the road were hard enough to follow on a clear day with a dry surface.

Hill climbed on top of the crate encasing the mirror, so he was high enough to signal to the front and rear. The final climb was a dozen miles, most of it on steep grades, the turns just wide enough that if the drivers took exactly the right line, they would keep the wheels of the trailer on the pavement and the mirror would clear the trees. Men walking alongside the lead tractor marked the edges of the pavement as the sleet began to stick in patches. There was no place to turn around. The drivers were reluctant even to risk stopping.

The sky was too gray to reveal the passage of time. When the reporters' cars stopped for photographs of the tractors and the trailer with the disk, men would have to get out and push to get them going again. While the reporters struggled, the tractors kept up their steady pace, shifting down to low-low, then back up through the range of gears to keep their speed steady on the changing grade. Men walking alongside counted the puffs of diesel exhaust for each shift, watching the three tractors synchronize their moves. After a mile it was hard to keep count.

At eleven o'clock in the morning, four hours ahead of schedule, the lead tractor rolled through the gate of the observatory grounds. Despite a dozen close calls, with much shouting and arm signals, there had been no real mishaps. A crowd was waiting at the dome, huddled in the wind and sleet, as the tractor pulled to a stop outside the big doors. One after another the cars that had followed the mirror up the mountain pulled up to the dome. The anxiety of the climb showed on every face. Someone said there was a pot of coffee on inside the dome, and everyone rushed in for warmth, hot coffee, and small talk. It was ten minutes before anyone inside realized that no one was outside with the mirror.

The trailer was exactly where they had left it. The wind and sleet were still blowing, and everyone was exhausted with the anxiety and fatigue of the trip up the mountain. There was talk of delaying the unloading for a day, but Hill and Rule had been told to watch expenditures, and demurrage for a tractor and trailer sitting on the mountain overnight was expensive. Byron had the big doorway opened and positioned men on either side of the trailer, and back at the doorway, to direct the driver. Lloyd Green, a Belyea driver with twenty-five years' experience, climbed up into the cab of the huge diesel and waved them all away. He pulled forward, leaned out of the window to look over his shoulder, and backed the huge trailer straight through the doorway. Like the highway patrolmen who later received souvenir photos to thank them for their help, Green would have stories for his children and grandchildren about the day he drove the most valuable cargo on earth up the mountain.

Max Mason, back in Pasadena, waited until he heard that the mirror was safely in the dome to call Warren Weaver at the Rockefeller Foundation. Mason passed on the reports he heard, letting Weaver share the good news. Only then did Mason break the bad news: The telescope project was broke.

34

Finishing Touches

The final journey of the mirror was front-page news. *Collier's*, *Life*, and *Time* all planned features on the telescope and, discovering that it was far from operation, sent their reporters to George Hall, in charge of publicity at Caltech, for material. Hall had been on the job long enough to know the weeklies loved nothing as much as a colorful personality, so he set up interviews with Fritz Zwicky and Edwin Hubble. Zwicky's thickly accented explanations that he would use the telescope to search for neutron stars and gravitational lenses sounded wacky, but Edwin Hubble was a reporter's dream.

Hubble had returned from his wartime service as chief ballistician at the Aberdeen Proving Ground with the Medal of Merit. He enthusiastically posed for pictures at the two-hundred-inch and at the forty-eight-inch Schmidt camera, and explained to the reporters, in his acquired English accent, that when the telescope was ready, he would extend his measurements of red shifts and counts of nebulae out to the "one-billion light years range of the 200-inch," and test the cosmology of an expanding universe and Einstein's geometry of space. Tantalizing quotes like, "Mathematical physicists believe (from Einstein's ubiquitous Relativity) that space is curved back upon itself, in a four-dimensional way," and photographs of the tweedy, handsome astronomer with his pipe, a bold adventurer preparing to solve the mysteries of the universe, sold magazines.

"It will be a historic night," *Time* wrote, "an extra-clear night, with the sky velvety black and the stars, though bright, twinkling hardly at all. Hubble will go into the observatory after dusk, rise to the big round telescope chamber in a push button elevator." Along with the interviews Hubble also gave speeches on the problems the two-hundred-inch telescope might solve. Mostly the problems he described were his own cosmological program, but he added that the telescope would determine, once and for all, whether there were canals on Mars.

The new wave of publicity raised hackles at Mount Wilson and

Caltech. Astronomers and engineers on the project had learned to live with minor inaccuracies. The "big round telescope chamber" was the prime-focus cage, six feet in diameter, cramped inside for a six-footer like Hubble. The telescope would never be used for observation of Mars, and the issue of canals had long been dismissed by most astronomers. What galled most of all was that Hubble had hoodwinked the press. Walter Adams, a quiet, gentlemanly sort who avoided publicity and nastiness, sent a handwritten complaint to Bowen, "because it is not material I should want to give to a stenographer."

Adams was troubled that the new round of news stories was so one-sided, that there was no mention of men like Max Mason, and instead it seemed to be

> a kind of glorification of two men, Hubble and Zwicky, the first of whom has done little work of the first order for twenty years, and the second hardly anything at any time.
>
> It's clear that Hubble cannot be relied upon to provide a fair or adequate description of the work of a large modern observatory to a journalist seeking information. He knows little about spectroscopy and what it is doing, and at the age of practically sixty is still eager for notoriety and has his press agent continuously at work. I judge he will never be able "to put away childish things."

The short shrift given to other astronomers was improper and just plain wrong to a fair-minded man like Adams. He was even more troubled by the impression Hubble had publicized that the two-hundred-inch telescope would *answer* the important questions of cosmology. "It is just possible," Adams wrote, "that the Hale telescope will not meet all our hopes in its penetration of space, or that even if it does, the gain will not be sufficient to answer many of the important cosmological questions." If the public was led to believe that "answers" about the size and shape of the universe were the sole purpose for the telescope, it might be considered a failure if it did not answer those questions— while the considerable contributions the telescope might make in dozens of other fields of astronomy were ignored.

Propriety wasn't the only reason for the confidentiality of Adams's letter. Harlow Shapley had begun another round of skeptical comments about the telescope, and the first tests of the completed telescope hadn't produced any good news to quash the rumors he started.

Before they could even test the telescope, the opticians had to turn the disk into a mirror. In 1928, when the telescope project began, mirrors in astronomical telescopes were coated with a thin layer of silver. In a household mirror the silvering is applied to the back of the mirror and is seen through the glass, which protects the delicate silvered layer. Viewing through the glass also adds distortions, intentional in an amusement park or a magnifying mirror, acceptable in a household mirror, and intolerable in a precision instrument. In an astronomical telescope the mirroring is applied to the front surface, so that nothing stands between the finely figured surface of the mirror and the light from distant objects. The reflective coating, perhaps one thousand atoms thick, *is* the telescope.

When the telescope is in use, that thin coating of the mirror is exposed to the elements, vulnerable to the corrosive effects of the atmosphere, the accumulation of dust, dripping oil from the telescope, an accidental drop of tools or equipment, rain, snow, hail, windblown debris, even a falling meteorite. The potential hazards are so great that it is tempting to never use the telescope.* When the two-hundred-inch telescope is not in use, the mirror and cell are protected by a diaphragm over the mirror. Leaves, like the blades of a camera shutter, are opened and closed by motors. In case of a power failure, auxiliary power is available to close the diaphragm and protect the mirror.

Silver had long been the material of choice for telescope mirrors. It was easy to apply to the mirror chemically, and immediately after it was applied, it would reflect almost 95 percent of the visible light that struck the mirror. But, as generations of English butlers have learned, silver tarnishes on exposure to air. After a relatively short period of use, the reflectivity of a silvered telescope mirror decreases to approximately 50 percent. In the ultraviolet spectrum—which, though invisible to the naked eye, is important to the astronomer—even a freshly deposited silver film reflects only 4 percent of the light to hit the mirror. John Anderson tried some experiments, coating silvered mirrors with silica and/or fluorite, and concluded it would not protect the silver.

The problems with silver coatings prompted Francis Pease to ponder the ideal mirror material, the imaginary substance mirrorite, which would have the "reflecting power of silver, the zero coefficient of expansion of Invar, the freedom from tarnish of stainless steel, and the lightness of magnalium." Even for the two-hundred-inch telescope, the Caltech geologists couldn't find mirrorite.

In 1932 John Strong discovered a mirror coating almost as good. Strong was trying to deposit a protective layer on rock-salt prisms, to prevent the surfaces from deteriorating on exposure to the air. He finally succeeded in depositing a thin aluminum film on the prisms in a high vacuum, then tried the same process on glass. By 1932 he had aluminized a twelve-inch-diameter telescope mirror. The new coating was nothing short of sensational. The aluminum film didn't need bur-

---

* Byron Hill is infamous for saying that the observatory would run just fine if they didn't have astronomers coming up there and messing with the equipment. Astronomers who tangled with Hill were certain he was not joking.

nishing, and instead of tarnishing on exposure to air like silver, the aluminum formed a hard, transparent oxide coating that *protected* the reflective surface. Dust could be wiped off the surface with a moist soft cloth or even washed off with soap and water—a treatment that would have removed a silver coating. The reflectivity was only 89 percent of visible light, slightly less than the initial reflectivity of silver, but the aluminum maintained its reflectivity even after continued exposure to the atmosphere and also reflected 85 percent of the ultraviolet light. Tests at the Lick Observatory showed that for stellar photography, the new coating reflected on average 50 percent more light to the photographic emulsion. It was like getting a new telescope.

The process was tricky. The aluminum had to be vaporized over the disk in a bell jar by heating it with tungsten coils. To produce a smooth, even layer on the mirror, the aluminum molecules evaporating from the coil had to travel to the surface of the disk in a straight line, which meant they could not hit another molecule. A collision-free path required that the bell jar be evacuated to a vacuum of one tenthousandth of a millimeter of mercury, in a contained space so free of leaks that if it were evacuated and sealed off, it wouldn't reach onehalf of atmospheric pressure for fifteen years. Strong's process required that the optician duplicate the near emptiness of outer space—no easy task in a laboratory.

George Hale, who kept his fingers on any technology that affected big telescopes, had eagerly followed Strong's work. When Strong aluminized the mirror of the Crossley reflector at the Lick Observatory, the telescope Heber Curtis had used for his surveys of "island universes," it improved the performance of the telescope so much that he was recruited to aluminize the mirror of the one-hundred-inch Hooker telescope on Mount Wilson. The results were spectacular. Before, the tiny companion star of the bright star Sirius had been difficult to photograph with the one-hundred-inch telescope, because the fine scratches that were inevitable on a silvered surface as it was burnished—no matter how fine the rouge used—would scatter the light. With Strong's new aluminum coating the companion star was easily resolved in plates. There was no question that they would try for an aluminum coating on the two-hundred-inch disk.

It had taken some tinkering to get Strong's temperamental process to work on a mirror as large as the one-hundred-inch. The coating was only one thousand atoms thick, less than one-thousandth of a millimeter. For optimum effectiveness the thickness had to be uniform to within 4 or 5 percent. The slightest contamination on the surface would cause the coating to fail. Strong constantly experimented with new techniques for cleaning the surface before depositing the aluminum.

In the spring of 1947 Anderson invited Strong to return to Caltech from Johns Hopkins University to supervise the aluminization of the two-hundred inch disk. Strong was delighted to work on the big telescope. He also had no illusions that it would be a tough job. Before he came to Pasadena he urged that whatever vacuum pumps Caltech planned to use, they should have an auxiliary pump system ready to supplement the original pumps.

The mirror cell of the two-hundred-inch disk, which held the support systems and the edge levers, complicated the plans for aluminizing the disk. The engineers had designed a wheeled platform, with screw elevators on each corner that could raise and lower the mirror and cell to the telescope. The platform served both as a carriage to remove and reinstall the disk in the telescope, and as the bottom half of the vacuum chamber. A steel bell was designed fit over the platform, with a rubber gasket fitted around the edge of the disk to isolate the high-vacuum area from the back of the mirror with its delicate support systems. Without the gasket, drawing down a vacuum to the level required by aluminizing would suck the oil and grease out of the bearings of the support assemblies, contaminating the vacuum and forcing a tricky relubrication of the mechanisms.

Strong had cleaned smaller disks by hand-burnishing with virgin chamois and extrafine rouge. The success rate was spotty. The worst contaminant, he discovered, was microscopic traces of oil from the human skin. With opticians hand-burnishing a two-hundred-inch disk, removing the traces of oil would be a Sisyphean task. So for the two-hundred-inch disk, he planned a new cleaning technique: he would first coat the surface of the disk with a "special fatty acid compound and precipitated chalk powder." The precipitated chalk would be wiped off with virgin felt pads, leaving the fatty acid on the surface of the disk. He would then burn the residue of fatty acid off with an oxygen glow, leaving a pristine surface for the aluminizing. Strong's procedure sounded as if it would work. The opticians eagerly awaited his arrival at Palomar with the special materials.

When Strong arrived, the "special fatty acid compound" turned out to be cases of Wildroot Cream Oil hair tonic. Radio listeners all across the United States knew what the magic ingredient was:

> *You'd better get Wildroot Cream Oil, Charlie;*
> *It keeps your hair in trim,*
> *Because it's non-alcoholic, Charlie;*
> *It's made with soothing lanolin.*

"In order to get glass clean," Strong told the opticians and astronomers, as he unpacked cases of hair tonic, "you first have to get it properly dirty."

Strong set to work on the two-hundred-inch disk with his Wildroot Cream Oil treatment. He and the opticians wiped the surface clean just before the overhead crane lowered the bell onto the base of the aluminizing chamber. When it was sealed in place, Byron Hill flipped a switch, and the oxygen glow of the heating coils burned the lanolin residue off the surface of the disk, along with every trace of human body oil. Then Hill flicked another set of switches to start the big oil-diffusion pumps to evacuate the bell to a high vacuum.

The pumps ran for days. Every hour someone would check the vacuum gauges and jot figures in the notebook in which Don Hendrix was recording each step of the aluminizing process. Progress was slow. Technicians drifted off to meals, naps, other work. The staff took turns checking the gauges and jotting notes in Hendrix's log. On the second day Ben Traxler, annoyed by the constant drone of the pumps and the funereal atmosphere of the watch over the process, wrote in the log that he had let air into the chamber to quiet the pumps. Hendrix saw the note and blew up. He had just been named optician for the two-hundred-inch mirror and wasn't ready for jokes.

After four days the pumps had pulled down the best vacuum they could achieve. Hendrix fired the tungsten coils that held the pure aluminum, releasing a rain of aluminum molecules. There were no windows into the chamber. They had to lift the bell jar to see the results.

The crane let off a soul-satisfying growl from the weight of the bell as Ben Traxler lifted it off. The disk was not soul satisfying. The surface was a blotchy mess.

The problem was obvious. The pumps hadn't achieved the required vacuum. Hendrix and Strong tried again, running the pumps longer. While the pumps labored, Strong ran around with a tin of shellac, trying to dope up the leaks in the bell jar. Again the process failed. They tried running the pumps for only twelve hours and firing the aluminum in a "dirty" vacuum. They tried firing the aluminum on one-quarter of the mirror at a time. Each time Byron Hill's notes recorded that the mirror emerged "dark in quarters of old firing. N.G."

Nothing they tried worked. The bell jar, with the tricky gasket around the edge of the mirror, leaked. Even without leaks high vacuums were a laboratory challenge. The early cyclotrons at Berkeley were covered with gobs of red sealing wax on the brass chamber to close leaks. There was no quick and dirty fix for this jar: Sealing wax wouldn't hold the vacuum the aluminization required. The only answer was pumps large enough to pull a high vacuum despite the leaks. The engineers searched the catalogs and found that no company stocked larger pumps. Even if they had found a supplier, there were no funds in the depleted telescope budget to buy larger pumps. Twenty years of work was at a standstill.

The bad news got to Max Mason, who made some phone calls. Under Mason's tenure as president, the Rockefeller Foundation had made the first grants supporting Lawrence's work on cyclotrons at Berkeley. Between his contacts in Berkeley and his friendship with Vannevar Bush, Mason got the old boys busy. One phone call led to another, and before long an official from the Manhattan Project located some huge vacuum pumps at Oak Ridge that weren't in use. The pumps were rushed to Palomar. The huge pumps were powerful enough to draw down the needed vacuum despite the leaks in the jar.

One more Wildroot Cream Oil treatment, and the aluminization succeeded.

First light was three nights before Christmas 1947. John Anderson led the delegation of astronomers and engineers who watched Byron Hill's workmen install the mirror. Anderson had been waiting twenty years for this moment. He had headed the project from the beginning, supervised eleven years of grinding and polishing the disk. Only a heart condition had forced him to relinquish the responsibility for final testing and figuring of the disk on the mountain. This was his moment.

Byron Hill's workmen raised the disk and cell into position with the electric screw hoists and the workmen went to work on the circle of bolts that held the mirror cell in position on the telescope tube. Suddenly a bang and a hideous squeal echoed through the dome. Eyes flashed toward Anderson. Had the disk cracked? Would Anderson's heart survive that?

No one moved. After a long silence Don Hendrix said, "You ever seen a one-million-dollar bolt snap?" The mirror hadn't cracked. The securing bolt hadn't snapped either, only squealed in protest. Anderson and his mirror were fine.

The telescope was far from finished. Bruce Rule was still working on portions of the drive mechanism. The mirror had a turned-up edge, and the support mechanisms were untested. Still, someone had to have a first look. The honor went to Anderson. There were no provisions for visual observations with the telescope, so he used a small reading glass as an eyepiece. Anderson sat in a lift chair that raised him up to the Cassegrain focus and gazed for a while at the Milky Way. When he finally came down from the eyepiece, everyone wanted to know what he had seen. He answered laconically, "Oh, some stars."

Anderson didn't say more. Others—Bowen, Porter, Brownie, and visiting Dutch astronomer J. H. Oort—followed him. When Hill took his turn, he was amazed. He had never seen so many stars in his life. They looked like pollen on a fish pond. Forty years later he remembered that after so many years of work, he felt "pretty good."

Byron Hill wasn't an astronomer or an optician. When Ira Bowen looked, he understood Anderson's silence. The images were worse than disappointing. The mirror appeared to have a staggering astigmatism of perhaps twenty wavelengths, which meant a very serious problem. Bowen ended the test session with a curt "What the hell!"

In the morning the workmen discovered that one of the three points of fixation for the mirror was out of position. The mirror cell had been forced back against an earthquake safety block.

For the next trial Edwin Hubble showed up. Hubble peeked through the eyepiece and pronounced the star images "good," just as the mirror was. He went back to Pasadena with his favorable report, and Anderson told Mason, "You can be as optimistic as you please about the mirror." Anderson's guarded words were halfway between circumlocution and euphemism.

While Hubble's pronouncement leaked to the public, Bowen did some tests with a Hartman screen, a full-size cover at the top of the tube, with regularly spaced apertures admitting light to selected regions of the mirror. The tests were disappointing, with zonal problems, unanticipated problems with the outer edge of the disk, problems with the alignment of the mirror, and a severe vibration in the mount that made testing difficult. Bowen knew then that it was going to be a long time and much work before the telescope would be ready. After Walter Adams's warnings about what the press would do with even a hint of bad news about the telescope, Bowen kept only handwritten records of his mirror tests. Even Mason's reports to Weaver at the Rockefeller Foundation were in handwritten letters, avoiding the dangers of talkative stenographers and carbon-copied memorandums.

It was a disappointing start.

Mason couldn't avoid the reports to the Rockefeller Foundation. As of January 1, 1948, the balance of funds available to the project was effectively zero.

For Mason, who had headed the Rockefeller Foundation for so many years, the bankruptcy of the project was a serious embarrassment. At the end of the war he had requested a supplementary grant of $250,000, assuring the foundation that with that additional sum in hand and the funds from the government for the use of optical and machine shop space and equipment during the war, Caltech would finish the telescope. Mason's request seemed reasonable, the explanation of the delays and extra expenditures of the war years made sense, and the board and officers of the foundation trusted their former president. The supplementary grant was promptly approved. Two years later, with much work still remaining to be completed, the $250,000 was gone.

Weaver, Mason's protégé at the Rockefeller Foundation, had been in charge of overseeing the telescope project since Mason left in 1936. He didn't welcome the task of reporting that the project he had watched over for so long had exhausted its funds.

What had gone wrong? Partly it was Murphy's law. The polishing of the mirror had taken much longer than even the most pessimistic expectations. Deflation had hit portions of the American economy, but the cost of retooling from war production inflated the prices of precision instruments and tools needed to complete the telescope by 50 percent. When a committee of astronomers convened in early fall of 1947 to compile a list of essential instrumentation for the telescope, their list of additional instruments needed at the different observing stations of the telescope, and supplementary measuring instruments needed for the reduction of observations, added another $36,500 of unanticipated costs. In round terms the minimum Caltech needed to finish the telescope was another $250,000 and even that sum would omit some instrumentation that would permit full utilization of the telescope.

Weaver had too much experience with scientists who were longer on ideas than on precise budgets to be surprised. The era of big science, ushered in by projects like the telescope, would see a new generation of scientists who accepted project proposals and detailed budgets as part of the work of science. But Palomar had begun as the project of gentlemen scientists who wrote their budgets in round numbers, like Hale's original budget of $6 million. "The project is a sort of scientific heroic poem," Weaver said in a handwritten memo to his boss R. B. Fosdick, the president of the Rockefeller Foundation. "And like all truly great poetry, it contains elements of both joy and sorrow." The joy was that the biggest telescope in the world was almost finished. The sorrow was that Max Mason's colleagues at Caltech had not been realistic about the cost and timetable for the telescope. They were dreamers, not accountants: "We must remember that a group with the scientific power of imagination to create this massive and precise instrument is not necessarily composed of persons who would also do the most systematic and orthodox job of accounting and budgeting." Mason himself was the perfect example: "Almost anyone else would have done a more careful and reliable and patient job on the estimates & the accounting: but if he [Mason] had not put his great immaginative [sic] ability to work on the problem of the elastic deformation of the mirror, it might very well have *never* been brought to a successful figure."

Fosdick and the board of the Rockefeller Foundation accepted Weaver's recommendation that the foundation grant an additional $300,000, $50,000 more than Mason had requested, to support the completion of the project and to ensure that there was no delay in getting the telescope functioning with a full scientific program. The grant would bring the total Rockefeller commitment to $6,550,000. From the time of the original grant, in 1928, the budget had increased by only 9 percent and had absorbed not only the unsuccessful efforts to develop fused quartz at General Electric but the cost of the 48-inch Schmidt camera, itself a major research instrument, the largest of its kind in the world. The final grant would be the first with conditions attached beyond the vague original specification that the telescope was to be as "perfect" as possible. Caltech was to agree that major construction would terminate on April 1, 1948, that Caltech and the Carnegie Institution would assume the regular operational budget of the observatory from that date, and that any expenditures from Rocke-feller funds after that date would be strictly accounted.

With the disk out of the optical shop, the shop was shut down. The big mirror would never return from the mountain, and future optical work on smaller mirrors and instruments could all be done at the Mount Wilson optical labs on Santa Barbara Street. Lick Observatory agreed to buy the 120-inch disk and grinding machine for $75,000, once it was demonstrated that the support mechanisms of the two-hundred-inch mirror actually worked. Because the 120-inch disk had originally been designed for use as a flat to test the two-hundred-inch disk, it wasn't thick enough to grind into a deep shape for a fast mir-ror. The Lick Observatory had to settle on a slow $f/5$ design for their big telescope, which would soon be the second largest in the world. The Griffith Observatory in Los Angeles bought the 40-inch-diameter plug that had been used in the center of the disk and put it on display. Across the country, Corning already had the first, flawed two-hundred-inch disk on display.

Marcus Brown wrote from his farm that he wanted to buy the grinding machine for the two-hundred-inch disk, for sentimental rea-sons, but couldn't afford it. Brownie, his task done, seemed lost. He dabbled briefly in a jewelry store in Pasadena, but his legs were gone, done in by too many hours at the machines. The parts of the machine weren't worth the cost of dismantling. It was sold for salvage.

Wickliffe Rose's original requirement that Caltech provide endow-ment for the maintenance and operation of the observatory had long been forgotten. Caltech did not have the $30,000 in annual mainte-nance costs that they were obligated to provide, an expense that was expected to increase, substantially, year by year. The president of Cal-tech and his fund-raising staff had their work cut out.

At the observatory Ira Bowen also had his work cut out.

The most immediate problem was the unexpected vibration in the mount. When it happened the telescope would tremble for as long as one minute, ruining any test in progress. Sometimes the vibration seemed to start spontaneously. It also started when an observer moved from one side of the prime-focus cage to the other. Byron Hill blamed the tiny round seat the designers had provided for the observers, so uncomfortable that an observer couldn't help but wiggle. He tore the old seat out and substituted the seat from an old hayrake that had been left around from the days when the Beech family farmed on the mountain. It was more comfortable—what observers call the tractor seat is still there—but the new seat didn't cure the vibration.

Bruce Rule brought engineers up from Pasadena to study the problem. They timed the period of the vibrations at 1.4 seconds. The oscillation of the telescope was large enough to be visible. Against a fixed mark on the horseshoe bearing, the structure moved as much as one-half millimeter—a tiny wiggle on the huge structure but enough to cripple the telescope.

It didn't take long to identify the culprit—the great oil bearing. The friction in the bearing was even lower than the optimistic engineers at Westinghouse had predicted. The same almost frictionless movement that let the five-hundred-ton telescope move from the weight of a milk bottle on one horn of the yoke also allowed the yoke to pick up a natural vibration from the torsion tube that connected the drive gear to the yoke. With virtually no friction to dampen the resonance of the system, the entire mount was vibrating like a tuning fork.

The cure was to *add* friction to the system. The machine shop came up with an assembly of rubber rollers that pressed on the outside of the yoke and drove a series of disks in an oil bath. The device added enough friction to damp the vibrations. The perfect machine had its first Rube Goldberg addition.

When the vibrations settled down, the rest of the mount of the telescope worked better than anyone hoped. Serrurier's trusses controlled the droop of the ends of the telescope tube so well that the intersection of the optical axis of the mirror and the corrector lens did not vary by more than 0.01 inch, well within the limits Ross had set for the lens.

The real question was the mirror. Ira Bowen started serious testing with star images—the ultimate test of an astronomical mirror. To make certain that atmospheric effects wouldn't distort the results, Bowen had to wait for nights of optimum seeing and choose stars that could be imaged with relatively short exposures. He reduced the initial data himself to be sure of the results. Later a team of women "computers" were hired to do the measurements of test plates and the calculations.

No one had ever attempted tests as tricky as these. The Foucault tests in the optics shop had demonstrated the detailed smoothness of surface. The disk was beautifully figured, one of the smoothest optical surfaces ever created. But the mirror of the two-hundred-inch telescope needed more than a fine optical surface to function effectively. The ribbed disk, deliberately designed so that no part of its surface was more than four inches thick, was flexible, a "floppy" sheet of glass at optical tolerances. The lab tests had confirmed that the mirror worked when it was vertical, as it would be when the telescope was pointed at the horizon. To function effectively wherever it was turned, the mirror needed the help of the support system, the complex mechanical devices that would compensate for sags in the shape of the mirror with lever- and cam-induced thrusts against the back of the mirror. As part of his lab-testing program, Anderson had introduced artificially large thrusts in some of the individual supports. The disk reacted exactly as the calculations predicted. But until the mirror was in the telescope and duplicating actual observing conditions, there was no way to know whether the supports would produce the exact compensating forces needed to maintain the figure of the mirror.

Despite Hubble's early pronouncement that the mirror was good as it stood, Bowen's tests showed the disk was elastic: The mirror was bending and flopping as the telescope moved, with resulting distortions to the focused images. Every adjustment he tried produced some form of distortion or astigmatism. The support systems weren't doing their job.

It didn't take Bowen and Rule long to figure out that the problem was their old enemy stiction, the initial friction of a mechanical assembly when it first begins to move. In the optical shop, where the mirror was tested in a fixed vertical position, the support mechanisms had functioned perfectly. In the telescope, where the supports had to respond quickly to the changing position of the mirror, the stiction in those complex systems turned out to be eight to ten times too high. The result was hysteresis: When the telescope moved, the disk would "remember" its previous shape, a few millionths of an inch off what it should be, enough that the mirror could not effectively focus all the light from faint, distant objects.

Rule, Hill, and Bowen tried different lubricants and adjustments to the support systems. In June 1948, after six months of testing, Bowen concluded that no tweaking of the support mechanisms would cure the problem. The supports had been designed with low-friction pivots, but the measured friction of the pivots varied from 0.5 to 1.5 percent. Bowen calculated that to get the mirror to work right the friction in the supports shouldn't be more than one-tenth of what they measured.

The only cure would be new support mechanisms. The estimated cost, even with the machine shop cannibalizing and economizing, was $18,000. There was no contingency fund in the budget. The final Rockefeller funds were already committed, and no new expenditures could be made after April 1948. The Carnegie Institution was responsible only for operating funds. If Caltech wanted to use its new telescope, it had to find some more money.

In 1946 Robert Millikan had turned the reins of Caltech over to the physicist Lee A. DuBridge, who had been a graduate student of Max Mason at the University of Wisconsin. His immediate plans focused on expanding the faculty, increasing the size of the campus, and adding new research fields in chemical biology, planetary science, geochemistry, low-energy nuclear physics, and astrophysics. With so many irons in the fire, DuBridge's fund-raising efforts were already stretched to the limits. Caltech did not have funds to spare for a twenty-year-old project. The equivalent of the salaries of two or three full professors, $18,000 was a large sum in a contracting postwar economy.

One of DuBridge's new hires was Jesse Greenstein, the new chairman of the astrophysics faculty. Greenstein had graduated from Harvard, where he had been charmed by Shapley's brilliance and sociability, and subjected to a heavy dose of Shapley's prejudice against big telescopes in general and the two-hundred-inch telescope and what Shapley deprecated as "West Coast" or "newspaper" astronomy in particular. After a stint working in a family business, Greenstein decided to go back to astronomy and ended up at Yerkes, which was then working on the optics of an eighty-two-inch telescope for the University of Texas. Greenstein discovered a gross error in the testing of the telescope and later used a nebular spectrograph, which incorporated a small Schmidt camera made by Don Hendrix at the Mount Wilson optical shop, on the big telescope. He became a big-telescope convert. Using data from the new spectrograph like David's sling and stones, Greenstein wrote an article that challenged Hubble's measurement of the color temperature of M31.

Greenstein had barely settled in at Caltech when he was told about the need for the new support systems. Greenstein saw his plans for a great astrophysics department stalled. He marched into DuBridge's office and persuaded a reluctant Caltech administration to come up with the $18,000 for the new supports. Before the telescope was finished, Caltech would have to put between $80,000 and $90,000 into the project.

Once the funds were committed, the astrophysics machine shop went into high gear for the summer to rebuild the supports.*

The newspapers, and George Hall at the Caltech publicity office, couldn't wait for the slow birth of the telescope. Except for a few astronomers and visiting bigwigs from the Rockefeller Foundation, the world hadn't seen the big telescope in action, and Caltech was eager to show it off. The gala dedication was scheduled for June 3, 1948. Byron Hill and the staff cleared equipment off the floor of the observatory and set up hundreds of folding chairs in rows under the telescope. The VIPs were scheduled to have a preliminary dinner at Mount Wilson on June 1 and then a private evening at Palomar on June 2—the ecumenical pairing of events was meant to demonstrate the friendly merger of the institutions—before the actual dedication on the afternoon of June 3.

---

* Bruce Rule, in an interview shortly before his death, claimed credit for the design of the new support systems. He also told a reporter that the support mechanisms, with more than a thousand parts each, were so complex that no one else really understood them. Rule's contributions to the telescope were extensive, but he did not design the supports; they contain closer to one hundred than one thousand parts and are not computers but simple though sophisticated and finely machined lever mechanisms. Hans Karoloff, a Finnish engineer at Caltech, did much of the reëngineering on the longitudinal portions of the support mechanisms. Draftsmen had plans for the new supports ready before Ira Bowen decided to have the supports rebuilt.

June 3 dawned bright and sunny. The invited crowd filled every chair in the dome. When Vannevar Bush spoke, a meadowlark flew into the dome through the open shutters, and for much of the following remarks by Robert Millikan, Raymond Fosdick of the Rockefeller Foundation, Max Mason, and James Page, the chairman of the board of Caltech, many in the audience watched the lark as it circled high in the dome.

Evelina Hale, George Hale's widow, was on the dais, and no one was really surprised when Lee DuBridge, the president of Caltech, announced that the telescope would be named the Hale telescope. The naming had been treated as a "state secret" at Caltech, but as far back as Hale's death, in 1938, the *New York Times* had urged that the telescope would be a suitable memorial. Some astronomers close to the project, like Walter Adams, had already begun calling it the Hale telescope.

When the formal speeches concluded, Ira Bowen described how the telescope worked. Bruce Rule stood at the night assistant's control panel. "Like the usual fond parents," Bowen said, "We hope that our new child will not follow the precedent of most infants and misbehave badly on this the first occasion that it has been put on display before a large group of distinguished visitors."

The telescope behaved. No loose parts or oil fell on the guests, and nothing creaked or groaned, as five hundred tons of machine majestically slewed back and forth. There were celebrities and dignitaries in the audience, along with Caltech trustees, distinguished scientists, engineers, and workmen. The actor Charles Laughton was there, wearing a rumpled raincoat. Some men in the audience had worked on the project for as long as fifteen or twenty years. Russell Porter, bitter that he had been slighted for so long, didn't come, but Marcus Brown was there, and George McCauley had driven across the country to see the complete telescope for the first time.

To give the visitors a peek at the fabled mirror, they were invited to go up to the balcony around the inside of the dome. When Rule rotated the dome, the motion of one thousand tons of steel and aluminum, even with eight hundred people on the balcony, was eerily smooth. A confused radio announcer, ignoring Bowen's explanations in favor of what he felt, announced that the floor with the telescope on it was rotating inside the dome.

Anne Price, George McCauley's daughter, had grown up with the telescope as part of her life. She could remember the evenings when her father worked late at the dining table, and the times he had run off to check on the disk after church on Sundays, the crises of the flood and earthquakes, and the high moment when the disk left Corning on its journey across the country. She knew, even as a child, that her father's part in the building of the telescope had been the centerpiece of his working career, as it had been for so many men. She and her husband were newlyweds. The trip to California was their honeymoon. That bright sunny afternoon she watched her father's masterpiece become part of America.

When the crowd left, Bowen went back to his test photographs. Every evening with clear weather, he exposed images of stars, using the full-size Hartman screen with four hundred holes. To search for irregularities in the surface between the diameters of the Hartman tests, Bowen set up a knife-edge and photographed the resulting patterns with a Leica camera.

It was the end of the summer before the new support mechanisms were ready. The redesign had included a new compound lever system, a lengthened lever arm, and sixteen new bearings. To make sure the supports were as friction-free as possible, the bearings were preselected from hundreds of samples of each type that Timken and other companies were willing to ship to Pasadena; in return the companies got the right to publicize the use of their bearings in the great telescope. The bearings were run at from 1/50 to 1/100 of their rated load. The initial friction in the supports dropped from 1.30 to 0.12 percent, theoretically low enough for the supports to do their job. A new series of tests showed that the mirror retained its shape, within the design limits, no matter which way it was turned.

With the support system working, the test photographs revealed the work still needed on the mirror: The outside edge was high, and the tests on stars revealed zones in the mirror which needed polishing. Don Hendrix estimated that without interruptions the work would take six months. Bowen thought the estimate reasonable. But the telescope was no longer a private project. Newspaper and magazine reporters had begun asking for results. The repeated answer that more testing was needed, along with some rumors that seem to have originated with Harlow Shapley, prompted gossip that the telescope was a bust.

To silence the rumors another first light was set up for January 1949, again without advance publicity. Edwin Hubble showed up with his favorite eyepiece from Mount Wilson. The eyepiece lacked an illuminating light, so Ben Traxler, then the senior electrician at Palomar, offered to install one. He took the eyepiece down to his workshop, drilled the barrel and carefully mounted a small light. When he was finished, he noticed brass shavings in the brass barrel of the eyepiece and wiped them out, wiping the crosshairs off the lens in the process. There wasn't time to return to Pasadena for another eyepiece before the observing session. Traxler wasn't an optician, but he remembered hearing that the opticians used spiderwebs for crosshairs. He ran back to the garage of his cottage, found a black widow spiderweb, and stole some threads to glue them in place. They weren't perfect— one thread staggered across the eyepiece like the trail of a drunk—but the eyepiece was usable. Improvisation was already a tradition at Palomar.

Carrying the hastily repaired eyepiece, Hubble ascended the elevator to the prime-focus cage and over the intercom gave the night assistant the coordinates for Coma Berenices. At 10:06 P.M. PST, on January 26, 1949, Hubble pulled the slide of the plate holder to expose photographic plate P.H.-1-H. (the P for Palomar, the first H for Hale [telescope], and the second H for the observer). The first image was of NGC 2261, a variable galactic nebula—a comet-shaped mass with a variable star, *R Monocerotis,* at the apex—that Hubble had discovered at Yerkes.*

The control mechanism did the guiding without intervention. When the plates were developed later that night, Hubble announced that the threshold images, the faintest specks that could be resolved, appeared to be nebulae rather than stars, and that some of the faint nebulae were "presumably at about twice the distance reached with the 100-inch." Exposures of only five or ten minutes, he later told audiences and reporters, recorded stars the one-hundred-inch telescope could not reach in a full night's exposure. "The 200-inch opens to exploration a volume of space about eight times greater than that previously accessible for study. . . . The region of space that we can now observe is so substantial that it may be a fair sample of the universe as a whole." While Hubble gave speeches to quell the rumors, Bowen and the opticians went back to work on the mirror.

Working in a cleared area of the floor of the dome, Don Hendrix and Mel Johnson stripped the aluminum coating off the disk and set up a portable polishing machine, with its hub in the center of the mirror and the outer edge resting on the rim of the disk. After each polishing session, Byron Hill and his crew would mount the mirror into the telescope, and Bowen would wait for a good night to take more test photographs. The disk was uncoated—the aluminizing procedure was far too complex to be used for each test—so he used only the reflectivity of the glass surface for his test photographs of stars. When he had another good set of test photographs, Hill's crew would remove the mirror from the telescope and Hendrix and Johnson would go to work again.

Hendrix started with a twelve-inch polishing tool on his machine and gradually worked his way down to tools the size of a half-dollar. As the edge was brought down, Bowen and Hendrix began to identify small zones of the mirror that needed correction. The zones were too

---

* Actually Bowen had exposed five plates a year earlier with the labels P.H.-1 through P.H.-5. Bowen, who cared little about publicity, let Hubble have the honor of the *official* first plate. Bowen's plates were stored at the Santa Barbara Street headquarters of the Carnegie Institution.

small to use the polishing machine, so Hendrix and Johnson would use handheld cork tools to remove five- or six-millionths of an inch of material. The observatory wasn't a "clean room" like the optics lab, but for the sake of the opticians' concentration and cleanliness, no one else was allowed on the observatory floor while Hendrix and Johnson worked.

Sometimes the weather or the seeing wasn't good enough for testing the mirror, and everyone would have to wait for days for another round of tests. When the remaining defects were too small for the cork tools, Hendrix and Johnson would polish the mirror with their thumbs. They would dip a watercolor brush in a slurry of water and polishing compound—a mixture called Barnesite—paint the zone with the brush, then polish with a few strokes of a naked thumb. The thumb was a good instrument; it didn't slip. Each stroke would remove one- or two-millionths of an inch of glass. After a few strokes, sometimes only one or two, they had to stop, because the heat of their thumbs and the pressure of the polishing would have heated and expanded the mirror. The whole polishing that day might have taken two or three minutes. The disk then had to be remounted on the telescope, an operation that took most of a day. The crews got practiced at removing and re-attaching fourteen and a half tons of mirror and mirror cell.

The final figuring of a telescope mirror is a battle between perfection and reality. The opticians begin talking fractions of a wavelength of light, distances that have little meaning when they are translated into inches or even millimeters. The opticians' eyes glisten as they dream of achieving what no other optical surface has achieved, a surface so smooth that if the two-hundred-inch mirror were the size of the continental United States, it would have no bumps on its surface higher than a few inches.

Several times a week Bowen would give the lunch group at his favorite sandwich shop his latest report on the status of the mirror. The astronomers were champing at the bit: There are never enough big telescopes, never enough hours of observation time. While the opticians polished, the research projects—especially projects that could *only* be completed on the two-hundred-inch telescope—piled up. The astronomers wanted, *needed,* the telescope. Atmospheric effects and defects in the photographic emulsions, they argued, would mask any further improvements in the mirror.

The working astronomers and astrophysicists weren't the only ones waiting impatiently for the telescope. News stories about the telescope had inspired a generation of aspiring scientists. Graduate students in astronomy came to Caltech, with its fledgling astrophysics department, in the hope of working on the biggest telescope in the world. Steven Weinberg, now a Nobel laureate in physics at the University of Texas, was in high school when he read that the "big telescope at Mt. Palomar was going to start operating soon. . . . I thought that as soon as they went from a 100-inch to a 200-inch telescope, then suddenly all the problems would be solved, and that would be really exciting. We would know whether the universe expands forever or collapses."

Across the top of the mountain, the forty-eight-inch Schmidt telescope was already in operation. Hendrix had taken the first official photograph, of M31 (Andromeda), on Septmber 24, 1948. The quality was good enough for the plate to later be included in Hubble's *Atlas of Galaxies*. On the night of July 19, 1949, Albert Wilson and Robert Harrington exposed the first two plates of a sky survey, sponsored by the National Geographic Society and Palomar Observatory and designed to photograph the entire observable sky from Palomar with both red- and blue-sensitive plates. The Schmidt camera would record stars and galaxies of even fainter magnitudes than those Hubble had urged when he thought the Schmidt telescope was for his own proposed program of galaxy counts, a research program that had been slighted in the observation plans for the two-hundred-inch telescope.

The survey was exacting. The emulsions from Eastman Kodak, on thin fourteen-inch square glass plates, had to be bent in the darkroom on a curved mandrel; many broke in the test. Each plate would take in a six-degree square chunk of the heavens. To get the focus right each night, the observer had first to expose a test plate with different focus settings, rush it downstairs to the darkroom by dumbwaiter, then examine the images under a microscope to determine the exact focus for the next plate. Eight or ten exposures would constitute a night's work. An airplane flying overhead during an exposure would ruin the plate. The plates were developed immediately; the microscopic examination of each plate for previously undiscovered asteroids, comets, and supernovas, might take weeks. Fritz Zwicky was accused of rushing over to the big Schmidt dome early in the morning to check the plates from the night before so he could be the first to follow up on any discoveries. The complete survey would take years. Collections of copies of the plates or printed photographic editions of the Palomar Sky Survey became a basic research tool for astronomers everywhere.

By the fall of 1949, Bowen's test results on the two-hundred-inch mirror were getting better. The Mount Wilson and Caltech astronomers had studied enough optics to read the test photographs, and they began to protest the prolonged final figuring. Hendrix and Johnson told Bowen they needed one more week with their thumbs and the Barnesite. When that was over, they wanted one more, and then another after that—a few more chances to reach perfection. Bowen was a physicist by training, closer to the astronomers than to the opticians, but he had been entrusted with a unique challenge. Like so many other men before him, he was caught up in the spell of the original grant language—to build a telescope "as perfect as possible."

He sided with Hendrix and Johnson. There was only one two-hundred-inch telescope. Bowen wasn't going to let it go until it was ready.

He kept testing. The design of the mirror cell allowed air to circulate freely around the outer edge of the mirror. After a substantial temperature change in the observatory, he discovered, the edge cooled quicker than the rest of the mirror, distorting the surface. Bowen had fans installed in the cell to circulate the air inside, and added aluminum foil insulation enclosed in heavy craft paper. Running the fans for an hour or two when there had been a major change in the temperature seemed to solve the problem.

By the end of September, Hendrix and Johnson were down to single-stroke polishing, a bare touch of the thumb with Barnesite and water. The additional figuring of the mirror had taken five months. Less than seven hours of that time had been spent actually polishing. The rest of the time was testing, removing, and reinstalling the mirror; reduction of the test results; and the long intervals between strokes of the opticians' thumbs. Hendrix and Johnson would have been content to continue another two years, but Bowen, under increasing pressure from the astronomers eager to use the telescope, began talking of "final" tests. The zonal problems had disappeared, and in most orientations of the telescope, the tests with the Hartman screen were almost perfect. The one remaining problem was a trace of astigmatism in certain elevations of the telescope. Hendrix estimated that it could take months to polish out the astigmatism and that polishing might not work, because the problem could be in the support mechanisms.

No one had the stomach for another round of rebuilding the support systems. Bowen calculated the forces needed to correct the astigmatism. It came out a few ounces of pull at selected points on the back of the disk. As an experiment he purchased four cheap fisherman's scales at a hardware store. When he got behind the mirror, he hesitated; his calculations had been exact, but he temporarily forgot whether he wanted to push or pull the mirror. Ben Traxler, watching, chided him: "Was it a plus sign or a minus?" Finally Bowen hooked the scales onto the weights of the axial supports at the quadrants of the mirror, northwest, southwest, northeast, and southeast. On the next set of tests the astigmatism disappeared. Bowen left the scales in place.

In October 1949 he told Hendrix to put a new coating of aluminum on the mirror. It was time to turn the telescope over to the astronomers.

35

Palomar Nights

The old road from Pasadena to Palomar ran past the campus of Pomona College in Claremont, through tree-shaded villages, orange groves, clusters of palm trees, and irrigated valleys, before reaching the desert mountain. The Pasadena astronomers had a running challenge for the fastest time for the 134 miles to the observatory; Jesse Greenstein, a master of back roads, claimed the record. The races didn't end until freeways and spreading development replaced the old California of orange groves and palms.

Today, only the climb to the peak from the bottom of the mountain is unchanged from the days when huge tractor-trailers hauled sections of the mounting and the great mirror up the mountain. The road traverses a rainbow spectrum, from the desert oranges and browns of the Native American villages at the base of the mountain, with their ramshackle fences, wandering cattle, and the tired machinery of hardscrabble farming; to the fine greens and tans and splashy wildflowers of the high meadows; and still higher, the dense, dark green of the evergreen forest.

The first glimpse of a telescope dome, from a curve in the road, is a surprise. The domes were originally painted with silver aluminum paint. Today they are covered with a brilliant magnesium paint that reflects the sun's heat, in an effort to improve the local seeing and the temperature stability of the optics. The dazzling white, glistening in the distance, is like a glimpse of the domes of Jerusalem by a pilgrim. Even for an astronomer jaded by hundreds of nights on big telescopes, that first sight is electrifying.

There's a long way to go from that first glimpse to the top of Palomar Mountain. The peak is a shallow glen between two long north-south ridges. Outcroppings of granite stand proud in the scrub brush, meadows, and big-cone spruce forest. On a good day the Pacific Ocean is a smear of blue to the southwest, thirty-five miles away.

There are almost always visitors in the daytime. The weekend of

Labor Day 1948, when the telescope was still in trials, three thousand tourists showed up. There's a small museum, with a mockup of the structure of the mirror and exhibits explaining and illustrating the work of the two-hundred-inch. From the museum a pathway leads to the main entrance of the two-hundred-inch dome, which opens into a glass-enclosed visitors' gallery. The old concrete dummy mirror is along the walk, just past the dome, but few notice it. In the afternoon tourists watch the astronomers and technicians readying instruments on the telescope. By dusk the gates are closed and the tourists go home. The mountain then belongs to the astronomers.

Observing time on the two-hundred-inch telescope is one of the world's rare commodities. The Allocation Committee, successor to the small group of astronomers and physicists who met at Hale's solar laboratory after the war to decide who would use the telescope, reviews a constant stream of applications. The guiding policy for the two-hundred-inch telescope has been to favor projects which could only be undertaken on the large telescope, and which promise a good chance of success. The tea leaves of their decisions are scrutinized with endless care. Some research topics and directions are fashionable. Some observers have a talent for pursuing the hot issues. With time on large telescopes severely limited, there is a bias toward positive results. A few observers get a lion's share of the valuable "dark time," when the moon is not up.

Time on big telescopes is too valuable to waste. Before World War II, Walter Adams kept the big telescopes at Mount Wilson operating every night of the year that weather permitted, including Christmas Eve, Christmas, and New Year's Eve. It was only during the war that he agreed that there would be no observing on Christmas Eve or Christmas Night, so the night assistants could be with their families. Observing is scheduled for every night of the year at Palomar except the occasional "engineering runs" that are reserved for maintenance work on the telescope, such as the periodic washing and realuminizing of the mirror or the installation of new instruments.

Like much of science astronomy today often calls for collective efforts. Teams of two or more astronomers and astrophysicists will work together on an observing run, sometimes joined by physicists, radio astronomers, computer experts, spectroscopists, and others. A shuttle several times a week brings astronomers, technicians, and graduate students from Pasadena to Palomar. Observers arrive from institutions across the country or across the world. The rental cars from San Diego, Los Angeles, or Orange County Airports appear in the afternoon, in time for a night of work. Chronic jetlag, aggravated by long nights and the shift from day- to nightwork, is an occupational hazard of astronomy

Supper in the Monastery is served before dark. There is a tradition of good, hearty food at the observatory—energy and warmth for the long nights—even though winter suppers can end up at an uncivilized hour to accommodate observers who want an early start. Observers with spare moments relax in the Monastery lounge, where the bookshelves are filled with mysteries, old issues of the yearbooks of Mount Wilson and Palomar Observatories, and recent scientific journals. Caltech is a school with a long tradition of practical jokes, and the joking sometimes extends to the observatory, where journal articles or books by rival astronomers will mysteriously appear open on tables in the Monastery.

Pool on cloudy nights is another longtime tradition. At Mount Wilson skill at pool was valued almost as highly as the famed observational skills of Walter Baade and Milton Humason. The favorite game was "cowboy," and the Saturday-morning games between John Anderson and Frank Ross, at the Santa Barbara Street labs, drew heavy betting. When Palomar opened for regular use, Olin Wilson asked for a pool table on the mountain. Ike Bowen reluctantly authorized the purchase of a used table, as long as it cost less than one hundred dollars. A collection from astronomers and staff bought cues and balls, and Wilson got one of the night assistants to help him recover the table with new felt.

Palomar is no longer the isolated peak Hussey surveyed in 1903. At dusk darkness comes suddenly to the mountain, but today the overhead canopy of stars is dimmed by the looms of light of Los Angeles to the northwest and San Diego to the southwest. The light pollution would bring chills to George Hale, who thought this remote mountain safe forever. San Diego years ago agreed to use low-pressure sodium illumination for outdoor public lighting; the distinct yellow light can be filtered from images and spectra. In 1993, supposedly as a crime-prevention measure, the San Diego City Council reversed the decision and authorized the use of high-pressure sodium lights. The light from Los Angeles is unbridled and has steadily diminished the effectiveness of the telescopes even as far away as Palomar.

But even with the troublesome light pollution of the distant cities, the heavens from Palomar are magic. No exterior lights burn at the observatory, so eyes can adjust quickly to the night sky. On a night with good seeing, the canopy overhead is ablaze with pinpoints of light. Familiar constellations are sometimes hard to find in the blanket of stars. The Milky Way can be so dense it is difficult to discern individual stars. The seeing at Palomar has never matched the best of the seeing at Mount Wilson, but the observatory has made a long effort to reduce locally generated atmospheric turbulence with new paint on the domes, insulation of the dome floor, and by removing or insulating internal sources of heat in offices, electrical equipment, oil tanks, pumps, and motors. Astronomers rate the seeing each night in arc seconds of resolution, the lower the better. The seeing the night before—on a good night it is below an arc second—is a sure subject of discussion at breakfast in the Monastery.

The shutters of the big dome are usually opened in late afternoon, to allow the telescope to adjust to the outside temperature. In early tests the dome insulation held the interior temperature to within 0.1° over twenty-four hours, but every night is different, and every procedure that can improve the local seeing helps. The open shutters on the dome at dusk are a beckoning invitation as the astronomers drive or walk from the Monastery to the telescope.

The walk is best. From a distance it is easy to mistake the dome of the two-hundred-inch telescope for the full moon rising over the crest of the mountains. The dome stands proud on a meadow, elegant in its simplicity, anchored by the simple banded base of Russell Porter's design. There are no frills. It is big enough to enclose a twelve-story building, but the proportions are visually comfortable. Even with a glimpse of the telescope visible through the shutter, it is hard to believe that the building is the housing for a scientific instrument and not a temple or monument.

Astronomers enter not through the porticoed entrance used by tourists, but at a simpler door into the lower level. The ground floor is a cluttered industrial workshop—tables and cabinets of tools, fork lifts, tanks of liquid nitrogen to cool electronic devices, barrels of Flying Horse Telescope Oil for the pressure bearings (when Mobil announced that they were discontinuing the product in favor of a synthetic oil, Palomar cached a big supply). The footings of the telescope stand out amidst the tanks and equipment—massive, simple girders, rising from four points. The joints are welded and bolted, a belt-and-suspenders precaution. The footings were later modified to add a safety brace, lest the telescope literally jump off its footings in an earthquake, but as the geologists predicted, the granite mountain has been spared major earthquakes. Around the edges of the lower level are storerooms and former darkrooms that have been converted for storage. Up a flight of steel stairs is the main floor of the observatory.

Even astronomers who have worked at other big telescopes are awed by the two-hundred. The arch of the huge interior space seems immense yet pleasing. The final dome dimensions—the width is equal to the height—were chosen to match the f-ratio of the telescope, but the balanced proportions recall the harmonious architectural magic of spaces designed for their effect, like the interior of Saint Paul's in London or the Pantheon in Rome. In the dim light of early evening, with a slice of the sky in the open shutters of the dome, the building feels like a cathedral.

Everyone has seen photographs of the two-hundred-inch telescope, but the scale of the machine, the sheer size of the massive horseshoe and the side tubes that lead down from the horseshoe to form the yoke, is more than the photographs convey, even the photographs that show tiny human figures next to the telescope. The apparent simplicity is striking. There are no frills, not a single ornament. The gray mounting is stark and smooth, without rivets or seams. It seems impossible that this huge machine, weighing twice as much as the Statue of Liberty, could move with the incredible precision to point at a star.

Observing on the two-hundred-inch telescope is a humbling experience. The machine carries a history, the legacy of achievement. This was the instrument that led the great twentieth-century voyage into cosmology.

It began when Walter Baade used the two-hundred-inch telescope to double the size of the universe.

As soon as the telescope was ready for astronomers, Baade turned it toward his favorite target, the Andromeda galaxy, hoping to end a long-standing dispute. Hubble had used Shapley's calibration of the period-luminosity relation for Cepheid variables to fix the distance of Andromeda at 750,000 light-years. If his distance scale was correct, the (*RR Lyrae*) variable stars of period less than one day in the central region of Andromeda should have had a photographic magnitude of 22, well within the range of the two-hundred-inch telescope with a thirty-minute exposure. In one of his few kind words for the two-hundred-inch telescope, Shapley predicted that it would resolve the *RR Lyrae* stars in Andromeda and thus give a final validation to positions he had held since the great debate of 1921. Baade, who had studied Andromeda for years, including his miraculous wartime resolution of stars in the nucleus with the Mount Wilson telescope, was convinced that simple extrapolation of the light scales of Cepheid variables was wrong, because it failed to distinguish between two different populations of stars. The test was whether the *RR Lyrae* variables in Andromeda could be detected at the predicted magnitude.

Baade was a master of the new machine as he had been of the old. His reputation as an observer was so formidable that other astronomers claimed that Bruce Rule would give the mirror supports of the telescope a special tuneup before Baade had a run on the telescope. In his earliest runs Baade turned the telescope toward Andromeda. "The very first exposures on M 31 taken with the 200-inch telescope," Baade recalled, "showed at once that something was wrong." Earlier tests had showed that the two-hundred-inch telescope would detect stars with a photographic magnitude of 22.4 in a thirty-minute exposure, but as he predicted, Baade could not detect the variables. Shapley responded with a last lick at the new telescope: Maybe, he said, the telescope wasn't good enough to detect the stars.

Baade then brought William Baum, an expert in photometry, up to Palomar to measure the light-gathering and resolving power of the telescope by setting up photometric sequences. Baum's tests agreed with the earlier calibrations. The telescope *could* detect magnitude 22.4 stars in a thirty-minute exposure. Separately, from a study by one of his doctoral students, Baade confirmed the absolute magnitude of *RR Lyrae* stars. That Baade couldn't detect these stars in the central region of Andromeda, after a careful survey, confirmed his division of the variables into two distinct populations. In Rome at the 1952 meeting of the International Astronomical Union, Baade announced his findings. Shapley's light curves for variable stars were too simple. By identifying two distinct populations of variable stars, Baade had corrected Hubble's scale, doubling the size of the universe. The first major discovery of the two-hundred-inch telescope was from what it couldn't see. Imagine, astronomers said to themselves, what the telescope will discover from what it can see.

Astronomy is an incremental science. Although reporters puff each report from a meeting of the AAS or the IAU into what sounds like a definitive proof of the big bang, black holes, or dark matter—a night, or even an entire observing run on a telescope, rarely produces a revolutionary discovery. Each night adds data, fragmentary glimpses and measurements of the reaches of universe that astrophysicists and cosmologists can use to build, modify, or undermine features of a constantly evolving model.

Yet amidst that steady accumulation of knowledge, the achievements of the two-hundred-inch telescope stand out as a history of twentieth-century astronomy. It was on the two-hundred that Baade's student Allan Sandage pursued his long search to refine the Hubble constant, the magic number that would define the age and size of the universe. It was on the two-hundred that Baade and others went beyond the geometry of space that Hubble had explored to identify distinct populations of stars and to explain the evolution of stars, the processes at work as stars were born and died. As the geometry and astrophysics came together, mostly from research done on the two-hundred-inch telescope, it became possible to age-date stars, to begin to understand the mysterious processes at work in the galaxies, and to discover large-scale structures in the universe.

With new instruments, tiny Schmidt cameras that Don Hendrix built with sapphire or diamond lenses, more sensitive photographic emulsions, phototubes and photomultipliers that could record light too faint for a photographic emulsion, corrective lenses with even broader fields than the Ross lens, finer spectrographic gratings, and tricky observational techniques that pushed the equipment to the limits—the reach of the telescope extended further than even the wildest optimists had dared to predict.

After Baade retired back to Germany, in 1958, Rudolph Minkowski took over some of the research Baade had pursued. Minkowski, a big hulk of an astronomer, was famed for ineptness around machines that seemed the perfect inverse of Baade's skill. Night assistants who drifted over to the two-hundred knew two sure ways to tell when Minkowski was on the telescope: the aroma of the smokey Lapsang souchong tea he brought to the dome with him, and the sounds— many not suitable for polite company—that came over the intercom from the prime-focus cage as Minkowski's body protested at the cramped quarters. Baade claimed that the hayrake seat in the prime focus, too big for his diminutive frame, had been shaped from a plaster cast of Minkowski's derriere.

Minkowski was persistent, and he got results. On his last run on the two-hundred-inch telescope, in the spring of 1960, he was determined to get a spectrum from an elusive and suspicious object, identified from the Cambridge compilation of radio sources as 3C 295. The object was so faint he couldn't see it to center it on the slit of the spectrograph. He would have to guide the telescope, for a whole night of exposure, on a dark area of sky where previous direct photographs had identified the object. His final run on the telescope was scheduled for four nights. The first two nights were too cloudy to observe. Minkowski moped around the Monastery, watching his chances slip away. The next night the weather cleared, and he held a faint guide star in the crosshairs of his guiding eyepiece for an entire night. He and Allan Sandage had previously taken "trial plates" to determine that the tiny area of blackness that hid 3C 295 would be on the slit of the spectrograph if the guide star was in the crosshairs. Minkowski gambled, and on the fourth night continued the exposure—trusting Sinclair Smith's wonderful control system to keep the telescope pointing to exactly the same spot. At midnight on the last night he rushed the slide down to the darkroom and developed it. The red shift he measured for 3C 295 was .46, meaning that the strange object was receding at 46 percent of the speed of light. It was the most distant and fastest-moving object yet discovered. Minkowski bounded into the Palomar library with a bottle of bourbon and three glasses, for himself, Sandage, and night assistant Robert Seares. Despite Byron Hill's rules, he was going to celebrate. The "lookback time" to 3C 295 was somewhere between one-third and one-half the age of the universe.

Not every experiment was a success. There were some legendary failures. Fritz Zwicky was allocated little time on the two-hundred-inch telescope. In one experiment he used the two-hundred-inch telescope and the "little" Schmidt telescope, a few hundred yards away, to study the effect of artificial meteors on the atmosphere. The experiment required Ben Traxler, Zwicky's favorite night assistant, to stand in the open shutter of the telescope, firing his .30-.30 carbine into the air while both the two-hundred-inch and Schmidt telescopes tried to record the path of the bullet. When the report from the rifle echoed over the mountaintop, astronomers accused Zwicky of trying to punch a hole in the atmosphere to improve the seeing. Experiments like that one weren't repeated.

Hubble suffered a heart attack in 1949 and was ill by the time the telescope was ready. In a photograph taken in the prime focus, he looks wan and pale. A bell was rigged near his bed in the Monastery, and the night assistants went out of their way to accommodate him on his few observing runs. When he could no longer observe, he picked a surrogate observer to carry on his work, Allan Sandage. As an undergraduate at the University of Illinois, Sandage had read about the telescope in David Woodbury's book and decided that he would go to California to study with Hubble, the most famous astronomer in the world, and somehow work on the most famous telescope in the world. As one of the first Ph.D. students in astrophysics at Caltech, Sandage learned from Walter Baade the skill of identifying a Cepheid star among the thousands of stars in an image of a distant galaxy and then using the remarkable abilities of a trained human eye to compare the brightness of the Cepheid with other stars so its light curve could be calibrated. Allan Sandage probably spent more hours of dark time on the two-hundred than any other observer during its early years, as he relentlessly pursued the goal of refining the Hubble constant that relates the red shifts of distant galaxies to their distance, and thus gives the size and age of the universe.

Sandage used the two-hundred to show that the most distant "stars" Baade had observed were actually H II (ionized hydrogen) clouds. That finding again doubled the scale of the universe. Later, on plates taken with the two-hundred-inch telescope, Sandage discovered a new class of stellar object. He called them "quasi-stellar radio sources," a name that was later shortened to "quasars." For a long time no one could explain these strange objects, until Maarten Schmidt, using spectra taken on the two-hundred, concluded that the light from the quasars was shifted so far toward the red end of the spectrum that they had to be moving at inconceivable speeds. Quasars were the most distant and most luminous objects ever recorded. The telescope, it seemed, was reaching to the very edge of the universe.

The race to find objects with ever greater red shifts was on. Reporters clamored for more exciting reports, and the Caltech publicity office was ever ready to satisfy the requests with a new breakthrough. "Those are the days when you'd come down from Palomar," Sandage recalled, "and everyone would expect you to come down with a pot of gold. Sometimes, almost always, it worked: new quasars, the biggest redshift, variability, are they galaxies or are they nearby?" The two-hundred had gone beyond even the most optimistic of hopes.

Even when larger telescopes came along, and when research on radio telescopes, neutrino detectors, and satellite-borne instruments compete for space in the journals and headlines in the science sections, those heady days of reaching for the edge anointed the two-hun-dred-inch telescope as the machine that opened the universe.

Astronomers usually show up at the dome around dusk. When the telescope first went into service, observers came down early to sensi-tize photographic plates in special hypering solutions. For some obser-vation programs the astronomer would have to work in absolute dark-ness to cut the glass emulsions into small squares and bake them in an atmosphere of dry nitrogen to increase their sensitivity. An invisible speck of dust in the wrong place on the emulsion would ruin a night's work. Before the run began, the observer would give the night assis-tant a handwritten or perhaps typed list of objects with their coordi-nates. The observing positions in the prime-focus cage and the swing-ing seat on the Cassegrain focus were connected by intercom to the night assistant's console, but the astronomer, once loaded into his seat, was alone for hours. For the prime focus the observer would ride a small open elevator up the inside of the shutter opening. The elevator would stop at the level of the prime focus, and the observer would step out, across a ten-inch gap, into the cage.

Once, when the prime-focus elevator broke before a scheduled observing run, the night assistant loaded a young, willing Allan Sandage into the cage by lowering the tube to its lower limit and let-ting Sandage climb up on a rickety ladder, like a carnival stunt man loading himself into the muzzle of a cannon. As the telescope slewed up, he was a prisoner in the cage. For cold winter nights the observers wore war-surplus electrically heated flight suits. For hours they would sit, cramped in a small tube, in subzero temperatures, without light, food, drink, or access to the toilet. As the suits wore out, sometimes the heating wires would chafe and be exposed. Bladder control in an electrical "hot" suit was at a new premium.

Yet for all the rigors of a cold night in a cramped cage, observing at the prime focus is a magical experience. The observer can glance over the edge of the cage, or down from the prime-focus elevator, and see directly into the mirror. The achievement of seventeen years of work—the masterpiece that McCauley, Brown, Anderson, Hendrix, Bowen, and uncounted other men created—is revealed in all its glory. Stars, *millions* of stars, seem to float above the disk. The focus of the disk creates an illusion: It seems as though there is nothing between the observer and the heavens. The great mirror has reached out and grabbed a chunk of the universe.

Few observers today use the prime focus. The last regular expo-sure of a photographic plate at Palomar was on September 29, 1989, almost forty years after Hubble exposed the "official" first plate at the prime focus on November 12, 1949. Photographic plates are still occa-sionally used on the two-hundred for late-epoch proper-motion mea-surements, but in the search for more sensitive detection devices, astronomers shifted to phototubes, photomultipliers, and solid-state light-sensitive devices for most imaging and spectra. The black-box devices of the newest imaging technology are phenomenally sensitive, but they have stolen the mystique of developing a plate in the darkrooms around the perimeter of the observatory and wondering what results would emerge from the developing and fixing baths. The move from the prime focus to the Cassegrain focus, and now to the heated observing room, has stolen the romance of lonely nights aloft *in* the telescope, with the heavens laid out beneath the astronomer in the great mirror. Astronomers eagerly trade romance to reach ever farther into the universe.

In the 1970s an experimenter at the Bell Labs working on a new memory device for computers discovered that the silicon chips he was using were sensitive to individual photons of light. More experiments disclosed that these CCDs, or charge-coupled devices, were phenomenally sensitive, recording up to 90 percent of the light that hit them, compared to the 3 percent that even the most sensitive photographic emulsion recorded. The detectors were tiny, a few hundred pixels in each direction, compared to the millions of grains on a photograph emulsion. But a grain of a photographic emulsion is black or white. Each pixel of a CCD can record hundreds or thousands of shades of gray, as the sensitive chip electronically integrates the photons falling on its surface. The military promptly developed CCDs as detectors for spy satellites. The astronomers had to wait, until an enterprising astronomer and tinkerer from Princeton and Caltech named Jim Gunn found sample chips at an electronics surplus shop that had bought up rejected CCDs that didn't meet military specifications.

Today CCD detectors are available even for amateur telescopes and video cameras. The electronic chips have revolutionized telescope research. Faint objects that could once be detected only in the two-hundred-inch telescope are now within the range of a one-meter telescope with a sensitive CCD detector.

In the basement shops in Robinson Hall, where instruments are built and maintained for the two-hundred-inch telescope, the electronic wizards stretch the capabilities of these new detectors. They combine supersensitive new CCD detectors with tiny Schmidt cameras and dispersion gratings. Another instrument uses fiber optics that can be manipulated so that in a single exposure, a large CCD can record spectrograms of hundreds of galaxies simultaneously, accomplishing in a fraction of an evening work that would have taken Hubble and Humason years. The new imaging cameras and spectrographs are so sensitive that the reach of the Hale telescope has doubled and doubled again, letting astronomers reach out to ever-more-distant galaxies and quasars, until they bump up against what seems a wall of detection, as if the telescope had finally reached the edge of the universe.

The CCDs feed their signals directly to a computer. The newest

instruments collect so much data, so rapidly, that for the next generation of detectors the coaxial cables that connect the instrument cage
on the telescope with the computers will have to be replaced with a
fiber optic pipe to provide adequate bandwidth for the data. With no
need for an observer to change photo plates or manually focus the
images for the plates and spectra, and with the image from a video
camera to guide the telescope, observations are now done from a
warm, brightly lit room on the mezzanine of the observatory. The
night assistant and the observers sit in swivel chairs, controlling the
telescope and the detectors from computer keyboards. Bags of Oreos
and ever-available coffee have replaced the treasured midnight breaks
when observers came down from their perch on the telescope to warm
themselves with hot coffee or tea and share their sausages and cheese
with the night assistant. A tape library that ranges from the Grateful
Dead and Led Zeppelin to Bach and Mozart has replaced the AM radio
the night assistants could pipe through the intercom to the prime
focus, that sometimes mistakenly got stuck on a southern station
broadcasting fundamentalist sermons for hours at a time.

The old control panel for the night assistant, developed from Sinclair Smith's plans, is still in place, and still works, although for most
use now, the motion of the telescope is controlled by a computer. At
his console in the data room, the night assistant has gauges to monitor
wind speed and direction, humidity, and the temperature at various
points on the observatory and the telescope. Instead of turning dials to
the observer's coordinates, the night assistant types them into a computer keyboard.

Yet for all the new detectors and other improvements to the telescope, and the substitution of the warm data room for the cold perch
in the observer's cage, much of the operation of the telescope hasn't
changed. The strictest rule at the observatory, from the days when Ben
Traxler became the first night assistant on the two-hundred-inch telescope, is that observers cannot move the telescope. The great machine
is too valuable to trust to an astronomer. The night assistants aren't
necessarily engineers. One started out as a barber before finding his
way to Palomar; another was a librarian. What they share is a respect
for the instrument entrusted to their care, fascination with life on a
lonely, beautiful mountain, and the skills to maneuver five hundred
tons of machine.

The old control panel for the night assistant, developed from Sinclair Smith's plans, is still in place, and still works, although for most
use now, the motion of the telescope is controlled by a computer. At
his console in the data room, the night assistant has gauges to monitor
wind speed and direction, humidity, and the temperature at various
points on the observatory and the telescope. Instead of turning dials to
the observer's coordinates, the night assistant types them into a computer keyboard.

It's a long-standing tradition for the night assistant to make a ritual, several times each night, of going up to the balcony around the
inside of the dome and then through the doorway to the balcony on
the outside. From there, five stories off the ground, he (or she—even
that has changed) can feel the humidity and view any threatening
weather or run a hand along the brass handrail of the balcony. If the
rail is wet with condensation, it is too damp to observe. It is then the
night assistant's responsibility to close the diaphragm over the mirror

until the air is drier or to shut the shutters of the dome if threatening weather appears.

Observers still go to the dome early in the evening, before dark, to calibrate equipment. Instead of hypering photographic emulsions, they refill the liquid nitrogen dewars (insulated flasks) on the CCD detectors, carrying a vacuum jug of supercold nitrogen up to the Cassegrain cage on the telescope. They get to the cage on a ladder that was modified from a war-surplus C-54 boarding platform. To make sure the ladder is out of the way when the telescope moves, it carries a plug that has to be secured in a wall outlet. Before the interlock system was installed, observers in the data room one night heard loud screeching during a run. When they finally came down to inspect, they discovered that the telescope had been dragging the heavy wheeled ladder with it as it slewed to different positions. There was no damage—a tribute to the robustness of the telescope—but no one likes to risk a priceless machine.

Before electronic instruments took over, the original Cassegrain cage, behind the mirror, had a swinging seat for the observer. An observation session, as the telescope slewed to new positions, felt like sitting in a gimballed drink holder on a boat.* The gimballed chair was replaced in 1965 with a new cage, large enough to hold the huge new electronic detectors that had replaced photographic plates. The refrigerator-size instruments are packed with auxiliary lenses, CCDs, dispersion gratings, cooling dewars, and a mass of wiring and electronics. There was a time when astronomers studied optics and built their own instruments. Now, the "wizards" who build and maintain the instruments in the basement warren of offices in Robinson Hall practice magic that most astronomers don't even try to understand. Fred Harris, who does much of the critical work on the CCDs, calls it a black art, part skill and experience, part luck. He eats a Mexican lunch "to steady [him]self" before he does critical work on a CCD, then does tricks with the sensitive chips that match Walter Baade or Milt Humason's tricks with their photographic plates. Connections have to be soldered to connectors thousandths of an inch apart. Extreme vacuums and baths in oxygen are the equivalents of the plate-hypering an early generation used to increase the sensitivity of their photographic emulsions. Harris is the first to admit that he sometimes doesn't know why the complex detectors choose or refuse to work. The early CCD detectors were mazes of wires and tubes for vacuum pumps, held together with miles of "Palomar glue," duct and strapping tape. After years of refinements, the instruments are reliable. The ones not in use wait in

---

* Some observers got nauseous, but as precarious as the swinging chair felt, it was a great relief to astronomers who had seen Russell Porter's original drawings of a seat suspended on cables, like a child's swing.

individual pens at the perimeter of the dome, labeled with names like "the four-shooter."

Astronomers rarely discuss religion in the Monastery, but privately even veterans of years of observing runs on the big telescope talk of the awe-inspiring mystery of deep space observations, the inescapable feeling that every night on a big telescope is a unique voyage to an uncharted corner of the heavens. It is a magic moment when the night assistant flips switches on his console to start the telescope. From within the bowels of the observatory, a low rumble starts. There is no vibration, just the sound of the pumps for the oil bearings. The night assistant may say nothing more than, "The telescope's ready," but it is hard to shake off the feeling of starting on a vast voyage. Under the canopy of sky the huge dome and telescope feel tiny. It seems the height of audacity to think that this machine, five hundred tons of glass and steel, can reach out to the edge of the cosmos, that from a mountaintop on a planet circling around an ordinary star, one of billions in a not very special galaxy, we are about to reach into the secret depths of the universe.

When the telescope is ready, the night assistant types the coordinates of the first target of the evening into his console and follows with the command "Go." As the telescope slews across the sky, images appear on the video display in front of the observer—stars, the fuzzy glow of a distant galaxy, the clear spiral of a near galaxy, the stellar gridlock of a cluster. Some observers talk of a new age of fully digital telescopes, when the observer won't need to go to the observatory at all and can sit comfortably at a workstation at his home institution, typing commands and waiting for sample images to appear on his display. If the observer is halfway around the globe, even the time changes of nighttime observing disappear. The wee hours at Palomar are the working day in Europe or Asia.

Most observers today still work at the telescope. Few can resist going outside, into the dome, to watch as the telescope slews from one target to another, marveling that so huge a machine can move so smoothly. From the balcony of the dome, an observer today feels the same deceptive sensation that captivated the workmen who first tried the dome, and the visitors on dedication day. The dome motion is so eerily smooth that it seems to stand still. Only a look outside convinces a visitor that the observatory underneath is not turning.

Today, the position of the dome is tracked with scanners that pick up barcodes placed around the dome. But the railroad carriages, the motor drives, and the ground rails are the same units that Byron Hill and his crew labored over more than a half-century ago. The phantom telescope that Sinclair Smith designed is still in place under the cabinet at the head of the stairs to the Coudé room. The tiny telescope tracks every motion of the telescope and dome. The probe that represents the telescope holds its place in the middle of the slot as the tiny dome mirrors the motions of the great overhead dome. The phantom telescope always brings a smile; it is a comprehensible machine, a far cry from the black boxes of contemporary technology.

Maintenance has kept away the ravages of age, but the telescope has the quirks of maturity. When the mirror is realuminized, volatile elements evaporate from the lubricants in the support mechanisms, leaving the bearings stiff. Observers discover the mechanical arthritis when the images the telescope produces degrade from pinpoints to blobs of light. It took years to diagnose the problem. Now the night assistants all know the prescription for stiff joints: exercise. When the images degrade, the night assistants slew the telescope across the maximum range of its movement, forcing the support weights awake. After a few "reps," the mechanisms limber up, and the images snap back to pinpoints.

Much of the telescope is unchanged. The edge supports that John Anderson introduced to correct the astigmatism required no service for forty years. The spring scales that Bowen put on the back of the mirror are still there, a reminder of human limits, like the scar an Arab craftsman puts in an artistic masterpiece lest Allah be offended by a human effort at perfection. The springs came out once, when someone thought they didn't *belong*. Observers immediately protested that the images of faint stars had changed from periods to commas. Bowen's springs went back on.

The two-hundred-inch telescope isn't perfect. There is a residual astigmatism in the mirror in declinations above sixty-five degrees that has resisted every effort to tune the mirror supports. Despite twelve years of grinding and polishing, the mirror is flawed, with tiny fractures in the surface that couldn't be ground or polished out. Minuscule holes have been drilled at the ends of the fractures to keep them from growing, and pitch fills the hairline cracks. A workman—no one remembers when—accidentally dropped a wrench on the mirror, dinging the surface. It hurts to see the scars, but the effect on the performance of the mirror is negligible. Starlight strikes all parts of the mirror at once. The scars cut down on the overall light-gathering ability of the mirror by a fraction of a percent, but the telescope won't miss a distant object because of the imperfections.*

Several times a year the mirror is removed from the telescope for a

---

* At the McDonald Observatory of the University of Texas, a disgruntled employee vented his anger by emptying a revolver into the surface of the eighty-two-inch primary mirror of their telescope. When the university tried to collect from the insurance company, the company paid only a prorated percentage of the cost of the mirror, on the grounds that the mirror had lost only a small percentage of its light-gathering ability.

wash. The procedure has been rehearsed and practiced—the technicians did it dozens of times during the final figuring—but the great mirror is still cradled like a newborn. Hardhats are required for the initial stages, when the mirror cell is unbolted from the telescope. Once the mirror is exposed, all hardhats come off to protect the mirror. The reason for the regulation is obvious, but the bare heads still seem an act of respect as the staff surrounds the mirror to wash it with Orvus soap, a fragrance-free industrial version of Ivory soap. The final rinse is with distilled water. For periodic realuminizing, the surface is stripped and cleaned with powerful solvents. A few bottles of Wildroot Cream Oil are still on the shelf of the cabinet that holds the chemicals, a reminder of the old days.

Astronomers, even veterans with hundreds of nights on the big telescopes, are guests at Palomar. They come up for an observing run, usually two to four nights, then go back to their home institutions. The staff, entrusted with the instruments, stays on—engineers, technicians, night assistants, groundskeepers, mechanics. There are job titles and specializations, but the mountain is a small world. Some jobs need every person on the mountain. The secretary to the superintendent sometimes tends the gift shop in the museum; she also helps wash the mirror of the two-hundred during maintenance runs.

Byron Hill was the first superintendent of the observatory. Hill's strong ideas didn't always make him popular, and a few of his stern rules, like the absolute ban of liquor on the mountain, were hard to enforce. One night assistant was famed for providing 150-proof rum to ease the frustration of cloudy nights. A groundskeeper clearing the brush around the dome of the forty-eight-inch Schmidt camera once discovered a cache of empty whiskey bottles that Humason had hidden during observing sessions. Hill was less tolerant of some foibles. Once he threw an English observer off the mountain for wearing shorts to lunch. Another observer was asked to leave the mountain when a maid reported that he had masturbated in his bed in the Monastery. Most astronomers believed Hill when he said, "We could run this mountain a hell of a lot better if the astronomers wouldn't come up here."

For those who stay year round, life on the mountain isn't easy. There's an elementary school at the observatory, but for high school, children usually live with relatives down in Escondido. Winters are hard. Only one regular staff member tries the daily commute to the mountain.

From the earliest days an unwritten policy at the observatory has required that the engineers, mechanics, and technicians who work at Palomar *not* be amateur astronomers lest they be distracted from their duties on the machines. But men and women who choose to spend their lives on remote mountaintops acquire an awe of the sky that much of America knew in a simpler era, before street lights, urban sprawl, and highways took away the dark skies and their wonders. For all the hardships and loneliness, the mountain can be compelling. When Byron Hill retired he moved to a trailer on a mountaintop near Sonora, with an unbroken view as far as Yosemite. Ben Traxler retired to a mountaintop in Northern California. The telescope makes romantics of cynics.

On one engineering run Fred Harris, the wizard of CCDs, was at Palomar working on a device at the prime focus, which was more crowded than usual because a portion of the observer's cage was taken up with a video display. Engineering runs are normally scheduled during the "light" of the month, when the moon is up, but Harris realized that an eclipse of the moon would darken the skies, if only briefly. He hurriedly phoned his girlfriend in San Diego, told her to drive up to the mountain, and took her up into the prime-focus cage with him. He said nothing until he had turned the telescope toward Lyra, the famous ring nebula. In an amateur telescope the ring nebula is tiny; on the two-hundred the beautiful, mysterious ring filled the screen of the display. There he asked his wife-to-be to marry him, offering her a ring like none on earth.

The success of the two-hundred was a model: The ribbed Pyrex mirror, massive equatorial mount, horseshoe bearing, Serrurier truss, passive supports to correct the shape of the mirror, oil pressure bearings, a prime-focus cage for the observer, "fast" optics with corrective lenses to broaden the good field—all design elements that had been pathbreakers at Palomar—set a norm from which few dared depart. For more than two decades after the telescope went into operation, the few large telescope projects, like the 120-inch at the Lick Observatory and an 84-inch telescope for Kitt Peak in Arizona in 1958, borrowed wholesale from the two-hundred. The Lick telescope used the 120-inch practice disk that had been cast before the two-hundred-inch disk. George McCauley came out of retirement in Corning to supervise the casting of a disk for Kitt Peak.

Many of the new telescopes had birthing problems. The Serrurier truss that had worked so well on the two-hundred-inch looked elegant on the longer tube of the $f/5$ telescope at Lick, but they were long enough to flex. Some suggested that the two-hundred-inch telescope had set an impossible standard, that a design that could only be built in the depression, when glass workers were paid fifty-four cents per hour and Westinghouse would build huge mounting structures for thirty-seven cents per pound, impeded new advances in telescope design.

Only gradually, in the 1970s, did new ideas appear. George Ritchey's optical design, the Ritchey-Chrétien telescope, which used a complex secondary mirror to produce a wide field at the Cassegrain focus, replaced the paraboloid mirror and correcting-lens combination of the 200-inch telescope in many new telescopes. In the Soviet Union,

Russian engineers, waging another battle in the war for space that had begun with Sputnik, built a six-meter telescope in the Caucasus, with a primary mirror a full meter larger than the two-hundred-inch. The Russian design eliminated the massive equatorial mount in favor of a simple alt-azimuth mount: an oil pressure bearing turns the telescope on a table; the tube of the telescope rises and falls in a short fork. The complex motions to translate the altitude and azimuth motions to match the sidereal motion of the stars are controlled by a computer.

By the late 1970s a new wave of telescope building was on. Four-meter telescopes went up in Arizona and Australia. Some design points from the two-hundred were inescapable. Mark Serrurier kept a list of the telescopes that used his truss in their tube designs; it is a listing of every large telescope built since the two-hundred. Designers tried lighter telescope tubes, pivoting the tube in the horseshoe bearing and dispensing with the massive yoke. Pyrex disks and support systems became simpler as glass foundries copied McCauley's newest procedures, "sagging" mirrors from chunks of pure borosilicate glass.

Some innovations failed. The Russian six-meter telescope has never rivaled the performance of the two-hundred. The first try at a mirror developed cracks on the grinding machine in Leningrad (St. Petersburg). A second mirror, designed as a monolithic mass, was poorly figured and subject to thermal shocks. The relatively slow $f/4$ optics required a long tube and a big dome with poor thermal characteristics. The location in the Caucasus had few nights of good weather or seeing. Even with a third mirror, figured to higher standards than the earlier efforts, the telescope is not a productive instrument.

It was only in the 1980s and 1990s that new telescopes came along with designs that finally broke away from the standards set by the two-hundred. Ideas that had been considered and set aside fifty years before reappeared in new guises. Corning began producing mirror blanks from fused quartz, using high-temperature casting techniques that finally solved the problems that had defeated GE. Designs appeared with superthin meniscus mirrors, supported by computer-controlled plungers; George Hale had once hoped to do the same thing with a cushion of air or water. Honeycomb mirrors, like those Ritchey had tried to fabricate in France, provided light weight and quick thermal response. Mirrors were cast in spinning ovens, resulting in a dished shape that cuts the time needed for rough-grinding of the mirror to a fraction of the years required on the two-hundred. Hale had asked McCauley to do the same trick.

At the new Keck Observatory in Hawaii, a telescope with a primary mirror twice the diameter of the two-hundred is now in service. The mirror is made of individual segments, each constantly reshaped by computer-controlled supports. On some telescopes, including the venerable sixty-inch reflector on Mount Wilson, experiments are under way with "adaptive optics," computer-controlled optics that use continuous measurements of artificial stars created with a laser to adapt the optics of the telescope to the changing atmospheric conditions, so that bad or marginal seeing no longer limits the telescope. When the systems are refined, the quality of images at the telescope will rival those of a telescope freed from atmospheric limitations, as if it were in outer space.

The Hubble Space Telescope goes a step further, putting a 96-inch telescope, with a variety of detectors and instrumentation, in orbit above the atmosphere. It was a frightfully expensive project, and with the many delays, it rivaled the two-hundred-inch in the total years it took from conception to launch. The infamous error in the primary mirror of the space telescope would have popped out to John Anderson and Marcus Brown on one of their Saturday tests. As a measure of scale, which needs some adjustment for inflation, the cost of the repair mission for the Hubble Space Telescope was almost exactly one hundred times the total cost of designing and building the Hale telescope.

The two-hundred-inch telescope is no longer the biggest working telescope in the world. The mirror is no longer the most perfect optical surface ever polished. Stressed-lap polishing, using a machine with a polishing surface that continuously adjusts itself as it turns over the disk, enables opticians to grind deeper, faster mirrors, to finer tolerances, in a fraction of the time. Instead of the twelve years it took to figure the two-hundred-inch, a large mirror can be figured in less than two years. Alt-azimuth mounts have become the mark of newer computers, permitting a simpler, lighter, and more compact instrument. New dome designs, some looking like futuristic angular architectural experiments, offer better thermal adjustment and insulation for the telescopes inside.

But the two-hundred endures. If some newer telescopes are bigger, and in sites with superior seeing and without the light pollution of Los Angeles and San Diego, for some work the two-hundred is peerless. The scarcity of observing time on big telescopes puts a premium on reliability and versatility, and the two-hundred is the "rock" of big telescopes. Newer telescopes with their alt-azimuth mountings look more modern than the massive mount of the two-hundred, but they have the disadvantage that to match the sidereal motion of the heavens, instruments have to be rotated constantly, requiring complex software and slip-ring connections. In the interest of lightness, which generally translates to a saving in cost, some new telescopes have limited space and rigidity to support big instruments. Although the size and weight of modern instrumentation wasn't contemplated in the original design, the two-hundred was overbuilt, constructed to last a century. Pease, Porter, Serrurier, Kroon, and others produced a tube and mount rigid enough to maintain the critical optical alignment with two tons of instruments hung behind the mirror. That stability and reliability, and the long record of successful observations from Palomar, make the telescope an ideal testbed for new instrumentation and a prime candidate for new innovations, like adaptive optics.

The history and lore of the telescope are never far away, but when the telescope has slewed to position on the coordinates of a tiny region of the heavens, and distant objects begin to snap into focus on the video displays in the data room, all else seems to disappear. The voyage has begun.

It takes a few moments to match the gray dots and blobs on the video display to the myriad objects on the Palomar Sky Survey plates, from the forty-eight-inch Schmidt camera, that observers frequently use to identify their targets. For a moment it is difficult to match the image on the video display with the plate, then suddenly the tiny bite of sky that the two-hundred has brought into the data room fits perfectly. With the buttons on a handheld paddle, the observer moves the telescope incrementally, placing a remote object—a galaxy or quasar billions of light-years away—on the slit of the spectrograph. A few keys punched into a computer keyboard and the sensitive CCD detector goes to work, sucking up photons launched across the heavens billions of years ago. From the warmth of the data room it seems almost too easy. After ten or twenty minutes, the computer can present a quick sampling of the data, enough for the astronomer to know if he needs another try to get what he wanted.

The apparent ease of the operation is deceptive. The sample of data on the video display is a fragment of the data the detector has gathered. In the few minutes of each exposure, the CCD integrates tens, even hundreds, of megabytes of data. The real findings will wait until the information accumulated on the computer disks at Palomar is brought back on tape for reduction and image analysis on big number-crunching computers. Observers leaving the mountain carry their tapes by hand.

Often the observer will take two spectra, favoring the blue and red ends of the spectrum. Imaging is often done with several exposures, like a photographer bracketing his shutter speeds to make sure he gets a usable photo. Then the night assistant gets a new set of coordinates, and the telescope moves on, seeking another object, another point of data. As huge as the telescope seems, as voluminous as the data from an evening's work, a mirror two hundred inches in diameter is a tiny window on the universe. Forty years of observing is a blink in the eons of time. The astronomer and astrophysicist are forever condemned to bring order and understanding out of a paucity of data, momentary glimpses of the depths of the universe.

Night after night, whenever the weather is clear enough, the research on the big telescope goes on, too important to be stopped by war or politics. In the late 1970s, after the Carnegie Institution poured money and energy into a new observatory at Las Campanas, in Chile, old rivalries and tensions came to a boil at Mount Wilson and Palomar, Santa Barbara Street and California Street. Finally, in 1979, Maarten Schmidt, then director of the joint observatories, presided over a divorce. Caltech kept Palomar and the Big Bear Solar Observatory. The Carnegie Institution kept Mount Wilson and Las Campanas. The Telescope Allocation Committee survived the split for a decade; until 1990 the telescopes of Mount Wilson, Palomar, and Las Campanas were all listed on the pink sheets that announce who gets observing time on the telescopes. But it was an ugly divorce. Friendships among astronomers were rent; some would never be repaired. Allan Sandage, who had been one of the most expert and productive observers on the two-hundred-inch telescope, refused to set foot on the mountain again.

As the Carnegie Institution shifted its attention to the new telescopes at Las Campanas, it announced that it could no longer afford to operate the instruments at Mount Wilson. The one-hundred-inch telescope was mothballed. On its last night of operation, in 1985, night assistants and old friends of the telescope gathered to read portions of the poem Noyes had written to celebrate first light, sixty-eight years before. An effort to reopen the telescope under the operation of a foundation foundered. A second effort has been successful, and the new Mount Wilson Institute will soon bring the one-hundred-inch Hooker telescope into full operation.

The maintenance costs of the Palomar Observatory, and the cost of new instruments, are daunting. Bequests help. A generous bequest gave a name to the forty-eight-inch Schmidt telescope, now known as the Oschin telescope, and paid for a new corrector plate, sensitive into the infrared. Recently, a portion of the time on the two-hundred-inch telescope has been allocated to astronomers from Cornell University and the Carnegie Institution (which will drop out of the program at the end of 1994), in return for contributions to the upkeep and instrumentation of the telescope.

Sixty-six years after George Hale began the project in 1928, the two-hundred-inch telescope keeps working. Bigger instruments have taken the place it long held as the largest working telescope, but the two-hundred remains a premier research instrument. In 1991, more than forty years after the telescope entered service, Jim Gunn and Maarten Schmidt measured a red shift of 4.9 on an object they discovered with the two-hundred-inch telescope, making it the fastest and most distant object ever observed.

That discovery, like so many others, found its way to the science reporters. For a brief moment the name "Palomar" and the machine that had so many times captivated a nation was once again in the news. For men and women old enough to remember the journey of the disk across the country or the move of the mirror to the mountaintop, or for another generation that had read and heard the stories of the technological wonder, the name was a reminder of an earlier era, a time when American engineers, scientists, and workmen dared to build a machine that would challenge the mysteries of the universe.

In the early morning, as dawn comes to the mountain and the sky grows too light to continue, the night assistant closes the diaphragm over the mirror and the shutters in the dome. The pumps are suddenly quiet as the telescope shuts down. The observers, eager for the quiet sanctuary of the Monastery after a long night in the data room, leave by the inconspicuous door on the side of the dome to walk or drive back to the Monastery. Most cannot resist a glance back at the great dome and a quiet smile of gratitude for the privilege of a night voyaging into the unknown.

Notes on Sources

Much of this book is based on archival material and interviews. In the notes that follow, I have used the following abbreviations for archival citations:

RF	Rockefeller Foundation Archives, Rockefeller Archive Center, Pocantico Hills, New York
GEB	General Education Board Archives, Rockefeller Archive Center, Pocantico Hills, New York
IEB	International Education Board Archives, Rockefeller Archive Center, Pocantico Hills, New York
CIT	California Institute of Technology Archives, Pasadena, California. The George Hale Archives are also available in an excellent microfilm edition, edited by Daniel Kevles.
GE	General Electric Archives, Hall of History Foundation, Schnectedy, New York
Corning	Corning Glass Works Archives, Corning, New York
Corning Museum	Corning Museum of Glass, Corning, New York
Serrurier	Mark Serrurier Papers, courtesy of Naomi Serrurier, Pasadena, California
Rule	Bruce Rule Papers, courtesy of Carol L. Roth, Redwood City, California
Hagley	Hagley Museum and Library, Wilmington, Delaware
Huntington	Mount Wilson Archives, Huntington Museum, San Marino, California
Westinghouse	Westinghouse Electric Archives, Pittsburgh, Pennsylvania

I have also drawn on papers and documents provided by Ben Traxler, Byron Hill, Mel Johnson, Allan Sandage, Rein Kroon, Olin Wilson, Robert Thicksten, Sylvia Marshall, and the Astrophysics Library at Caltech. My notes, the material given to me by participants, copies of audio- and videocassettes, and the photocopied material from various archives that I used for this book will eventually be donated to the

423 # NOTES ON SOURCES

Astrophysics Library of the California Institute of Technology for the use of future researchers.

1 April 1921

Eddington on Einstein: "Forty Years of Astronomy," *Background to Modern Science* (Cambridge University Press, 1938), pp. 140–42. The quote on Chinese astronomers is from Joseph Needham, *Science and Civilisation in China*, vol. 3 (Cambridge University Press, 1959). The physicist before Congress is quoted in D. Kevles, *The Physicists* (Vintage, 1979), p. 96. Carnegie is quoted in Helen Wright, *Explorer of the Universe: A Biography of George Ellery Hale* (Dutton, 1966), p. 309.

Shapley describes the train ride in *Through Rugged Ways to the Stars* (Scribner's, 1969).

2 Washington

On women "computers," see Margaret W. Rossiter, *Women Scientists in America: Struggles and Strategies to 1940* (Johns Hopkins University Press, 1982), p. 53, and John Lankford and Rickey L. Slavings, "Gender and Science: Women in American Astronomy, 1859–1940," *Physics Today*, March 1990, p. 60. Owen Gingerich, "Faintness means Farness," *The Great Copernicus Chase*, (Cambridge University Press, 1992), pp. 213–24, is a superb essay on the importance of this concept.

On Shapley's work with Cepheids: Owen Gingerich and Barbara Welther, "Harlow Shapley and the Cepheids," *The Great Copernicus Chase*, pp. 238–45. The actual remarks at the symposium were not transcribed. Later, the two participants exchanged, polished, and published their papers in *Bulletin of the National Research Council of the National Academy of Sciences 2* (1921), pp. 171–217. The basic positions they took did not change in the revisions. Portions of the symposium are quoted in Kenneth R. Lang and Owen Gingerich, *A Source Book in Astronomy and Astrophysics, 1900–1975* (Harvard University Press, 1979).

Shapley on van Maanen is quoted in Richard Berendzen, Richard Hart, and Daniel Seeley, *Man Discovers the Galaxies* (Science History Publications, 1976), p. 116. Hale on Shapley is quoted in Wright, *Explorer of the Universe*, pp. 325–26.

3 The Worrier

Hale wrote a brief memoir of his youth, *Some Personal Recollections* [undated], CIT/Hale 92. The London trip is described in Burton Holmes, "Boyhood Memories of George Ellery Hale," *The Griffith Observer* 9:9 (September 1947), p. 106. On Hale's MIT years, see Edward C. Pickering to George E. Hale, 27 February 1888, CIT/Hale 33.

On Lick: Helen Wright, *James Lick's Monument* (Cambridge University Press, 1987), p. 8; M. W. Shinn, "The Lick Observatory" *Overland Monthly*, 1892, p. 2, cited in Henry C. King, *The History of the Telescope* (Dover, 1955). Lick's will is quoted in "An Extract from the Will of James Lick" in Harlow Shapley, ed., *A Source Book in Astronomy* (McGraw-Hill, 1929), pp. 316–17.

The report on the proposed Southern California answer to the Lick telescope is in *Scientific American*, October 1990, p. 16. On the Crossley telescope, see George Hale, *The Study of Stellar Evolution* (University of Chicago Press, 1908), p. 45.

On Root and the Carnegie Institution: Philip C. Jessup, *Elihu Root* (Dodd,

Mead, 1938), vol. 2, p. 489. Carnegie commented on Lick in *The Gospel of Wealth and Other Timely Essays* (Doubleday, 1933).

Hussey's report on Palomar: "Report by W. J. Hussey on Certain Possible Sites for Astronomical Work in California and Arizona," *Appendix A to Report of Committee on Observatories*, Carnegie Institution of Washington, 1903, IEB I-22-322. Also, George E. Hale, "Observatory of the California Institute," *Astrophysical Journal* 82 (September 1935), p. 125.

Ritchey: Donald E. Osterbrock, *Pauper & Prince: Ritchey, Hale, & Big American Telescopes* (Tucson, 1993). On the precautions for grinding the sixty-inch mirror: George W. Ritchey, "The Two-Foot Reflecting Telescope of the Yerkes Observatory," *Astrophysical Journal* 14 (1901), pp. 218–20; Dinsmore Alter, and Clarence H. Cleminshaw, *Palomar Observatory* (Griffith Observatory, Los Angeles, n.d.), p. 5. The original grinding machine has been restored.

4 The Whirligus

On Hooker and the one-hundred-inch telescope: George Ellery Hale, *Signals from the Stars* (London: Scribner's, 1932), p. 16. There is a collection of photographs of the Hooker home at the Huntington Museum in San Marino, California. Also, A. H. Joy, *Publications of the Astronomical Society of the Pacific* 39 (1927), p. 14.

On the American phenomenon of nervous exhaustion: Tom Lutz, *American Nervousness, 1903: An Anecdotal History* (Cornell University Press, 1991). Ptsosis: Osterbrock, *Pauper & Prince*, p. 128.

The decision to go back to the original one-hundred-inch disk: *Mt. Wilson Observatory Yearbook* 8 (1909), p. 179; Carnegie to Hale, 27 November 1931. Quoted in Helen Wright, Joan N. Warnow, and Charles Weiner, *The Legacy of George Ellery Hale* (MIT Press, 1972), p. 63.

Ritchey's arrogance: Osterbrock, *Pauper & Prince*, p. 154, points out that there is no correspondence with the signature Adams claimed Ritchey used, but plenty of instances of "Officer in Charge, Ordnance Department, Mount Wilson Observatory" or "Production Officer, Mount Wilson Observatory." Ritchey on the disk is quoted in Hale, "The Astrophysical Observatory of the California Institute of Technology: The 200-inch Reflector," supplement to *Nature* 137: 3458 (1934), p. 224.

5 First Light

Adams on first light with the one-hundred-inch telescope: *Publications of the Astronomical Society of the Pacific* 59 (1947). Also, "Autobiographical Notes" [undated], CIT/Hale 92.

6 Waiting

On Hale in retirement: F. H. Seares, "George Ellery Hale: the Scientist Afield," *Isis* 30:2 (May 1939), p. 264.

Views of Shapley are quoted in C. Whitney, *The Discovery of Our Galaxy* (Angus & Robertson, 1972), p. 218, and in Adams to Hale, 10 December 1917, CIT/Hale 30. Shapley and Adams may also not have gotten along because they quarreled over the war: Shapley was not as hard on the Germans as Adams. See Robert W. Smith, *The Expanding Universe* (Cambridge University Press, 1982), p. 93.

Much of the biographical work on Hubble has been muddled by the hagiographic diaries kept by his wife. Donald E. Osterbrock, Joel A. Gwinn, and Ronald S. Brashear have corrected the record from archival sources at the

Huntington Library. See "Edwin Hubble and the Expanding Universe," *Scientific American*, July 1993, pp. 84–89.

Hubble's dissertation is "Photographic Investigations of Faint Nebulae." Its publication was delayed three years because of his war service. He wrote of resolving "condensations" in Messier 33 in "A Spiral Nebula as a Stellar System," *Astrophysical Journal* 63 (1926), pp. 236–74

Hubble's search for Cepheids: Hubble to Shapley, 24 August 1924, quoted in Smith, *The Expanding Universe*, p. 125. Hubble, "NGC6822, a Remote Stellar System," *Astrophysical Journal* 62 (1925), p. 432.

Hubble's paper at the AAS: Stebbins to Hubble, 16 February 1925, AAS Archives. Quoted in Gingerich, *The Great Copernicus Chase*, pp. 236–37.

Early interest in the three-hundred-inch telescope: Markel to W. S. Adams, 30 September 1926, 1 October 1926. A. W. Christian to W. S. Adams, 13 September 1926, Huntington/Adams 66.1176. H. Shapley to W. S. Adams, 11 January 1927, Huntington/Adams 61.1077. A copy of the original drawing is in CIT/Hale 33. The model is in the attic of the Carnegie Observatories on Santa Barbara Street.

Einstein on the cosmological constant: "Cosmological Considerations on the General Theory of Relativity," *The Principle of Relativity*, trans. W. Perrett and G. B. Jeffrey (Dover, 1952), p. 188.

Humason is quoted in John Kord Lagemann, "The Men of Palomar," *Collier's*, May 7, 1949, p. 66.

Hubble on the revolution in astronomy: Hubble to Shapley, 15 May 1929, quoted in Smith, *The Expanding Universe*, p. 197. Edwin Hubble, "A Relation Between Distance and Radial Velocity among Nebulae," *Proceedings of the National Academy of Sciences* 15 (1927), pp. 169–73. Hale on relativity is quoted in Smith, *The Expanding Universe*, p. 173.

Hubble on future research: *The Realm of the Nebulae* (Yale University Press, 1936), p. 117.

7 Old Boys

Rose is quoted in Raymond B. Fosdick, *Adventure in Giving: the Story of the General Education Board* (Harper & Row, 1962), p. 229. His trip to Europe is described in Robert Jungk, *Brighter than a Thousand Suns* (Harcourt Brace Jovanovich, 1962), p. 21. The changes in foundation policy are documented in General Education Board, *Review and Final Report, 1902–1964* (New York, 1964), p. 47.

The meeting in Rose's office: Hale to Adams, 15 March 1928, quoted in Osterbrock, *Pauper & Prince*, p. 225. Rose's correspondence with Hale is in CIT, Hale Papers, Box 35. The *Harper's* article is *Harper's Magazine* 15 (1928) p. 639. Sending a copy to Rose: Hale to Lee Foster Hartman, Harper & Brothers, 15 February 1928, CIT/Hale 8.

Hale described the application process in "Autobiographical Notes," CIT/Hale 92. The notes are not dated, but from internal references would seem to have been written in 1934 or 1935. See also Hale to Rose, 14 February 1928; Rose to Hale, 21 February 1928; Rose to Hale, 13 April 1928, CIT/Hale 35. The pencil draft, dated and sent 16 April 1928, is in CIT/Hale 35.

Thorkelson's trip to Mount Wilson: Thorkelson Diary, 1 October 1926, IEB 1-21-312. Rose's trip, Rose Diary, IEB 1-21-312.

The two-hundred vs. the three-hundred: Wickliffe Rose Diary, 10 April 1928, IEB 1-21-1032. Rose's institutional preferences: Rose Diary, 12 April 1928, IEB 1-21-1032.

Endorsements of the proposal: Thorkelson to Adams, 23 April 1928. Rose to Arthur Day, 23 April 1928, IEB 1-21-312. Gano Dunn to Hale, 20 April 1928; Michelson to Hale, 26 April 1928, CIT/Hale 35.

8 The Politics of Money

Rockefeller on serving on the IEB is quoted in Fosdick, *Adventure in Giving*, p. 227. On Merriam: John Merriam to George Hale, 3 April 1928, CIT/Hale 3. Warren Weaver Diary, 31 January 1937, GEB 1-4-612-6473. Merriam's letter to George Pritchett is quoted in Helen Wright, *Palomar: The World's Largest Telescope* (Macmillan, 1952), p. 55. Proposal dead: George Hale to Wickliffe Rose, telegram, 28 April 1928; Rose Diary, 28 April 1928, IEB 1-21-312.

Damage control: Hale to Henry Robinson, telegram, 4 May 1928, CIT/Hale 30. Hale to Robinson, 16 October 1928, CIT/Hale. Pencil draft: Merriam and Root to Rose, 5 May 1928, CIT/Hale 30.

Merriam gives in: Thorkelson diary, 10 May 1928, IEB 1-21-312. Memo, Warren Weaver to Fosdick, 24 March 1929, GEB 1-4-612-6474. Merriam and Rose exchanged cordial though stiffly formal letters to confirm their new understanding. John Merriam to Wickliffe Rose, 3 May 28, IEB 1-21-312.

Merriam's sham poll of the board: 11 May 1928, CIT/Hale 30. Rockefeller's blessing: George W. Gray, *Education on an International Scale* (Greenwood Press, 1978).

9 Elation

Startup problems: Gano Dunn to Wickliffe Rose, 20 April 1928, CIT/Hale 35. Rose to Hale, 26 May 1928, IEB I-21-312. Thorkelson Diary, 11 June 1928, IEB I-21-313. Hale to Root, 14 June 1928, CIT/Hale 30.

Porter: Hale to Albert Ingalls, 10 August 1928, CIT/Hale 8. Berton C. Willard, *Russell W. Porter* (Bond Wheelwright Company, 1976), p. 177. The Anderson/Pease story is told in David O. Woodbury, *The Glass Giant of Palomar*. Woodbury often embellishes his tales, but he was a close friend of Porter, so this anecdote seems trustworthy. See also Willard, *Russell W. Porter*, p. 175f.

Problems with the one-hundred: Hale to Max Mason, 21 November 1928, IEB I-21-314. Anderson's design memo: CIT/Anderson 4.6.

Early negotiation with Thomson: Hale to George Ritchey, 5 March 1904, 31 March 1904, CIT/Hale 30. Deborah J. Mills, "George Willis Ritchey and the Development of Celestial Photography," *American Scientist*, March 1966, p. 73. Elihu Thomson, "The 200-in. Telescope," *GE Review* 33: 3 (March 1930), p. 138f. Pease to Hale, 25 March 1925, CIT/Hale 33. Ellis budget: 30 March 1928, CIT/Anderson 4.2. Gerard Swope to Henry Robinson, 6 July 1928, IEB I-21-314.

Startup rumors and the press: Adams to Hale, 8 September 1928; Adams to Robinson, 17 September 1928, Huntington/Adams 67.1189. Correspondence after the announcement, including Johnston to Hale, 20 December 1928, is in CIT/Hale 8.

10 Beginnings

Shapley's opposition: Shapley to Adams, 11 January 1927. Huntington/Adams 61.1077. *The Diary of H. L. Mencken* (Alfred A. Knopf, 1989).

Sensitivity to Mount Wilson light pollution: Hale to Adams, 15 March 1928, quoted in Osterbrock, *Prince & Pauper*, p. 225. Site dispute: Arnett Diary, 24 September 1928; Thorkelson to George E. Vincent, 7 September 1928, IEB I-21-314. Hale on Mount Wilson, quoted in Wright, *Palomar*.

Anderson's equipment: J. A. Anderson, "The Astrophysical Observatory of the California Institute of Technology," *Royal Astronomical Society of Canada Journal* (1942), p. 179–81. Anderson to Hale, 8 November 1928; Hale to Anderson, 7 December 1928, CIT/Hale 3. Hale to Anderson, 20 July 1928, 23 July

1928, CIT/Hale 3. Hale's defense of Southern California: William H. Pickering to Hale, 17 November 1928. Hale to Pickering, 10 December 1928, CIT/Anderson 1.18. Arnett and the outside committee: Arnett Diary, 26 September 1928, IEB I-21-314. Trevor Diary, 24 September 1928, IEB I-21-314. "Summary of History to end of 1938" [internal Rockefeller Foundation memo]. GEB 1.4.612.6474.1103.1, p. 1f. Arnett to Hale, 26 September 1928, IEB I-21-314. Hale to Arnett, 22 September 1928, CIT/Anderson 1.20. Hale to Arnett, 26 September 1928 and 29 September 1928, IEB I-21-314. Arnett to GEH, 4 October 1928. GEH to Arnett, 4 October 1928, IEB I-21-314.

The meeting in NY: Arnett Diary, 10 October 1928, IEB I-21-314. Hale's health: Hale to Merriam, 21 December 1928, CIT/Hale 8. Hale to Pickering, 10 December 1928, CIT/Anderson 1.18. A memo summarizing the status of the project in 1928 is in IEB I-21-313.

On Pasadena in 1928, see Carey McWilliams, *Southern California Country: An Island on the Land* (Books for Libraries Press, 1970).

11 Hope

Early work at GE: John Winthrop Hammond, "Building a Looking-Glass to Mirror Unknown Stars," GE internal publication [undated], p. 71f. Early budget disputes, A. E. Ellis to Anderson, 21 February 1929. CIT/Anderson 4.2. The earliest proposal for a ribbed mirror is in Thomson to Hale, 16 July 1928, CIT/Anderson 4.6.

Hale memos: Memorandum to Anderson and Pease, 16 September 1928, CIT/Hale 3. Memo on Committees, 21 January 1929, CIT/Anderson 1.17. Hale to Anderson, 29 June 1928, CIT/Anderson 1.5. J. E. Ross to Anderson, 8 November 1928, 19 November 1928, CIT/Anderson. Hale to Swasey, 4 September 1928, CIT/Anderson 2.1. Hale describes the early decisions in "Building the 200 Inch Telescope," *Signals from the Stars*, p. 100. Anderson to H. J. Thorkelson, 22 October 1929, CIT/Anderson 1.22. The possibility of building the mounting in Southern California is in a Pease memo, 11 December 1928, CIT/Anderson 6.19.

12 Depression

Ammonia as a fuel: Ellis to Anderson, 14 August 1929, CIT/Anderson 4.2. Problems with fused quartz, Anderson to Ellis, 1 July 1929; Ellis to Anderson, 6 July 1929, 16 July 1929, CIT/Anderson 4.2. Anderson to Ellis, 14 November 1929, CIT/Anderson 4.2.

Ritchey's designs: G. W. Ritchey, "L'Evolution de l'Astrophotographie et les Grands Télescopes de l'Avenir," *Publié sous les Auspices de la Société Astronomique de France* (1929), plate 33. Adams, Anderson, Pease, Seares to Hale, 3 July 1928; Adams to Thorkelson, 28 August 1928, CIT/Anderson. Gérard de Vaucauleurs, "George W. Ritchey and the Dina Laboratory," *Sky & Telescope* 85:1 (January 1993), pp. 98–100. Anderson to H. J. Thorkelson, 22 October 1929, CIT/Anderson 1.22.

Robinson's endowment funds lost: Brierly memo of meeting with A. H. Fleming, 19 December 1930, IEB.

Ideal material: Anderson to Ellis, 14 November 1929, CIT/Anderson 4.2. Ellis to Anderson, 14 August 1929, CIT/Anderson 4.2. Hale to Porter, 19 August 1929, CIT/Anderson 1.19. Hale to H. M. Robinson, 18 November 1929, CIT/Hale 30.

Hale's mounting ideas: Memo to Anderson, 19 July 1928, CIT/Hale 3. On the ribbed design: Thomson memo, 24 November 1928, GE. Anderson to Thomson, 22 December 1928; Thomson to Anderson, 28 December 1928, CIT/Anderson 4.2.

13 Orderly Progress

The early Corning disks: McCauley to Sullivan et al., 11 September 1929, Corning. Also, McCauley, "Glass Mirrors at Corning . . . " [unpublished manuscript], Corning. G. V. McCauley, "Some Engineering Problems Encountered in Making a 200-Inch Telescope Disk," talk at the Annual Meeting, American Ceramic Society, Buffalo, N.Y. (February 1935).

Thomson's talk was later published in *GE Review* 33:3 (March 1930), pp. 137–40.

Sixty-inch mirror preparations: Porter to Anderson, 15 August 1930, CIT/Anderson 1.19. Ellis to Anderson, 4 March 1930, IEB 1-21-316. The new budget is in an Ellis memo, 5 June 1930, CIT/Hale 30 (Robinson Papers).

Thomson and Ellis plan: Porter to Anderson, 15 August 1930, CIT/Anderson 1.19. New disasters: Ellis to Anderson, 4 December 1930, CIT/Anderson 4.3.

14 Change of Guard

Will Rogers on Einstein is in a letter to the editor, *Los Angeles Times*, March 6, 1931. Mrs. Einstein is quoted in Berendzen, Hart, and Seeley, *Man Discovers the Galaxies*, p. 200.

Hubble's new research: Hale to Max Mason, 25 February 1931, CIT/Anderson 1.20. Humason to Hale, 9 March 1931, CIT/Hale 3. Einstein wavering on Λ: Eddington, "Forty Years of Astronomy," p. 128. Einstein to Millikan, 1 August 1931, Millikan Papers, quoted in Judith R. Goodstein, *Millikan's School* (W. W. Norton, 1992), p. 101.

Zwicky: Paul Wild, "Fritz Zwicky," *Morphological Cosmology*, proceedings of the Eleventh Cracow Cosmological School, Poland, August 22–31, 1988, p. 392.

John Anderson, "Sinclair Smith," *Publications of the Astronomical Society of the Pacific* (August 1938), p. 12.

Adams on Bowen: Adams to Hale, 30 April 1928, CIT/Hale 35.

Mason and Hale on relativity: Hale to Max Mason, 8 November 1929. Max Mason to Hale, 12 November 1929, IEB I-21-315. The same year that Einstein had written Hale, Millikan was working on an experiment to disprove another Einstein prediction.

Hale on progress with GE: Hale to Max Mason, 25 February 1931, CIT/Anderson 1.20.

The optics lab: Hale to Anderson, 21 March 1931, CIT/Anderson 1.6. Porter to Burrell, 8 April 1930, CIT/Anderson 2.1. Anderson to Hale, 20 December 1929, CIT/Anderson 4.6.

Shapley on the quartz program: Shapley to Adams, 31 March 1931, CIT/Anderson 4.4. Thomson, McManus to Thomson, 30 March 1931, GE. Memorandum, 16 March 1931; Ellis to Thomson, 17 March 1931, GE.

Hale on Pyrex: Hale to Mason, 30 March 1931, CIT/Anderson 1.6. Hale to Anderson [undated, with reference to a letter from O. A. Gage at Corning of 25 March 1931], CIT/Hale 3. Memo on conference in New York City, 8 October 1931, Corning.

Renegotiations with Thomson: Hale to Thomson, 30 March 1931, GE. Ellis to Hale, 22 May 1931, GE. Hale to Anderson, 8 December 1930, CIT/Hale 3. Robinson, Millikan, Noyes, Hale, 29 May 1931, GE. Ellis to C. E. Eveleth, 19 June 1931, GE. Ellis to Hale, 18 September 1931, GE. Hale to Ellis, 10 September 1931, GE.

15 New Light

The old boys on Pyrex: Rev. Anson Phelps Stokes to Warren Weaver, 2 March 1932, IEB. A. L. Day to J. C. Hostetter, 11 May 1931, Corning 7.9.1.1. Walter Adams to H. A. Spoehr, 21 April 1931, IEB.

Hale goes east: Hale to Arnett, 5 June 1931, IEB. Ellis to Swope, 28 September 1931; Ellis to Eveleth, 27 July 1931; Ellis to F. A. Storz [undated], GE. Max Mason Diary, 30 September 1931, IEB.

Meeting in New York City with Corning: Max Mason Diary, 8 October 1931, IEB 1-21-318. Hostetter, Memo on conference in New York City, 8 October 1931, Corning. Hostetter to Falck et al., 15 October 1931, Corning 7.9.1.1.

GE publicity: Ellis to Hale, 9 November 1931, GE.

McCauley: "Glass for the Lens Maker" [unpublished, 18 November 1940] McCauley Papers, Corning Museum. "Corning Glass Works and Astronomical Telescopes" [unpublished], Corning, November 1965, p. 8. G. V. McCauley, "Electrically Controlled Cooling of a 200-Inch Telescope Disk" [unpublished], 25 March 1936, Corning. Hostetter to Sullivan et al., 17 November 1931, Corning. Max Mason to Hale, 5 February 1932, IEB. O. A. Gage to Hale, 25 April 1932, Corning 7.9.1.1. Day to Sullivan, Gage, McCauley et al., 12 January 1932, Corning.

On the shape of the disk and molding a concave face: Day to Hale, 24 February 1932; Hale to Day, 29 February 1932, Corning. In 1992 Steward Observatory, at the University of Arizona, succeeded in casting a 6.5-meter telescope mirror in a rotating oven, so that the surface of the disk assumed a concave shape as the glass cooled.

Fortune magazine snooping: Max Mason to Dwight Macdonald, 16 May 1932, IEB.

McCauley, "Corning Glass Works and Astronomical Telescopes," p. 22. McCauley, "Some Engineering Problems," is the best description of the details of the disk-casting process and equipment.

16 Good News

Hostetter asserting himself: Hale to Trevor Arnett, 26 June 1933, GEB 1-4-611-6470.

17 "The Greatest Item of Interest ... in Twenty-Five years"

Invitations to the casting: Pease to Adams, 31 October 1934, Huntington/Adams. Memo from McCauley, 12 February 1934, Corning. An amateur filmmaker did a grainy silent movie of the preparations and the casting and later tried to sell it as a documentary. There is a copy in the Corning Museum Library. The warning to McCauley is from my interview with Jerry Wright at Corning, 22 August 1990.

18 Salvaging Hopes

Changing the name of Palomar: Catherine M. Wood, *Palomar: From Tepee to Telescope* (San Diego, 1937), pp. 52–53.

Design changes: Minutes, Advisory Committee Meeting, 29 January 1929, CIT/Astrophysics 3.7. Hale to W. J. Luyten, 20 February 1931, GE. Hale memo to Anderson and Pease, 16 September 1928, CIT/Hale 3. Hale memo to Committee on Design of 200-inch Telescope Mounting, 21 January 1929, CIT/Anderson 6.19. C. S. McDowell, "Building the 200-Inch Telescope," *Journal of the Franklin Institute*, 224: 6 (December 1937), p. 684.

Sinclair Smith: Anderson to Hale, 24 August 1925, CIT/Hale 3.

Support-system bearings: Anderson and Porter, "The 200-Inch Telescope," *The Telescope*, March–April 1940, p. 35.

Scandal at Mount Wilson: Osterbrock, *Pauper & Prince*, pp. 274–75. Osterbrock thinks Herbert was blackmailing George Hale. John Merriam: Merriam to Adams, 20 March 1934, Huntington/Adams. Max Mason memo, 30 January 1934, GEB. Max Mason memo, 20 March 1934; Hale to Fred Wright, 13 April

1934, Huntington/Adams. Fred Wright to Hale, 17 April 1934, CIT/Anderson 1.7. Root to Hale, 19 May 1934, CIT/Hale.

19 Revelation

The second pour: Memo, WHC, 3 December 1934, Corning. Hostetter to Hale, 27 November 1934, CIT/Hale. F. S. Kriger to W. H. Curtis, 21 December 1934, Corning.

Construction bids: Ferguson to Hale, 3 October 1934, CIT/Hale. McDowell, Max Mason memo, 18 October 1934, GEB. McDowell to Serrurier et al., 9 December 1935, CIT/Rule 2.9. McDowell to Hale, 25 March 1936, CIT/Hale. McDowell to Anderson, 4 January 1935, 27 January 1935, CIT/Anderson 7.2.

Palomar work camp: George Mendenhall to McDowell, 18 March 1936, Hale/Anderson 1.9.

20 Swept Away

Shipping preferences: Hale to McCauley [telegram], 21 December 1933, Corning. McCauley to Hale, 26 January 1934, CIT/Hale.

21 The Journey

The optics lab: McCauley, "Glass for Palomar" [unpublished], Corning, 26 March 1947, Corning. SKF Industries advertisement, *Western Machinery & Steel World*, August 1937, p. 9.

Public unveiling of the disk: Hostetter to Hale, 2 December 1935; Hale to Hostetter, 4 December 1935, CIT/Hale. *The Evening Leader*, 8 December 1935.

Transport schemes: Argonaut to Anderson, 6 October 1932; Wiley to Anderson, 10 May 1934, CIT/Anderson 4.9. McDowell to Hostetter, 1 October 1935, Corning.

The journey is documented in clippings in the Corning Archives. See also Dennis de Cicco, "The Journey of the 200-inch Mirror," *Sky & Telescope*, April 1986, p. 347. A copy of the Burlington brochure is in the Corning Museum Library. McCauley to Hale, 26 March 1936, CIT/Anderson 1.7.

22 On the Roll

On the mountain: Hale to Max Mason, 1 May 1933, CIT/Anderson 1.20.

Byron Hill: McDowell to Hill, CIT/Anderson 1.9.

Baade and the idea of the Schmidt telescope: Anderson to Hale, 5 October 1931, CIT/Hale 3. Zwicky and the little Schmidt: Roland Müller, *Fritz Zwicky, Leben und Werk des grossen Schweizer Astrophysikers, Raketnforschers und Morphologen (1898–1974)* (Verlag Baeschlin, 1986), p. 147. Fritz Zwicky, "The Early History of the Faint Blue Star Program," *First Conference on Faint Blue Stars*, Strasbourg, August 1964 (*The Observatory*, University of Minnesota, 1965), p. 4. I am indebted to Allan Sandage for bringing this reference to my attention. In the winter of 1935, when Zwicky went back to Hamburg, Bernhard Schmidt was drinking cognac so heavily that, as one of his colleagues put it, "Death took the polishing tool from his hands."

Design work: Anderson to Hale, 10 March 1933, CIT/Anderson. Memo, McDowell to Hale, 9 March 1936, CIT/Anderson. Mark Serrurier, "Structural Features of the 200-Inch Telescope for Mt. Palomar Observatory," *Civil Engineering*, August 1938, pp. 524–25.

Welded construction: Memo from W. A. Kirkland, 3 April 1935 CIT/Rule 10.14. McDowell to Max Mason, 25 November 1935 [with enclosed memorandum on "Development of Telescope Design"], GEB. McDowell to Cmdr. E. D. Almy, USN, 3 April 1935, CIT/Rule 10.14. Hale to McDowell, 30 September

1935, CIT/Anderson 1.10. Frank Fredericks and N. L. Mochel, "Mounting of the 200-Inch Telescope: A Welded Structure," *Metals and Alloys* (November 1939), p. 336.

Rein Kroon: Reinhout P. Kroon, "What's Past is Prologue: a Personal History of Engineering," Symposium at the Towne School of the University of Pennsylvania, 7 October 1970. Kroon later worked on the Westinghouse design team that developed the first American jet engines. Emerson to McDowell, 18 July 1935, CIT/Rule 11.2. Froebel to McDowell, 9 October 1935, CIT/Rule 11.4. Engineering Committee Minutes, 24 January 1936, Huntington. R. P. Kroon, "Unique Bearings Support Yoke of 200 Inch Telescope" [undated, provided by Mr. Kroon]. Hale to McDowell, 30 September 1935, CIT/Anderson 1.10.

23 The Endless Task

Optics lab: Construction Committee Minutes, 1 October 1937, Huntington Library. David O. Woodbury, "Man Bites Glass," *This Week*, 10 September 1939, p. 4. I've generally not relied on Woodbury, because his facts are often wrong and his inventiveness ran afoul of people who were there. Anderson confirmed that the biography of Marcus Brown in this article was essentially correct. See Anderson to Emily Butler of Scott, Foresman & Co., 21 May 1946, CIT/Anderson 1.12.

Max Mason: Hale to Anderson, 26 April 1932, CIT/Anderson. Hale to Max Mason, 26 April 1932, IEB 1-22-319. Progress report, 1 June 1936, Huntingon/Adams 67.1198.

24 Crisis

Another disk: Max Mason to Amory B. Houghton, 11 January 1937, CIT/Astrophysics 2.1. McCauley memo, 26 January 1937, Corning. Warren Weaver Diary, 29–30 January 1937, RF. Warren Weaver Diary, re phone call with McDowell, 18 December 1936, RF.

The big Schmidt: R. B. Fosdick to Max Mason, 21 May 1937, RF. Zwicky believed that an additional $500,000 was awarded by the Rockefeller Foundation for the Schimdt camera because of his success with the little Schmidt. See Fritz Zwicky, "The Early History of the Faint Blue Star Program," p. 5.

Mason and Weaver: Max Mason to Warren Weaver, 19 November 1936; Warren Weaver to Max Mason, 3 December 1936, RF.

25 Big Machines

McDowell and the 1/10 scale model: McDowell to Max Mason, 17 January 1935, GEB. Mark Serrurier, "Structural Features of the 200-Inch Telescope for Mt. Palomar Observatory," *Civil Engineering*, August 1938, p. 525. McDowell, "Final Report on the 200-Inch Telescope Project," January 1939, p. 4, CIT/Astrophysics 2.14.

Westinghouse negotiations: Hale to McDowell, 31 January 1936, CIT/Anderson 1.7. Serrurier to McDowell, 13 March 1936; G. W. Sherburne to McDowell, 12 March 1936, CIT/Anderson 1.10. G. H. Froebel to McDowell, 27 March 1936, CIT/Anderson 7.4. Norman L. Mochel, "Welding and Annealing the Telescope Parts" [unpublished], Westinghouse [courtesy of Rein Kroon], p. 2. McDowell to Jess Ormondroyd, 19 March 1936, CIT/Anderson 7.4. Press Release, News Department, Westinghouse, East Pittsburgh [undated], Corning.

Milling the horseshoe: "Sunbonnet for a Bearing," internal Westinghouse publication [undated], Westinghouse.

McDowell: McDowell to Max Mason, 27 June 1935, GEB. Max Mason to Warren Weaver, 30 March 1937, GEB.

The design of the dome: Porter memo, 20 November 1936, CIT/Rule 3.8. Engineering Committee Minutes, 1 July 1936, Huntington/Adams 67.1199. Pease Memo, 7 April 1936, Huntington/Adams.

26 Fine Points

Disk progress: Warren Weaver Diary, 30 January 1937, RF. Max Mason to Warren Weaver, 1 June 1937, CIT/Astrophysics 2.10.

Hubble's survey plans: "The Distribution of Extra-Galactic Nebulae," *Astrophysical Journal* 79 (1934), pp. 8–76.

Palomar wives poll: Max Mason to J. J. Johnson, 29 March 1938; Zwicky to Mason, 9 April 1938; Baade to Mason, 6 April 1938; Humason to Mason, 2 April 1938; Adams to Mason, 31 March 1938, CIT/Astrophysics 3.8.

Monastery: Hale to McDowell, 22 April 1936, CIT/Anderson 1.7.

27 Passing the Torch

Dark matter: Sinclair Smith, "The Mass of the Virgo Cluster," *Astrophysical Journal* 83 (January 1936), pp. 23–30. F. Zwicky, "On the Masses of Nebulae and of Clusters of Nebulae," *Astrophysical Journal* 86 (October 1937), pp. 217–46. Control systems: Robert McMath to Walter Adams, 1 April 1938, Huntington/Adams 45.783. Minutes of the Construction Committee, 3/25/38, Serrurier. Memo of Robert R. McMath, CIT/Advisory Committee.

Construction: Max Mason to Hale, 5 December 1934. Hale to Max Mason, 12 December 1934, CIT/Hale. Construction Committee Minutes, 28 October 1937, Construction Committee minutes, 19 January 1937, Serrurier. Smith and McDowell, Drive and Control of the 200-Inch Telescope (Memo, 14 May 1937), Rule.

Pease: Engineering Committee minutes, 8 April 1936, Huntington/Adams.

Hale: Hale to Max Mason, 8 October 1935, 13 November 1935, GEB. Hale to McDowell, 30 September 1935, CIT/Anderson 1.10. Dedicating the telescope to Hale: See H. S. Mudd to Millikan, 5 May 1938, CIT/Observatory Council Correspondence.

The bust: Anderson to Thorkelson (Kohler), 8 July 1942, CIT/Anderson 1.22.

Electrical systems: Irwin, "Report on the Electrical Power Supply for the Astrophysical Observatory on Palomar Mountain," 13 February 1936, CIT/Anderson 1.9.

Shipment: Fredericks to McDowell, 9 November 1938, CIT/Rule 16.

Mobil Oil: "The Greatest 'Eye' in the World," *Socony-Vacuum News* 14: 5 (December 1948); George H. White, "The World's Biggest 'Eye' Floats on MobilOil," *Doings in General,* October 1948. J. Ormondroyd, "The Two Hundred Inch Telescope Mounting," Address at the Final Assembly of the 200-Inch Telescope Tube, Westinghouse South Philadelphia Works, 30 April 1937 [provided by Mr. Kroon].

Control systems after Smith's death: McDowell to Bush, 10 April 1937, 10 September 1937, 10 May 1938, 19 May 1938, 3 June 1938; Bush to McDowell, 24 May 1938; McDowell to Hannibal Ford, 31 May 1938; Poitras to McDowell, 28 July 1938, Rule.

28 Testing

First tests: Construction Committee minutes, 28 October 1938, Serrurier. Anderson to Max Mason, 16 September 1938, CIT/Anderson 1.17.

Supports: Construction Committee Minutes, 26 January 1940, Huntington Library/Adams 68.1207.

McDowell: Max Mason to Hale, 27 June 1935; Hale to Max Mason, 11 March 1936, GEB. McDowell, *Final Report on the 200-Inch Telescope Project,* CIT/Astrophysics 2.14. Confidential Report, Walter Adams to John Merriam, July 1936, Huntington/Adams. Warren Weaver memo, after a meeting at Caltech, 5 March 1936, GEB. Warren Weaver diary, 27 September 1938, GEB.

Woodbury: Anderson to David Woodbury, 2 March 1938, CIT/Anderson 2.2. Adams to Woodbury, 3 January 1939, 19 January 1939; Edward Dodd to Adams, 16 January 1939; Adams to Dodd, 23 January 1939; Huntington/Adams 72.1300. Warren Weaver diary, 4 March 1939, RF. Warren Weaver to Shapley, 17 February 1939, RF. "Comments on Draft of The Glass Giant of Palomar" [undated], CIT/Anderson 2.2.

29 Almost

Completion dates: AMJ interview with FBH re Pasadena trip, 1–18 April 1940, GEB. Merriam: Warren Weaver diary, 31 January 1937, GEB. Warren Weaver diary, 7 April 1937, GEB. Bush: Vannevar Bush to Warren Weaver, 20 March 1939; Warren Weaver to Fosdick, 24 March 1939, GEB. Warren Weaver diary, 17 March 1939, GEB. Warren Weaver diary, May 8/9, 1939, GEB.

Corning: Decker to Houghton, 26 January 1937; Turbett to Decker, 28 April 1938; Sullivan to Decker, 16 March 1939, Corning.

Millikan: Warren Weaver diary, 12 February 1941, GEB. Warren Weaver diary, 21 March 1940, GEB.

The big Schmidt: R. B. Fosdick to Max Mason, 21 May 1937; Warren Weaver diary, 4 March 1938, GEB.

30 Impossible Circumstances

Optics: John Anderson, "Optics of the 200-Inch Telescope," 23 June 1948, CIT/Anderson.

Gears: Bruce Rule, "Engineering Aspects of the 200-Inch Telescope," June 1948, CIT. McDowell, "Final Report," p. 6.

Shops: "Shops at Caltech," *Engineering & Science*, June 1948, p. 13. War at Caltech, John H. Rubel, "The Committee," *Engineering & Science*, Summer 1992, pp. 29–35. Samuel W. Fernberger, Technical Aide, D-2, to Warren Weaver, 28 September 1942, RF. Paul Wild, "Fritz Zwicky," p. 394. Mason to R. B. Fosdick, 22 December 1942, RF.

Mount Wilson during the war: Minkowski, "The Crab Nebula," *Astrophysical Journal* 96 (1942), pp. 199–213. Baade, "The Resolution of Messier 32, NGC 205, and the Central Region of the Andromeda Nebula," *Astrophysical Journal* 100 (1944), p. 137. "NGC 147 and NGC 185: Two New Members of the Local Group of Galaxies," *Astrophysical Journal* 100 (1944), p. 147. Allan Sandage, "The Population Concept, Globular Clusters, Subdwarfs, Ages, and the Collapse of the Galaxy," *Ann. Rev. Astron. Astrophys.* 24 (1986), pp. 425–28.

31 Endless, Damnable War

Shapley: Bush to Adams, 23 October 1944; Adams to Bush, 28 October 1944, Huntington/Adams 61.1084. Hubble: "Confidential Report of Adams to Merriam," July 1936, Huntington/Adams. Adams to Bush, 9 January 1945. Huntington/Adams 68.120. Bush to Herbert Hoover, 14 June 1945, Huntington/Adams. Hubble to Bowen, 16 October 1945, Huntington/Hubble 408. Greenstein is quoted in Heinz R. Pagels, *Perfect Symmetry* (Bantam, 1986), p. 98.

32 Starting Anew

Starting again: Warren Weaver to Max Mason, 26 September 1945, RF. Anderson to Dr. E. D. Tillyer, 21 October 1946, CIT/Anderson 1.23. Finishing the disk: Max Mason to R. B. Fosdick, 15 June 1947, RF. Max Mason to Warren Weaver, 1 October 1947, RF.

33 Delicate Cargoes

Moving the mirror to Palomar: Byron Hill to Rule, 3 September 1947; Rule to Jack Belyea, 22 September 1947, CIT/Rule 9.4. Rule to Jack Belyea, 3 November 1947, CIT/Anderson 4.10. Robert S. Richardson, "The 200-Inch Mirror Goes to Palomar Mountain" (unpublished ms., Astrophysics Library, Caltech).

34 Finishing Touches

Hubble and the reporters: *Time*, 9 February 1948, pp. 56–65. "The 200-inch Telescope and Some Problems It May Solve," *Alexander F. Morrison Lecture* (Pasadena, 8 April 1947), Huntington/Hubble 66. Adams to Bowen [undated], Huntington/Bowen 1.1. A handwritten note identifies the date as [c. 1946], but the letter was almost certainly written in late 1947 or early 1948.

Coating experiments: Anderson to Robert G. Aitkey, Lick Observatory, 3 August 1933, CIT/Anderson 3.14. Pease on mirrorite, quoted in Edward G. Pendray, *Men, Mirrors and Stars* (Funk & Wagnalls, 1935), p. 276. Strong: Hale to Mason, 25 March 1935, RF. John Strong to J. A. Anderson, 6 March 1947, CIT/Anderson 1.21. Strong on getting it dirty first: quoted in Richard Preston, *First Light* (New York, 1987), p. 46. John Strong, "The Evaporation Process and Its Application to the Aluminizing of Large Telescope Mirrors," *Astrophysical Journal* 83:5 (June 1936).

Mirroring trials: Byron Hill's notebook, 11 October 1949, Thicksten Papers, Palomar. Bolt snap, quoted in Preston, *First Light*, p. 47. Mason to Weaver, 30 December 1947, RF.

Budget poetry: Warren Weaver to R. B. Fosdick, 12 December 1947, RF.

Brown's plans: Brown to Bruce Rule, 4 March 1950, CIT/CIT 1.

Endowment: Max Mason to Sherman, Steely, and Wright, 6 March 1946, CIT/Observatory Council Minutes.

Vibration: Max Mason to Warren Weaver, 8 January 1947, RF. I. S. Bowen, "The 200-Inch Hale Telescope," *Telescopes*, ed. Gerald P. Kuiper and Barbara M. Middlehurst (Chicago University Press, 1966), p. 2.

Tuning the mirror: Byron Hill notebook, 21 December 1947, Thicksten Papers, Palomar. Mason to Warren Weaver, 28 June 1948, from Nantucket, RF.

Lever supports: Bruce Rule, "Lever Support Systems," Paper delivered at a symposium on Support and Testing of Large Astronomical Mirrors, Tucson, 4–6 December 1966 (KPNO and University of Arizona Press, 1968), p. 93.

Final budgets: Max Mason to Warren Weaver, 26 November 1948, RF. Warren Weaver diary, 8–9 January 1949, RF.

Dedication: Lee DuBridge to R. B. Fosdick, 1 March 1948; Max Mason to R. B. Fosdick [undated, probably May 1948], RF.

Second *First Light*: Hubble, "First Photographs with the 200-Inch Hale Reflector" [unpublished] (14 May 1949), Huntington/Hubble. Later published in *Publications of the Astronomical Society of the Pacific* 61: 360. Jane Dietrich, "First Lights," *Engineering & Science* 54:2 (Winter 1991), p. 9. Hubble, "First Photographs with the 200-inch Hale Reflector" [unpublished], Huntington.

Final polishing: Caltech press release, 22 October 1949, RF. I. S. Bowen, "Final Adjustments and Tests of the Hale Telescope."

Weinberg is quoted in Alan Lightman and Roberta Brawer, *Origins: the Lives and Worlds of Modern Cosmologists* (Harvard University Press, 1990), p. 452.

Hubble on the sky survey: Handwritten memo [undated, ca. 1947], Huntington/Hubble 14.

Final touches: Byron Hill notebook, 24 February 1949, Thicksten Papers, Palomar.

35 Palomar Nights

Baade and the distance scale: "A Revision of the Extra-Galactic Distance Scale," *Transactions of the International Astronomical Union* 8 (1952), pp. 397–98.

Support mechanism: I. S. Bowen, "Optical Tests and Adjustments of the 200-inch Hale telescope," *Support and Testing of Large Astronomical Mirrors*, Symposium, Tucson, 4–6 December 1966 (KPNO and University of Arizona Press, 1968), p. 101.

Acknowledgments

I am indebted to many men and women who mined their memories, libraries, file cabinets, scrapbooks, office shelves, and back closets for material on the building of the two-hundred-inch telescope. Often memories from half a century back seemed as clear as yesterday—testimony to the incredible impression the enterprise made on so many.

In California I drew on the memories of Ben Traxler; Mel Johnson; Sylvia, Bill, and Mary Marshall; and Byron Hill, who also provided audio and video tapes, documents, and photographs from their private collections. Jim McCauley, Anne Price, Walter Smith, and John Hoxie brought back a Corning that is no more. Rein Kroon not only recounted details of his work in Pasadena and at Westinghouse but prepared a collection of otherwise unavailable documents from his private papers. I am truly sorry that Ben Traxler and Rein Kroon did not live to see this book completed. I would have enjoyed their reactions to the story, and the book would undoubtedly have profited from their critical readings.

In Pasadena, Jesse Greenstein, Horace Babcock, Olin Wilson, and Allan Sandage evoked the excitement of the anticipation and early days of the telescope. Larry Blackeé, Bill McLellan, and Earle Emery shared memories of their years of working on the telescope, and pointed out treasures hidden in the domes at Palomar. Fred Harris, Hal Petrie, and Robert Brucato fielded questions and shared the contemporary excitement of the basement of Robinson Hall. Christine Shirley graciously allowed me to tour Hale's solar lab, with Horace Babcock as a guide.

During my stays at Palomar, Bob Thicksten, Will McKinley, Jeff Phinney, Dana and Bruce Cuney, John Henning, Paul van Ligten, Jean Mueller, Merle Street, Luz Lara, and Gerry Neugebauer shared a sense of the continued excitement of the telescope and the mountain. Willem Baan and John Salzer graciously opened the observing room to a visitor during their run on the two-hundred. I would particularly like to thank Robert Brucato, Bob Thicksten, Hal Petrie, and the Palomar staff for putting up with my questions during the removal and washing of the mirror and the first-ever removal and replating of the edge supports.

In Pasadena and Redwood City, Olga Rule, Carol Roth, and Naomi Serrurier shared memories of Bruce Rule and Mark Serrurier, along with private papers from their collections. Richard Preston, in Princeton, was generous with his notes and with tapes he made in his own interview of Byron Hill. Neal Matthews of the *San Diego Reader* shared the notes he made during a visit to Palomar. William Niering, of Connecticut College, helped with botanical fine points.

I would also like to acknowledge the many archivists and librarians who assisted my research. John Anderson and Ruth Shoemaker at the Hall of History at General Electric in Schenectady, Ron Brashear of the manuscript department of the Huntington Library, Michelle Cotton at the Corning Glass Works, Helen Knudsen at the Caltech Astrophysics Library, Melissa Smith and Emily Oakhill at the Rockefeller Archive Center, Virginia Wright at the library of the Corning Museum of Glass, Charles Ruch at Westinghouse, the staff at the Hagley Museum and Library, and Carol Buge and the current and past staffs of the Institute Archives at Caltech not only addressed dozens of requests for obscure documents and photographs but provided suggestions and surprises from their own expert knowledge.

I owe special debts to Paul Routley, now of the U.S. Naval Observatory, who introduced me to the wonders of astronomy and the Hale telescope; to Edward Burlingame for his faith and encouragement on this project; to Jesse Greenstein, Daniel Kevles, and Allan Sandage for their thoughtful reading of the manuscript of this book; and to my wife and son for putting up with many years of Palomar anecdotes.

For the errors that remain, I claim full credit.

Index

450 ✳ INDEX